# Genomics of Plant-Associated Fungi and Oomycetes: Dicot Pathogens

Ralph A. Dean · Ann Lichens-Park
Chittaranjan Kole
Editors

# Genomics of Plant-Associated Fungi and Oomycetes: Dicot Pathogens

Springer

*Editors*
Ralph A. Dean
Department of Plant Pathology
Center for Integrated
Fungal Research
North Carolina State University
Raleigh, NC
USA

Chittaranjan Kole
Bidhan Chandra Krishi Viswavidyalaya
Mohanpur
West Bengal
India

Ann Lichens-Park
United States Department
of Agriculture
National Institute of Food
and Agriculture
Washington, DC
USA

ISBN 978-3-662-44055-1 ISBN 978-3-662-44056-8 (eBook)
DOI 10.1007/978-3-662-44056-8
Springer Heidelberg New York Dordrecht London

Library of Congress Control Number: 2014946389



Printed on acid-free paper

Springer is part of Springer Science+Business Media (www.springer.com)

# Preface

The turn of the century not only ushered in a new millennium, but the age of microbial genomics with the genome sequence of the first plant pathogen *Xylella fastidiosa,* the bacterium that causes citrus variegated chlorosis. At that time, knowledge of genomes of agriculturally significant microbes was extremely limited. In an effort to improve the situation, the United States Department of Agriculture (USDA) offered a competitive grants program in 2000 to support the sequencing of agriculturally relevant microorganisms. In the following year, the United States National Science Foundation (NSF) partnered with USDA to jointly offer a competitive grants program that would support genome sequencing for more diverse microorganisms. The USDA/NSF Microbial Genome Sequencing Program was offered for 10 years. During that time, USDA supported the sequencing of a large number of agriculturally important microbes, including bacteria, viruses, fungi, oomycetes, and even a nematode. Some of the most expensive projects were jointly funded by USDA and NSF. Other sequencing projects were completed in collaboration with the United States Department of Energy's (DOE) Joint Genome Institute (JGI). In addition, NSF and DOE supported the genome sequencing of microbes of importance to other areas of science. The last year of the USDA/NSF Microbial Genome Sequencing Program was in 2009, by which time sequencing costs had decreased dramatically and sequencing speed had increased tremendously. USDA's support for microbial genomics had shifted toward functional analysis of the sequenced genomes. This book describes how the availability of the genome sequences of some agriculturally important plant-associated fungi and oomycetes, many of which were first sequenced with support from the USDA, has revolutionized our understanding of these eukaryotic microorganisms. This book also describes how knowledge derived from genomics can be translated into improved ways of managing microbes so as to increase the sustainability of agriculture in the United States and around the world.

The oomycetes were once classified as fungi due to their morphological similarities and because, like fungi, they feed on decaying as well as living plant material. It is now known that there are significant differences between oomycetes and fungi. In fact, they are not closely related at all. However, both these groups of microbes include pathogens that cause devastating losses to crop yields. Genome sequence information from fungi and oomycetes has laid the foundation for significant increases in knowledge of the lifestyles of these tiny eukaryotes, their dynamic evolution, and how they interact with plants. New approaches for functional analysis of the genomes are accelerating the progress toward novel understanding and improved management methods.

In this volume on "Genomics of Plant-Associated Fungi and Oomycetes: Dicot Pathogens," each chapter describes the genomic analysis of a genus, species, or group of related fungi or oomycetes. This volume focuses on the genera of fungi or oomycetes that include important pathogens of dicot plants, although some may also include pathogens that attack monocot plants. A companion volume, which we have also edited, focuses on the genera of fungi that include important pathogens of cereals and other monocot plants. A third volume (edited by Dr. Dennis Gross, Dr. Ann Lichens-Park and Dr. Chittaranjan Kole) describes the genomic analysis of plant-associated bacteria. Taken together, these three volumes illustrate some fundamental discoveries about these microbes with regard to the overall structure of their genomes, their lifestyles, and the molecular mechanisms that form the basis of their interactions with plants. Many of the genomes described exhibit considerable variation in DNA content, even among related species, illustrating selective invasion and expansion of repetitive genetic elements. Some genomes, such as those of *Phytophthora* species, show a large degree of conservation in gene content and colinearity (synteny) among related species. On the other hand, species of *Cochliobolus* and *Mycosphaerella* are more diverged but they exhibit mesosynteny, where gene content is conserved within chromosomes but gene order is not. In other genera, gene conservation is minimal. Some fungal genomes, such as *Pyrenophora tritici-repentis* are unstable and dynamic with large differences in genome size and chromosome number within a population. A number of chapters show that fungal genes, DNA segments (*Verticillium*) and even chromosomes (*Fusarium*, *Alternaria* and *Mycosphearella*) can move within species, genera, or even across kingdoms. In many instances, these events affect pathogenicity and host range.

Biotrophic fungi feed on living organisms. Genomics has provided new insights about fungal lifestyles, such as obligate biotrophy. Mildews, rusts, and other fungi that are obligate biotrophs have lost many genes involved in primary metabolism (for example, nitrogen, sulfate, and amino acid biosynthesis) and also secondary metabolism. Indeed, *Blumaria* and other powdery mildews have only half of the gene content of related fungi. Necrotrophic fungi live on dead plant material. Genome structural and functional analysis has revealed the necrotrophic lifestyle to be more sophisticated than once thought. Necrotrophs possess effectors, typically gene products that affect the development of diseases on host plants. Moreover, necrotrophs do not contain an excess of genes for degradation of plant material. Regulation of these degradative genes, rather than expansion of them, may be the key to the necrotrophic lifestyle.

In oomycetes, genomic analysis has led to the discovery of large families of effectors, such as the RXLR protein family and Crinklers. Effectors are predominantly found in genomic regions that are rich in rapidly evolving, repetitive DNA, which is subject to positive selection. Interestingly, true fungi appear to lack analogous families of effectors.

Genomics has led to practical advances and the understanding needed to implement fair and effective policies. Beginning in 2013, all fungi must have a single name as determined during the Nomenclature Session at the Botanical Congress in Melbourne. It is no longer acceptable to use the anomorph or teleomorph names. Genome sequences, which are the foundation of modern classification, have clarified the species concept in some instances but in other cases the "One Fungus One Name" concept has prompted much debate and controversy among the mycologists. A name has important federal and global implications. For example, clear nomenclature is needed for effective quarantine policies.

Fungal diagnostics has been greatly advanced through genomic technologies. The ability to accurately distinguish between closely related pathotypes is another requirement for effective quarantine policies. This can be very important for tracking devastating plant pathogens, such as the Ug99 pathogen that causes wheat stem rust.

Genomic studies of populations enable accurate reconstruction of previous disease epidemics, notably that of *Phytophthora infestans*, the oomycete pathogen that caused the notorious Irish potato famine. Genomic information enables predictions about the spread and evolution of new races. Knowledge about changes in race structure, including fungicide sensitivity, provides insights that influence fungicide use and contribute to improved plant breeding and cultivar release.

We wish to express our thanks to all of the authors and co-authors who contributed to the chapters in this volume. They have done a tremendous job, clearly describing the novel findings and exciting advances enabled by genomics with regard to the microbes addressed in their chapters. We also wish to specifically thank some current and former employees of USDA, NSF, and DOE whose support has been invaluable to the success of the microbial genomics program and to the existence of this volume. These people are Dr. Sonny Ramaswammy, Dr. Colien Hefferan, Ms. Betty Lou Gilliland, Ms. Erin Daly, Mr. Edward Nwaba, Dr. Deborah Sheely, Ms. Cynthia Montgomery, Dr. Michael Fitzner, Dr. Daniel Jones, Ms. Pushpa Kathir, Dr. Anna Palmisano, Dr. Mark Poth, Dr. Maryanna Henkart, Dr. Daniel Drell, and all of the USDA and NSF Program Officers and Staff who worked with Dr. Ann Lichens-Park while the Microbial Genome Sequencing Program was offered. Space limitations prevent us from describing the roles played by each of these individuals but their contributions were significant and we are immensely grateful to all of them.

Dr. Ralph A. Dean
North Carolina State University
Raleigh, NC, USA

Dr. Ann Lichens-Park
United States Department of Agriculture
National Institute of Food and Agriculture
Washington, DC, USA

Dr. Chittaranjan Kole
Bidhan Chandra Krishi Viswavidyalaya
West Bengal
India

# Contents

# Genomics of *Sclerotinia sclerotiorum* 1

Jeffrey A. Rollins, Christina A. Cuomo,
Martin B. Dickman, and Linda M. Kohn

## 1.1 Introduction

*Sclerotinia sclerotiorum* (Lib.) de Bary is a ubiquitous phytopathogenic fungus that causes unrestricted lesion development and tissue maceration of its hosts. This necrotrophic fungus has one of the broadest host ranges of any known phytopathogen. A compiled list of *S. sclerotiorum* hosts includes 408 different plant species from 278 genera encompassing 75 plant families (Boland and Hall 1994). Nearly all dicotyledonous plants assayed are susceptible to *S. sclerotiorum* infection and despite intensive efforts, adequate levels of genetic resistance have not been attained. Susceptible crops of importance to commerce and food security include soybean, dry bean, green bean, sunflower, canola, tomato, green pepper, lettuce, and numerous other vegetable and fiber crops.

In the United States alone, annual losses of crops from diseases caused by *S. sclerotiorum* are in the multimillion dollar range (http://www.whitemoldresearch.com). It is believed that this fungus occurs somewhere in almost every country (Steadman 1979). On the basis of the significance of this organism as an important plant pathogen, the US Congress has made appropriations for a National *Sclerotinia* Initiative, managed by the US Department of Agriculture, to foster research on biology and control of this pathogen. In 2004, we were awarded a grant from the USDA's National Research Initiative Competitive Grants Program via the USDA/NSF microbial genome-sequencing program to obtain the complete genome sequence of this fungus.

The key rationale for undertaking this effort included:

- There has been, and still is, only limited genetic resistance to *S. sclerotiorum* available in crop plants. Cultural practices have proven inadequate for effective durable, long-term disease control.
- The host range is exceptionally large, in contrast to most plant pathogens, which tend to exhibit narrower specialization. Why and how is *S. sclerotiorum* so successful in infecting such a diverse range of broad leaf hosts?
- *S. sclerotiorum* represents a large, poorly known group of Ascomycetes—the Helotiales and associated Rhytismatales. Sequence information will thus impact other important

J. A. Rollins
Department of Plant Pathology, University of Florida, Gainesville, FL 32611, USA

C. A. Cuomo
Broad Institute of MIT and Harvard, Cambridge, MA 02142, USA

M. B. Dickman (✉)
Department of Plant Pathology and Microbiology, Institute for Plant Genomics and Biotechnology, Texas A&M University, College Station, TX 77843, USA
e-mail: mbdickman@tamu.edu

L. M. Kohn
Department of Ecology and Evolutionary Biology, University of Toronto Mississauga, Mississauga, ON L5L 1C6, Canada

R. A. Dean et al. (eds.), *Genomics of Plant-Associated Fungi and Oomycetes: Dicot Pathogens*,
DOI: 10.1007/978-3-662-44056-8_1, © Springer-Verlag Berlin Heidelberg 2014

**Table 1.1** Functions of oxalic acid in pathogenesis

| Function | Reference |
|---|---|
| Acts synergistically with pectolytic enzymes through pH and $Ca^{2+}$ chelation | Bateman and Beer (1965) |
| Chelates calcium and other divalent cations | Bateman and Beer (1965), Maxwell and Lumsden (1970) |
| Acid toxicity | Bateman and Beer (1965), Marciano (1983) |
| Vascular plugging by oxalate crystals | Lumsden and Dow (1973) |
| Signal molecule—pH dependent and independent | Rollins and Dickman (2001), Kim et al. (2008) |
| Guard cell misregulation | Guimarães and Stotz (2004) |
| Down-regulates the host oxidative burst | Cessna et al. (2000) |
| Induces ROS in the host and likely in the fungus | Williams et al. (2011) |
| Elicits host apoptotic-like programmed cell death and suppresses host autophagy | Kim et al. (2008), Kabbage et al. (2013) |
| Modulates the host redox environment | Williams et al. (2011) |

genera and will serve as a link in comparative fungal genomics.

- Population biology of *S. sclerotiorum* is among the best characterized of fungal pathosystems, facilitating rational choice of epidemiologically significant genotypes for sequencing and comparative genomics.
- As with several other important soilborne pathogens, *S. sclerotiorum* forms sclerotia, highly melanized vegetative propagules that survive long periods in soil and that germinate to produce either infective hyphae or spores that function as inoculum. As such, it is a model for sclerotial development and live-cell biology.

Pathogenesis of *S. sclerotiorum* is complex and not well understood. The fungus produces a wide array of degradative, lytic enzymes (e.g., endo, exo-pectinases, cellulase, hemicellulase, protease), which are believed to facilitate colonization and host cell wall degradation (Marciano et al. 1983; Riou et al. 1991). In addition to degradative enzymes, the role of oxalic acid (OA) in pathogenicity of this fungus has been intensively studied (Godoy et al. 1990; Dickman and Mitra 1992; Rollins and Dickman 1998, 2001; Cessna et al. 2000). Studies with $OA^{-}$ mutants strongly suggest that OA is an essential pathogenicity determinant in *S. sclerotiorum* (Godoy et al. 1990). Oxalate secretion has been demonstrated to enhance *Sclerotinia* virulence in several (and ever increasing) ways (see Table 1.1).

Given that *S. sclerotiorum* is an aggressive necrotroph that excels at killing cells, significant research activity has focused on means by which host cells die during infection. Programmed cell death is characterized by a cascade of tightly controlled events that culminate in the orchestrated death of the cell. In multicellular organisms, autophagy and apoptosis are recognized as two principal means by which these genetically determined cell deaths occur (Dickman and Fluhr 2013). During plant-microbe interactions, cell death programs can mediate both resistant and susceptible events. We have shown that via OA, *S. sclerotiorum* hijacks host pathways and induces cell death in host plant tissue resulting in hallmark apoptotic cell death in a time- and dose-dependent manner (Kim et al. 2008). Oxalic acid deficient fungal mutants are severely attenuated in virulence and trigger a restricted cell death phenotype in the host. This unexpectedly includes markers consistent with the plant hypersensitive response (which is generally associated with biotrophic/hemi-biotrophic defense) including callose deposition and a pronounced oxidative burst, suggesting that in the case of the mutant (but not wild type), the plant can recognize and respond defensively. Using a combination of electron and fluorescence microscopy, chemical effectors and reverse genetics, we have demonstrated that this constrained cell death is autophagic (Kabbage et al. 2013). Inhibition of autophagy rescued the attenuated virulence mutant phenotype. These findings suggest that plant autophagy is a host defense response in this necrotrophic fungus/plant interaction and point toward a novel function associated with OA, namely the suppression of autophagy. These data suggest that not all cell deaths are equivalent, and though programmed cell death occurs in both disease and resistance, the outcome is predicated on whether host or pathogen is in control of the cell death machinery. Based on these data, we suggest that it is not cell death per se that dictates the outcome of certain plant-*Sclerotinia* interactions, (and likely other pathogenic fungi), but the manner by which cell death occurs that is crucial. From the perspective of the pathogen, the annotated genome provides a valuable resource for understanding the mechanisms by which *S. sclerotiorum* both promotes and suppresses host cell death.

Despite the variety of pathogenicity-related mechanisms involved (some of which appear to be indiscriminate of host), accumulating evidence indicates that necrotrophic plant pathogens interact with their hosts in a manner much more subtle than originally considered and that signaling between the necrotroph and the host plays a significant role in the lifestyle of these pathogens (van Kan 2006). In this chapter, we discuss how the application of genome data can facilitate a better understanding of these subtleties, which may share characteristics with biotrophic/hemibiotrophic interactions (Dickman and Fluhr 2013).

## 1.2 Genome Structure

The *Sclerotinia sclerotiorum* genome project utilized the '1980' isolate originally identified by Dr. Jim Steadman from bean culls in western Nebraska, USA. Before sequencing, this isolate was cultured from a single ascospore, and hyphae from this culture were used to prepare DNA for sequencing. The assembled 8X chromosome-anchored genome sequence was constructed by a shotgun sequencing strategy, utilizing Sanger sequencing chemistry and optical mapping (Amselem et al. 2011). The sequencing, assembly, and annotation were performed by the Broad Institute of Harvard and MIT from a USDA-funded project awarded to Drs. Christina Cuomo (Broad Institute), Martin Dickman (Texas A&M University), Linda Kohn (University of Toronto Mississauga) and Jeffrey Rollins (University of Florida). The project coordination at the Broad Institute was directed by Dr. Christina Cuomo. Feature highlights of the genome are discussed below and can be viewed in Table 1.2.

The 38 Mb genome assembly is anchored onto 16 chromosomes, one of which terminates with tandem copies of the rDNA locus. The 37 assembled scaffolds cover 96 % of the optical map, with 2–3 large scaffolds aligned to each chromosome. The largest uncovered regions correspond to the gaps between the scaffolds, which likely correspond to centromeric locations. The centromeres of filamentous fungi consist of islands of repetitive elements, which can be difficult to assemble using shotgun data and are often missing from whole genome shotgun assemblies. By contrast, the *S. sclerotiorum* scaffolds extend to the telomeric ends of the map suggesting subtelomeres are not highly repetitive in this species. Raw sequence trace files can be downloaded from the National Center for Biotechnology Information (NCBI) trace archives and the assembled genome was deposited as 37 genomic and one mitochondrial scaffold as project accession AAGT00000000 in the DDBJ/EMBL/GenBank databases. The annotated genome is hosted by the Broad Institute and can be accessed at http://www.broadinstitute.org//scientific-community/data. The website provides datasets for download, a graphical genome map interface, search algorithms for identifying genes based on sequence or annotated features, and browsing capabilities to view genes based on conserved domains or function.

Expressed Sequence Tags (ESTs) derived from a diverse collection of cDNA libraries were utilized in genome annotation and gene verification. These reads are mapped to the genome and viewable through the graphical genome browser. These ESTs were derived from the

**Table 1.2** Feature highlights of the *Sclerotinia sclerotiorum* genome

| Feature | Nuclear | Mitochondrial |
|---|---|---|
| size (bp) | 38.33 Mb | 128.85 kb |
| Number of chromosomes | 16 | 1 |
| Number of Scaffolds | 37 | 1 |
| Scaffold N50 | 1.55 Mb | 128.85 kb |
| Number of Contigs | 679 | 1 |
| Contig N50 (Kb) | 122.55 | 128.85 |
| Total G + C content (%) | 41.84 | 30.9 |
| Exon G + C content (%) | 45.88 | 30.22 |
| Intron G + C content (%) | 39.94 | 32.85 |
| Intergenic G + C content (%) | 39.17 | 31.07 |
| Number of high confidence genes | 11,865 | 45 |
| Exons | 35,199 | 66 |
| Mean gene length (bp) | 1,502 | 1,583 |
| Percent coding (%) | 38.59 | 35.3 |
| Median intron length (bp) | 78 | -ND- |
| Mean intergenic length (bp) | 1,723 | 1,996 |
| Repetitive sequences (%) | 7.7 | -ND- |
| tRNA genes | 191 | 33 |
| rRNA genes | 50 | 2 |

Summary data from "*Sclerotinia sclerotiorum* Sequencing Project, Broad Institute of Harvard and MIT (http://www.broadinstitute.org/)"
*-ND-* not determined

major developmental stages of the life cycle including the earliest hyphal aggregation stage of sclerotium development (library G781 (center_project field in NCBI trace repository record); 17,533 sequences); mycelia grown at acidic pH treated 4 h at neutral pH (G786; 18,885 sequences); apothecia during expanding disc development (G787; 21,333 sequences); and compound appressoria (infection cushion library G866; 1,532 sequences). Additionally EST sequences from infected plants, *Brassica napus* (G865; 5,107 sequences) and tomato (G2118; 13,144 sequences) and oxidative stress conditions associated with host colonization (G2128, 13,621 sequences) are mapped to the genome and are available for download.

Repetitive elements occupy 7 % of the genome with no apparent hotspots along chromosomes. This is a 10X higher proportion of repetitive sequence than the 0.7 % observed for the T4 strain of the related species *Botrytis cinerea* (Amselem et al. 2011). This expansion includes diverse types of transposable elements. LTR elements are most frequently observed, with large numbers of TIR, LINE, unknown, and MITE elements also found. As many of the LTR Gypsy and Copia retroelements appear excised, with only solo LTR present, this may have mitigated the overall impact of TE proliferation on genome size. However, repetitive elements appear to have rearranged genome structure. While most of the genome of *S. sclerotiorum* is found in syntenic regions with *B. cinerea*, breaks in synteny are often associated with repetitive elements.

## 1.3 Life Cycle

The genome sequence has provided an important resource for understanding the developmental and pathogenic lifecycle of *S. sclerotiorum* through direct annotation of conserved regulator genes as well as via gene expression and protein accumulation analyses. A brief overview of the *S. sclerotiorum* life cycle is provided here to emphasize and compare several unique aspects of the lifecycle relative to other sequenced fungi and to delineate where studies relating the genome to the pathogen's biology may be focused for disease control (Fig. 1.1).

Despite the dramatic symptoms associated with *S. sclerotiorum* diseases, it is instructive to note that this fungus spends the majority of its lifecycle quiescently within the soil; it exists as a sleeping menace in the unassuming form of a sclerotium. Although a seemingly unremarkable form of existence, the sclerotium's simplicity masks a complexity of form and function essential to the pathogen's survival and dissemination. Initially, it serves as the primary survival structure marking the end of the disease cycle. Subsequently, the sclerotium serves as the structural and nutritional base of apothecial and myceliogenic germination, signaling the start of a new disease cycle. A multitude of physiological, microscopic as well as molecular genetic studies have documented the structural organization and make up of sclerotia (Kohn and Grenville 1989; Willetts and Bullock 1992), revealed exogenous and endogenous regulators of sclerotium development (Willetts and Bullock 1992; Rollins and Dickman 1998; Harel et al. 2006; Erental et al. 2007; Patsoukis and Georgiou 2007), and characterized environmental parameters associated with survival and germination (Coley-Smith and Cooke 1971; Hao et al. 2003; Erental et al. 2008). The key genes underlying the makeup and processes associated with sclerotia, however, remain mostly undiscovered and uncharacterized.

Myceliogenic germination of sclerotia is generally associated with the availability of exogenous nutrients and is epidemiologically significant in the disease cycle on a small number of hosts, including carrots and sunflower. In other species, e.g., *S. minor* and *Sclerotium cepivorum*, demonstrated by phylogenetic analyses of multilocus sequence data to be very closely related or congeneric with *Sclerotinia* (Carbone and Kohn 1993; Holst-Jensen et al. 1998; Xu et al. 2010; Andrew et al. 2012), myceliogenic germination is the primary route of infection and apothecia are rarely encountered in the field. Sclerotia develop phialospores on the outer surface which likely

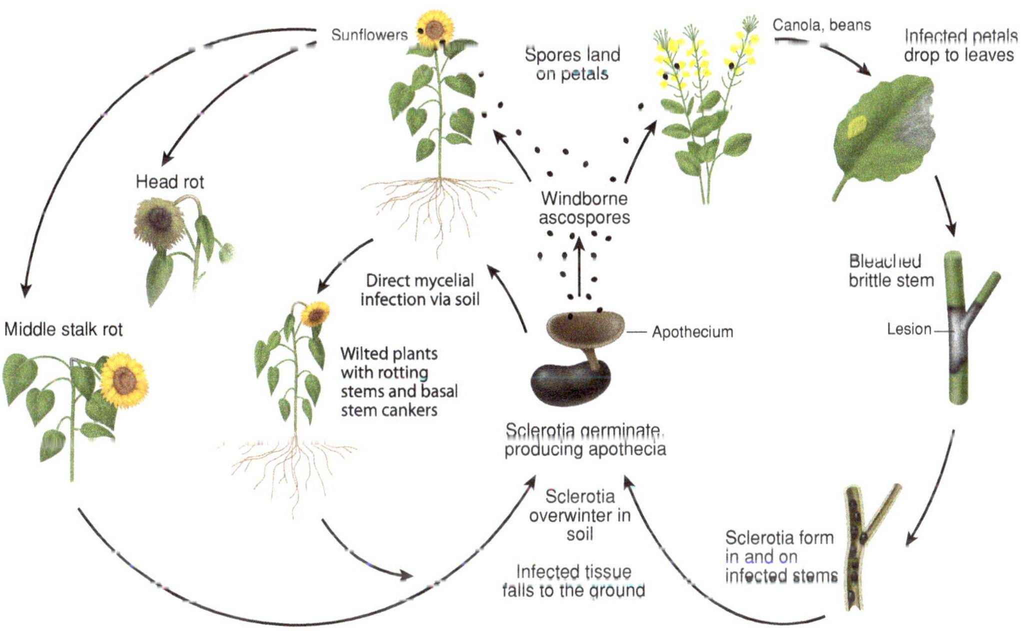


**Fig. 1.1** *Sclerotinia sclerotiorum* life cycle

function as spermatia in fertilization similar to related heterothallic Sclerotiniaceae species. Under appropriate environmental conditions, sclerotia carpogenically germinate to produce stipitate apothecial fruiting bodies. Carpogenic germination often coincides with canopy closure and flowering but seasonality of germination is dependent on the local climate and the adaptation of the local pathogen populations. Each sclerotium may produce multiple apothecia and although susceptible to predation and microbial degradation, has the potential for multiple years of carpogenic germination. Each apothecium produces millions of ascospores (Steadman 1983) that forcibly discharge (Roper et al. 2010) providing a means of local infection and spread within and between fields of susceptible crops. The multiyear survival of sclerotia and their movement with soil and harvested seed ensures their persistence and spread within and between agriculture fields. The movement of a single viable sclerotium can represent the inoculum potential equivalent of 10 million or more infectious ascospores (Steadman 1983).

Though the lifecycle of *S. sclerotiorum* is anchored by the sclerotium, it is also noted for its conspicuous lack of a propagative, asexual sporulation stage. Copious conidiation characterizes other closely related genera of the Sclerotiniaceae including *Botryotinia* spp. with *Botrytis* anamorphs and *Monilinia* spp. with *Monilia* anamorphs. Conidia in these and many other genera are important epidemiologically and influence disease management strategies. Comparative sequence analysis of the *S. sclerotiorum* and *B. cinerea* genomes using known orthologous conidiation-associated genes conserved between *A. nidulans* and *N. crassa* indicate that these conserved genes are also present in *S. sclerotiorum* (Amselem et al. 2011). The absence of conidia in *Sclerotinia* spp. is of both evolutionary and practical interest. The lack of air-dispersed conidial inoculum would seem to be costly from the standpoint of fitness. However, these species appear to suffer no fitness penalty with only the production of asexual sclerotia and sexual ascospores to sustain the lifecycle and initiate the disease cycle.

The dependence on ascospores for aerial dispersal and infection also comes with a unique caveat; in particular, ascospores are unable to directly establish infections on healthy host tissues. Ascospores can germinate and even penetrate host tissues in the absence of exogenous nutrients; however, compatible infections are established only when senescent or dead tissues have been colonized saprotrophically prior to infection of healthy tissues (Abawi et al. 1975; Sutton and Deverall 1983). In a field situation, this is generally accomplished by the colonization of senescent flower blossoms and leaves or damaged host tissues that are in direct contact with healthy tissue.

Following saprotrophic colonization, hyphae that encounter healthy tissues form appressoria. Generally appressoria are compound structures consisting of two or more modified hyphal tips. Commonly tens to hundreds of bifurcated modified hyphal tips develop to form these infection structures and in the latter case they are macroscopically observable and termed 'infection cushions'. The level of complexity is influenced by a number of factors including nutritional status and inductive surface hardness (Tariq and Jeffries 1984, 1986). The penetration process appears to be facilitated by enzymatic digestion (Jones 1976; Lumsden 1979) and a number of cytological studies using various hosts have been conducted to better understand the infection process. These studies have revealed a common strategy in which the plant cuticle is breached by multiple tips of the compound appressorium, penetration hyphae generally form bulbous subcuticular hyphae and then filamentous subcuticular infection hyphae and secondary hyphae that ramify to colonize inter- and intracellularly, with rapid host cell death and necrotic symptom development (Jamaux et al. 1995; Lumsden and Dow 1973; Lumsden and Wergin 1980). As no mitosporic conidial stage is produced by *S. sclerotiorum*, diseases are monocyclic. Under conducive conditions, the infection spreads from the point of infection throughout the plant and from infected plant to healthy plant. Sclerotia are formed on the surface and within hollowed stems and fruits of colonized host tissue thus completing the lifecycle.

With the exception of infectious hyphae types produced following host penetration, all of the multicellular developmental stages of *S. sclerotiorum* are readily produced in the laboratory under axenic conditions. This ability to induce and manipulate development under controlled conditions provides an exceptional experimental system for applying genomic tools to understand the genes, gene products and metabolic processes that regulate their development and function. Studies aided by the genome have revealed (1) the proteome associated with sclerotium development and secreted proteins in culture (Liang et al. 2010; Yajima and Kav 2006), (2) the classification of genes unique to sclerotium-producing species conserved across taxonomic boundaries (Amselem et al. 2011) and (3) the identity of highly accumulating sclerotial-specific proteins (Li and Rollins 2009). Transcription can be monitored by RNA-Seq, including a number of projects initiated by independent research programs, and by using publically available Agilent microarrays representing all nonredundant predicted genes and orphan EST sequences (Agilent Technologies Design ID 015292). The application of these technologies and analyses of results from the studies will bring broad insight into the transcriptome of *S. sclerotiorum* during interactions with its hosts for disease development and with its environment for life cycle completion.

An example illustrating how the genome sequence improved our understanding of life cycle regulation has come from the characterization of the mating type locus. The underlying regulation of sexual compatibility and development in fungi is controlled by the mating type locus. In *S. sclerotiorum*, a single mating type locus encodes both an alpha box (alpha1) protein and the mating type high mobility group (HMG) protein. This arrangement is consistent with those observed in other self-fertile, homothallic mating systems in the Pezizomycotina (Amselem et al. 2011). This so-called "fused" MAT locus contrasts with the heterothallic *B. cinerea* MAT locus, which consists of two idiomorphs, one encoding the alpha1 protein and the other encoding the HMG mating type protein. In both species, two lineage-specific proteins are also encoded at the MAT locus: two at the Ss-MAT locus, and one at each of the two Bc-MAT idiomorphs.

A unique feature, discovered from the comparative genomic analysis of *S. sclerotiorum* and *B. cinerea*, is that in *B. cinerea*, redundant, truncated copies of the alpha1 and the HMG genes are present at the opposing idiomorphs. That is, a truncated copy of the HMG-encoding gene exists along with the full-length alpha1-encoding gene at the MAT1-1 idiomorph; correspondingly, a truncated copy of the alpha1-encoding gene exists along with the full-length copy of the HMG-encoding gene at the MAT1-2 idiomorph. The truncated copies appear to be nonfunctional, lacking start codons and functional domains. It has been proposed that two simple inversion and deletion events could explain the conversion from the MAT locus of a *S. sclerotiorum*-like homothallic ancestor to the configuration of functional and disabled gene fragments in the MAT locus of *B. cinerea* (Amselem et al. 2011). This was the first evidence of an evolutionary transition from homothallism to heterothallism, although evidence for the reverse pattern from heterothallism to heterothallism in nature based on such rearrangements had been advanced (reviewed in Amselem et al. 2011).

Further structural variation was recently discovered (Chitrampalam et al. 2013) in a study of MAT loci among a sample of 283 isolates of *S. sclerotiorum* from multiple hosts and locations. Two MAT alleles were identified, one with a 3.6-kb inversion (inv+), the other without the inversion (inv−) as exemplified by the genome isolate. The inversion affects three of the four MAT genes, *mat1-1-1*, *mat1-2-1*, and *mat1-2-4* and is truncated at the 3′ end in *mat1-1-1*. Isolates are either *inv*+ or *inv*− and both are self fertile. The inversion is flanked by a 250 bp repeat sequence and it occurs during each meiosis such that tetrads segregate 1:1 for $inv^{+}$ and $inv^{-}$, suggesting maintenance of equal proportions of the two alleles. The MAT inversion correlates with MAT gene expression. Altered

MAT gene expression did not influence sexual development or mycelial/vegetative compatibility groupings in this study. The authors suggest that comparable inversion regions may be implicated in mating type switching in filamentous ascomycetes, including other species of *Sclerotinia*. These studies facilitated by the genome sequence have uncovered a remarkably dynamic MAT locus within the Sclerotiniaceae both at the evolutionary and cellular levels. New insights into the mechanisms and selection for mating specificities are likely to be revealed by further comparative genomic analyses.

## 1.4 Pathogenesis

Investigation of the genome has provided some clues to the identification of pathogenicity determinants related primarily to OA metabolism, ROS dynamics, and the regulation of these factors. *S. sclerotiorum*, which regulates ROS dynamics during pathogenic development has a similar repertoire of genes involved in oxidative stress (both quantitatively and qualitatively) similar to the model saprophyte *Neurospora crassa*. As well, evaluation of genes presumably associated with cell signaling has revealed the usual suspects. One observation of note, however, is that during infection *S. sclerotiorum*, induces ROS, yet, also inhibits oxidative stress. These opposing observations were difficult to reconcile. From studies using redox-regulated GFP in a time course of infection, we now know (Williams et al. 2011) that *S. sclerotiorum* initially inhibits ROS (dampening the host response) and then following what we believe is establishment, transitions to the induction of ROS, cell death, and disease.

Understanding how *S. sclerotiorum* dynamically modulates the host environment in regard to ROS and OA accumulation is certain to provide new insights into pathogenicity as well as host resistance. Relevant to OA metabolism, the gene encoding oxaloacetate acetylhydrolase (*oah*) was identified in the genome. The encoded enzyme activity is hypothesized to catalyze the final step in OA biosynthesis and experimental evidence including gene deletion mutants in *B. cinerea* and *S. sclerotiorum* supports this role (Stefanato et al. 2008; Liang et al. 2013). Additionally, genes with predicted, secreted oxalate decarboxylase activity were also identified (*odc1* and *odc2*) suggesting that *S. sclerotiorum* dynamically regulates OA levels in its environment which may be pertinent to both development and infection. Oxalic acid is known to directly and indirectly affect ROS levels during infection and host colonization, and several associated candidate activities and their genes have been characterized from the genome. These include NADPH oxidase (*nox1*, *nox2*), superoxide dismutase (*sod1*) catalase (*cat1*), glutathione metabolism genes (gama glutamyl transpeptidase (*ggt1*, and others), pH signal regulators (*pac1*), and ROS resistance regulators (*yap1*). Indeed, knockout and knockdown mutants in most of these genes result in attenuated virulence and/or developmental phenotypes (Kim et al. 2007, 2011; Li et al. 2012; Liang et al. 2013; Liang and Rollins 2013; Rollins 2003; Veluchamy et al. 2012; Xu and Chen 2013).

An understanding of how signals from OA, pH, and ROS accumulation are integrated and transduced to effect virulence and development is beginning to come to light. One example of this is found in the highly conserved homolog of ERK-type mitogen-activated protein kinases (MAPKs) from *S. sclerotiorum* (Smk1). This gene was identified and demonstrated to be required for sclerotial development (Chen et al. 2004). Transcription of *smk1* and MAPK enzyme activity was induced dramatically during sclerotiogenesis, especially during the production of sclerotial initials. When PD98059 (a specific inhibitor of MAPK activation) was applied to differentiating cultures or when antisense expression of *smk1* was induced, sclerotial maturation was impaired. The *smk1* transcript levels were highest under acidic pH conditions, suggesting that Smk1 regulates sclerotial development via a low pH-dependent signaling pathway, involving the accumulation of OA. Indeed, OA treatment induced sclerotia formation, which is consistent with disease development. Addition of cyclic AMP (cAMP) inhibited *smk1* transcription, MAPK activation, and sclerotial development. Thus, *S. sclerotiorum* can coordinate environmental signals (such as pH) to trigger a signaling pathway mediated by Smk1 to induce sclerotia formation, and this pathway is negatively regulated by cAMP.

The mechanism(s) by which cAMP inhibits MAPK, however, are not clear. Expression of a dominant negative form of the prototypical small G Protein, Ras an upstream activator of the MAPK pathway, also inhibited sclerotial development and MAPK activation, suggesting a conserved Ras/MAPK pathway is required for sclerotial development. Evidence from bacterial toxins that specifically inhibit the activity of selected small GTPases suggested that either Rap-1 or Ras is involved in cAMP action. The Rap-1 inhibitor, GGTI-298, restored MAPK activation in the presence of cAMP, further suggesting that Rap-1 is responsible for cAMP-dependent MAPK inhibition. Importantly, inhibition of Rap-1 restored sclerotial development blocked by cAMP. These results suggest a novel mechanism involving the requirement of Ras/MAPK pathway for sclerotial development that is negatively regulated by a PKA-independent cAMP signaling pathway. Cross-talk between these two pathways is mediated by Rap-1. Further characterization of these processes including the spatial and temporal regulation of these activities during infection, their interconnected regulation, and their roles throughout the life cycle is certain to refine our understanding of development and necrotrophic pathogenicity in *S. sclerotiorum* (Chen and Dickman 2004).

## 1.5 Outstanding Questions

In addition to understanding the genes and regulatory circuitry involved in the transduction of exogenous and endogenous signals in *S. sclerotiorum*, outstanding questions that may be addressed with genome-wide transcriptome, proteome, and functional genomic analyses include: (A) determining what genes, gene products and pathways maintain sclerotium quiescence. Such knowledge may lead to strategies for prematurely breaking quiescence when a suitable host is unavailable; (B) determining what genes, gene products, and metabolic pathways are responsible for protecting sclerotia from microbial degradation and insect predation; (C) understanding how sclerotial tissue is programmed to form apothecia and what genes regulate the timing and numbers of carpogenic (apothecial) germinations; (D) determining what genes, gene products, and metabolites are produced by the compound appressoria and how the temporal and spatial dynamics of this production relates to plant infection; (E) understanding the basis for sensing surface hardness as it relates to appressorium developmental complexity; (F) exploring the genetic and metabolomics basis of nutrient requirements for infection; (G) determining if specialized hyphae are produced during infection and if transcript profiles for subcuticular hyphae, infection hyphae and ramifying hyphae reflect a specialization in pathogenic function; (H) understanding how the fungus coordinates and balances ROS production and; and (I) revealing how the fungus regulates cell death decisions to induce apoptosis-like cell death and how autophagic cell death is repressed. These areas and others relating to the developmental and pathogenic forms of *S. sclerotiorum* will provide new understandings of fungal biology and offer new strategies for interfering with these processes to break the cycle of disease development.

## 1.6 Comparative Genomics

By comparing the predicted proteins of *S. sclerotiorum* to those in *B. cinerea* and other fungi, unique and shared features of these necrotrophic pathogens have been identified (Amselem et al. 2011). Most of the 11,860 protein coding genes predicted for *S. sclerotiorum* are shared with *B. cinerea*; 8,609 proteins share an average of 84 % identity between these species. In comparison to seven other fungi, selected to include other necrotrophs, biotrophs, and saprobes, *S. sclerotiorum* and *B. cinerea* are enriched for two glycosyl hydrolase families (GH71, involved in alpha-1,3-glucan degradation and GH28, a class of polygalacturonases involved in pectin degradation), a GMC oxidoreductase, a GPD-fucose protein O-fucosytransferase, and a putative amidoligase. These expansions were not found in the only other Leotiomycete in the comparison, *Blumeria graminis* f.sp. *hordei*, a biotrophic plant pathogen; however, the reduction in total gene count for *B. graminis* f.sp. *hordei* to only 5,854 total may bias against finding such expansions. By examining the genes specific to *S. sclerotiorum* and *B. cinerea*, and without homology to those in other fungi, cytochrome p450 and transcription factors were found to predominate, suggesting this subset may have diverged by more rapid evolution. Gene families depleted in *S. sclerotiorum* and *B. cinerea* compared to other fungi include two glycosyl hydrolase families (GH2 and GH43), subtilases, zinc finger domains, and terpene synthases.

Further annotation of the carbohydrate active enzymes (CAZymes) and growth profiling suggests that *S. sclerotiorum* and *B. cinerea* have specialized to degrade pectin (Amselem et al. 2011). Comparing the CAZyme content of these species to the other necrotrophic, biotrophic, and saprobic fungi revealed a smaller total number of CAZymes in *S. sclerotiorum* and *B. cinerea* than in most other plant pathogens. However, the total number of pectin degrading enzymes, including GH28 noted above, is high in these two necrotrophs. This specialization is reflected in the higher growth for these species on pectin than on other substrates. Pectin-related CAZymes are a major class of genes up-regulated on growth on sunflower cotyledons for both *S. sclerotiorum* and *B. cinerea*. Other large classes of cellulose-, xylan-, and mannan-degrading enzymes are also significantly up-regulated, suggesting that a broad group of enzyme activities could be used by these fungi during infection. Thus, while the complete cell wall degrading enzyme (CWDE) profiles are similar in comparison with saprotrophs; specific subsets/families of CWDEs may account/contribute to particular nectrotrophic niches. Gene conservation patterns correlated with the appearance of a necrotrophic lifestyle. To examine shared features of necrotrophic pathogens, gene families expanded in *S. sclerotiorum*, *B. cinerea*, and the distantly related necrotrophic species *P. teres* and *P. nodorum* were examined. This search identified only heterokaryon incompatibility proteins. In addition, comparison of the full set of plant CWDEs revealed that *S. sclerotiorum* and *B. cinerea* contain fewer total enzymes than *P. teres* or *P. nodorum* (Amselem et al. 2011). The lack of common expansions of particular families of degradative enzymes which target plant cell walls or of protective enzymes to mitigate plant defense responses suggests necrotrophy can be established with different subsets of genes, or may be more at the level of gene regulation.

Looking in more depth within the Sclerotiniaceae, an evolutionary lineage related by recent common ancestry, do species share a set of genes associated with necrotrophy, i.e., a common toolbox? With a toolbox shared across the Sclerotiniaceae, we would expect a common set of genes or gene families with expression modulated in time and magnitude. Modulation of expression might or might not differ across these species. In a preliminary investigation, Andrew et al. (2012) pursued this hypothesis. To start, a phylogeny of 24 species that was inferred from combined sequences of three housekeeping genes, *cal*, *hsp60*, and *g3ph*, was found to be consistent with previously published phylogenies based on other loci (reviewed in Amselem et al. 2011) with four main lineages. Two lineages were necrotrophic, including both host generalists and host specialists (*Sclerotinia/Dumontinia/Sclerotium cepivorum*, and *Botryotinia/Botrytis*). The third lineage included facultatively biotrophic/host specialists (*Myriosclerotinia*). The fourth lineage included biotrophic/necrotrophic/host specialists (*Monilinia*) and an obligately biotrophic/host specialist (*Ciborinia whetzelii*). The conservative interpretation of this phylogeny was that the common ancestor of the Sclerotiniaceae was a necrotroph, with biotrophy the more recent, derived state in one or two shifts from necrotrophy. These shifts were in the evolution of *Myriosclerotinia* from an ancestor in the *Botryotinia/Botrytis* lineage and in the evolution of *Monilinia*, *Ciboria*, and *Ciborinia* from a common ancestor in the *Sclerotinia* and *Botrytis* lineages. In site-specific likelihood analyses of a small set of pathogenicity-associated genes, there was evidence for purifying selection on all 8 genes and positive selection on sites within 5 of 8 genes. Purifying or positive selection are evidence that these genes convey fitness in adaptation to the plant-associated life style of the Sclerotiniaceae. Constitutive production of OA in bromophenol blue indicator plates showed early and abundant production by all taxa sampled from the necrotrophic *Botrytis* and *Sclerotinia/Dumontina/Sclerotium* lineages. Within these two lineages, early and abundant OA production characterized both host generalists and those specialized more broadly for plant groups such as *S. cepivorum* on Alliums or *Sclerotinia trifoliorum* on forage legumes, or specifically for single plant species. The exceptions here were each of two strains of *Dumontinia ulmariae* with late and low production contrasting with the expected early and abundant production in the congeneric *Dumontinia tuberosa*. In contrast, late or low production of OA, or no production over the time course of the experiment, was observed for the biotrophic/host specialist lineages: *Monilinia/Ciboria/Ciborinia* and *Myriosclerotinia*. This polarity of either early and abundant or late and low production was borne out as induced expression in a second experiment *in planta* on *Arabidopsis thaliana*. Here, normalized *oah* transcript copy numbers were compared through a course of eight sampling times: 1, 2, 4, 8, 12, 24, 48, and 72 h after inoculation. Interpreting from all time points, *S. sclerotiorum* and *B. cinerea* had 10–300 times greater *oah* transcript accumulation than necrotrophic or biotrophic host specialists. In these assays, the low transcript accumulation group included the more specialized *Botrytis* and *Sclerotinia* species, notably including *S. homoeocarpa*. The advantage of early, abundant *oah* expression may allow the generalists to exploit their wide host range, while specialists and biotrophs deploy expression of *oah* and other genes differently. From this study, the authors hypothesize that within the Sclerotiniaceae interactions with plants are modulated by genes in a common toolbox by means of differences in timing and magnitude of gene expression. The toolbox shares its common features via one common evolutionary origin from the necrotrophic ancestor. These ideas now await comparative genomic and expression studies taking full advantage of whole genome screens for patterns of selection.

## 1.7 Population Biology

Ideally, population biology captures genomics, transcriptomics, and proteomics in the context of evolutionary time and space. It is in this context we find the conceptual framework and analytical tools for the following research goals: (1) determining the boundaries of populations (groups of individuals that have interbred or are interbreeding); (2) testing hypotheses about genetic and evolutionary mechanisms (e.g., hybridization, introgression, mutation, adaptation, selection, genetic drift, impact of sexual and asexual reproduction, gene flow/migration); and (3) finding the distribution of phenotypes such as avirulence factors, chemotypes, and fungicide response. All of this can be accomplished at the spatial scales of interest: from global, to regional, to a crop, to a single plant.

Descriptive and evolutionary aspects of population structure in *S. sclerotiorum* have been reviewed by Malvárez et al. (2007). Though much has been done, several important questions await answers. Does *S. sclerotiorum* have fully functional meiosis in which two parents form a normal dikaryon that undergoes recombination and segregation? There is as yet no cytological evidence. Following from this, what is the relative contribution of clonality and outcrossing with recombination to population structure on different geographical scales? Is high standing genetic variability most accurately explained by outcrossed sexual reproduction or by agronomic practices, climate and lack of selection? Does indirect evidence of recombination describe contemporary or retrospective events? What is the mechanism of vegetative/mycelial incompatibility and how does it evolve?

A series of studies on over 3,000 isolates over 15 years in several laboratories has provided evidence of a pattern of epidemic clonality characterized by a small number of clones appearing at high frequencies and a large number of clones or genotypes at low frequencies. This pattern has been reported from north-central and eastern United States and adjacent Canada as well as New Zealand, Australia, Argentina, and Europe (Ekins et al. 2011; Kohn 1995; Carbone et al. 1999; Carbone and Kohn 2001a, b; Carpenter et al. 1999; Cubeta et al. 1997; Durman et al. 2003; Kohli et al. 1992; Kull et al. 2003; Phillips et al. 2002).

In the early studies, clone boundaries were determined by vegetative compatibility grouping (in this system termed mycelial compatibility groups, MCGs; Kohn et al.1990) and genomic DNA fingerprinting using a probe containing a 4.5 kb repeated dispersed element of nuclear DNA from *S. sclerotiorum* (Kohn et al. 1991). The repeated dispersed nuclear element was later shown to be a fragment of a non-LTR retrotransposon (LM Kohn, unpublished). Members of a clone are vegetatively compatible (but incompatible with other clones) and have a fingerprint of several hybridizing bands, among which one or 2–3 are unique to the clone. Decoupling of mycelial compatibility group and DNA fingerprint, consistent with recombination, was first observed in a Norwegian population of wild buttercup, *Ranunculus ficaria*. This was in contrast to coupling of MCG and fingerprint consistent with clonality in Norwegian agricultural populations and other agricultural populations in North America, (Kohn 1995). Interestingly, a discrete population distinct from agriculture of another wild buttercup, *R. acris*, has recently been reported in the UK (Clarkson et al. 2013).

Later studies, in addition to MCGs, utilized multilocus DNA sequences. Multilocus DNA sequences supplanted fingerprinting methods owing to (1) their ease in implementation and accurate scoring; (2) their rich potential trove of sequence polymorphisms, especially across longer sequence spans; and (3) the potential for finding evidence of recombination or differential rates of evolution among different loci. These attributes of multilocus studies are exemplified in studies of *S. sclerotiorum* that included the 4 kb intergenic spacer of the nuclear ribosomal repeat (IGS), with 55 polymorphic sites for four populations studied in North America, plus the two populations in California lettuce and Washington pea-lentil studied later. Prominent recombination blocks—sites of inter-block but not intra-block recombination—were evident in the IGS sequences from the southeastern US, Norwegian wild buttercup, California lettuce, and Washington pea-lentil populations (Carbone and Kohn 2001a, b; Malvárez et al. 2007). The discovery of recombination blocks indicated that *S. sclerotiorum* has had a history of recombination even though epidemics associated with asexual clonal reproduction have marked recent agricultural history in monoculture crops in some regions. Other loci used to infer population history in these studies were translation factor 1-alpha (EF-1a), calmodulin (CAL), chitin synthase (CHS), and an anonymous nuclear region from *S. sclerotiorum* (44.11). In these studies, the use of multiple loci strengthened the statistical power of the analyses by providing abundant markers, which in turn strengthened the support of conclusions regarding the clonal structure, population boundaries, and patterns of migration among populations.

Haplotype distributions have been used to determine population boundaries in many species and have been applied to *S. sclerotiorum*. Haplotypes are the basic unit in multilocus population studies. A haplotype is a sequence of a locus with a unique suite of polymorphisms as determined in comparison to a reference sequence; one or more strains will have the same haplotype in a population sample. Some haplotypes are widely distributed; in a recent study of *S. sclerotiorum* on various crops and wild buttercup, 3 of 14 IGS haplotypes from the United Kingdom matched those reported from North America, New Zealand, or Norway, while 8 of the 14 were found only in the UK (Clarkson et al. 2013).

Microsatellites are another tool, most useful at the local crop or regional scale (Sirjusingh and Kohn 2001), and have been applied recently in

Brazil (Gomes et al. 2011; Litholdo et al. 2011), Iran (Barari et al. 2013; Karimi et al. 2012; Hemmati et al. 2009), Turkey (Yanar and Onaran 2011), India (Madal and Dubey 2012), China (Attanayake et al. 2012, 2013; Li et al. 2009), England (Clarkson et al. 2013), with further study in Australia (Sexton et al. 2006) and Washington state (Attanayake et al. 2012). Five observations emerge: (1) These studies differ enormously in design, methodology, and execution; they are difficult to compare. Here, larger samples and positive controls, such as clone tester strains from North America, could be helpful in facilitating communication among research groups and anchoring results over time. (2) Clonality is not the rule everywhere. In areas without harsh winters, or without crop rotation, or with high crop diversity, there appears to be the high level of genetic diversity comparable to what Malvárez et al. (2007) reported from California lettuce with evidence of clonality lacking. (3) Genetic exchange and recombination are certainly possible but not all of the evidence is acceptable. For example, the lack of correspondence among the genetic markers and phenotypic traits is not evidence of outcrossing as intraclonal mutation is well established, such as in somaclonal variation in populations derived from plant tissue or experimental evolution of asexual yeast lines (Evans 1989). (4) There is little or no correlation between genotypes determined with neutral markers and pathogenicity or virulence; phenotypic variation is observed but is independent of these genotypes. (5) A strain with variation in the ITS as compared to the *S. sclerotiorum* reference genome sequence (strain 1980) is likely to be a new species. Workers must be careful to determine that all strains are conspecific; there are new species of *Sclerotinia* yet to be discovered. Confirmation of species identity of strains is a prerequisite for execution of a quality population analysis. There are two reasons why: (1) a population by definition is a group of con-specific individuals with the potential to interbreed and (2) owing to their expected high mutation rate, microsatellites should be deployed only for the species from which they were developed. This last point is mainly due to the difficulty of interpreting null alleles when the test group is not con-specific.

From well-designed studies with sufficient sampling and sequence data, we can use powerful statistical methods to infer the evolutionary past. For example, Malvárez et al. (2007) were able to infer that Washington pea-lentils shared a common origin with Ontario lettuce (part of a larger North central/Eastern North American population on many hosts), but that the hypervariable California lettuce population was older with a different origin. We speculate that the pathogen may have moved to California from South America on beans with the Gold Rush, when many people migrated from Uruguay and other countries to California just as California agriculture was diversifying from livestock to crops. In another study on canola, as a relatively newly introduced crop in the Southeastern US, Phillips et al. (2002) were able to determine an association between symptoms, indigenous pathogen haplotypes and newly arrived haplotypes.

While analyses of population structure may seem both descriptive and retrospective, experimental evolution of populations allows us to test predictive hypotheses. Kohn et al. (2007) took 12 strains of *S. sclerotiorum* through 400 serial transfers over 1 year at two temperatures, 22 and 30 °C. The strains represented five clones from bean, canola, soybean or pea, from Nebraska, Ontario, and Washington. Given that a clone is a mitotic lineage, the experiment considered the potential elevation of mutation rate at a stressful temperature. One of the isolates, a member of the most frequently recovered clone across the region, was isolated following 2 years of drought; it both grew and infected within the normal temperature range for *S. sclerotiorum*, but unlike other strains in the same clone, also infected at 28 °C, above the normal range. During the course of 400 serial transfers, no mutations were detected among any of the markers: MCGs, DNA fingerprints, alleles at 7 microsatellite loci, and alleles at 56 AFLP loci. By contrast, all of these markers were variable in field populations where populations are much larger and influenced by a variety of environmental processes. The drought strain was stable in its ability to grow at 30 and 22 °C, indicating that *S. sclerotiorum* is capable of adapting to environmental stochasticity and global climate change.

With the *S. sclerotiorum* genome in hand, the distribution and evolutionary trajectories of phenotypic determinants within populations should be more finely resolvable that they have been to date. These would include determinants of pathogenicity and virulence, plasticity under environmental extremes, and fungicide tolerance. It is now possible to develop a panel of single nucleotide polymorphisms (SNPs) across the genome and to screen more widely and more deeply, not just for markers toward elucidating population structure, but also for signs of selection or more specifically for variation in genes of interest. It should be possible to identify "divergence genes" in a genealogy of haplotypes representing different populations.

## 1.8 Conclusions

The genome sequence of *S. sclerotiorum* represents a strong first step in developing a more complete understanding of the complexities of the necrotrophic pathogen life style. Admittedly, the generation and initial analysis of the *Sclerotinia* genome, has not provided complete answers to underlying mechanisms responsible for pathogenic development but provides a framework to build upon. We have come to a fuller understanding of the roles of OA during pathogenesis and have genetically demonstrated its requirement for systemic host colonization. Concomitantly, we have learned that the subtleties of the host-pathogen interaction are much more nuanced than previously imagined. Understanding the underlying basis for this nuance may reveal the mechanisms of broad versus narrow host range specialization within the Sclerotiniaceae. To date, comparative analysis suggests that, with the exception of cutin degrading activities, necrotrophic pathogens do not encode a larger proportion of degradative enzymes. Instead, the regulation of gene expression, translation, protein modification or localization in the cell, may prove key to a necrotrophic lifestyle. Production of primary and secondary metabolites may also play an important role in interacting with the host plant. Secondary metabolite gene clusters are highly diverged between *S. sclerotiorum* and *B. cinerea*, with clusters unique to each species (Amselem et al. 2011). Characterizing the molecules produced by these clusters, their regulation, and roles in plant infection through metabolomics analyses may also contribute to our understanding of the specificity of plant host interactions. Moving forward, functional genomic approaches; comparative patho-biology and genome analyses; and a systematic characterization of population structure and dynamics is required. Such combinations of studies, all utilizing the resources of the genome sequence and annotation, are poised to reveal the basis for the complex nature of necrotrophic pathogenicity in *S. sclerotiorum*.

**Acknowledgments** Lifecycle illustration by Dave Mazierski, Biomedical Communications, Department of Biology, University of Toronto Mississauga.

## References

Abawi GS, Polach FJ, Molin WT (1975) Infection of bean by ascospores of *Whetzelinia sclerotiorum*. Phytopathology 65:673–678

Amselem J, Cuomo CA, van Kan JAL, Viaud M, Benito EP, Couloux A, Coutinho PM, de Vries RP, Dyer PS, Fillinger S, Fournier E, Gout L, Hahn M, Kohn L, Lapal N, Plummer KM, Pradier J-M, Quévillon E, Sharon A, Simon A, ten Have A, Tudzynski B, Tudzynski P, Wincker P, Andrew M, Anthouard V, Beever RE, Beffa R, Benoit I, Bouzid O, Brault B, Chen Z, Choquer M, Collémare J, Cotton P, Danchin EG, Da Silva C, Gautier A, Giraud C, Giraud T, Gonzalez C, Grossetete S, Güldener U, Henrissat B, Howlett BJ, Kodira C, Kretschmer M, Lappartient A, Leroch M, Levis C, Mauceli E, Neuvéglise C, Oeser B, Pearson M, Poulain J, Poussereau QH, Rascle C, Schumacher J, Ségurens B, Sexton A, Silva E, Sirven C, Soanes DM, Talbot NJ, Templeton M, Yandava C, Yarden O, Zeng Q, Rollins JA, Lebrun M-H, Dickman M (2011) Genomic analysis of the necrotrophic fungal pathogens *Sclerotinia sclerotiorum* and *Botrytis cinerea*. PLoS Genet 7(8):e1002230

Andrew M, Barua R, Short SM, Kohn LM (2012) Evidence for a common toolbox based on necrotrophy in a fungal lineage spanning necrotrophs, biotrophs, endophytes, host generalists and specialists. PLoS ONE 7(1):e29943

Attanayake RN, Carter PA, Jiang D, del Río-Mendoza L, Chen W (2013) *Sclerotinia sclerotiorum* populations infecting canola from China and the United States are genetically and phenotypically distinct. Phytopathology 103:750–761

Attanayake RN, Porter L, Johnson DA, Chen W (2012) Genetic and phenotypic diversity and random association of DNA markers of isolates of the fungal plant pathogen *Sclerotinia sclerotiorum* from soil on a fine scale. Soil Biol and Biochem 55:28–36

Barari H, Dalili SA, Rezaii SA (2013) Population genetic structure analysis of *Sclerotinia sclerotiorum* (Lib.) de Bary from different host plant species in Northern Iran. Arch Biol Sci 65:171–181

Bateman DF, Beer SV (1965) Simultaneous production and synergistic action of oxalic acid and polygalacturonase during pathogenesis by *Sclerotium rolfsii*. Phytopathology 58:204–211

Boland GJ, Hall R (1994) Index of plant hosts of *Sclerotinia sclerotiorum*. Can J Plant Pathol 16:93–108

Carbone I, Kohn LM (1993) Ribosomal DNA sequence divergence within internal transcribed spacer 1 of the Sclerotiniaceae. Mycologia 85:415–427

Carbone I, Anderson JB, Kohn LM (1999) Patterns of descent in clonal lineages and their multilocus fingerprints are resolved with combined gene genealogies. Evolution 53:11–21

Carbone I, Kohn LM (2001a) A microbial population-species interface: nested cladistic and coalescent inference with multilocus data. Mol Ecol 10:947–964

Carbone I, Kohn LM (2001b) Multilocus nested haplotype networks extended with DNA fingerprints show common origin and fine-scale, on-going genetic divergence in a wild microbial metapopulation. Mol Ecol 10:2409–2422

Carpenter MA, Frampton C, Stewart A (1999) Genetic variation in New Zealand populations of the plant pathogen *Sclerotinia sclerotiorum*. N Z J Crop Hortic Sci 27:13–21

Cessna SG, Sears VE, Low PS Dickman M (2000) Oxalic acid, a pathogenicity factor for *Sclerotinia sclerotiorum*, suppresses the oxidative burst of the host plant. Plant Cell 12:2191–2200

Chen C, Dickman MB (2004) Dominant active Rac and dominant negative Rac revert the dominant active Ras phenotype in *Colletotrichum trifolii* by distinct signaling pathways. Mol Microbiol 51:1493–1507

Chen C, Harel A, Gorovoits R, Yarden O, Dickman MB (2004) MAPK regulation of sclerotial development in *Sclerotinia sclerotiorum* is linked with pH and cAMP sensing. Mol Plant-Microbe Interact 17:404–413

Chitrampalam P, Inderbitzin P, Maruthachalam K, Wu B-M, Subbarao KV (2013) The *Sclerotinia sclerotiorum* mating type locus (MAT) contains a 3.6-kb region that is inverted in every meiotic generation. PLoS ONE 8:e56895

Clarkson JP, Coventry E, Kitchen J, Carter HE, Whipps JM (2013) Population structure of *Sclerotinia sclerotiorum* in crop and wild hosts in the UK. Plant Pathol 62:309–324

Coley-Smith JR, Cooke RC (1971) Survival and germination of fungal sclerotia. Annu Rev Phytopathol 9:65–92

Cubeta MA, Cody BR, Kohli Y, Kohn LM (1997) Clonality in *Sclerotinia sclerotiorum* on infected cabbage in eastern North Carolina. Phytopathology 87:1000–1004

Dickman MB, Mitra A. (1992) Arabidopsis as a model for studying resistance to *Sclerotinia infection*.Physiological and Molecular Plant Pathology 41:255–263.

Dickman M.B. and Fluhr R. (2013). Centrality of host cell death in plant-microbe interactions. Annu Rev-Phytopathol 51:25.1–25.28

Durman SB, Menendez AB, Godeas AM (2003) Mycelial compatibility groups in Buenos Aires field populations of *Sclerotinia sclerotiorum* (Sclerotiniaceae). Aust J Bot 51:421–427

Ekins MG, Hayden HL, Aitken EAB, Goulter KC (2011) Population structure of *Sclerotinia sclerotiorum* on sunflower in Australia. Australas Plant Path 40:99–108

Erental A, Dickman MB, Yarden O (2008) Sclerotial development in *Sclerotinia sclerotiorum*: awakening molecular analysis of a "Dormant" structure. Fungal Biol Rev 22:6–16

Erental A, Harel A, Yarden O (2007) Type 2A phosphoprotein phosphatase is required for asexual development and pathogenesis of *Sclerotinia sclerotiorum*. Mol Plant-Microbe Interact 20:944–954

Evans DA (1989) Somaclonal variation-genetic basis and breeding applications. Trends Genet 5:46–50

Godoy G, Steadman JR, Dickman MB, Dam R (1990) Use of mutants to demonstrate the role of oxalic acid inpathogenicity of *Sclerotinia sclerotiorum on Phaseolus vulgaris*. Physiol. Mol. Plant Pathol. 37:179–1991

Gomes EV, Do Nascimento LB, De Freitas MA, Nasser LCB, Petrofeza S (2011) Microsatellite markers reveal genetic variation within *Sclerotinia sclerotiorum* populations in irrigated dry bean crops in Brazil. J Phytopathol 159:94–99

Guimarães RL, Stotz HU (2004) Oxalate production by *Sclerotinia sclerotiorum* deregulates guard cells during infection. Plant Physiol 136:3703–3711

Hao JJ, Subbarao KV, Duniway JM (2003) Germination of *Sclerotinia minor* and *S. sclerotiorum* sclerotia under various soil moisture and temperature combinations. Phytopathology. 93:443–450

Harel A, Bercovich S, Yarden O (2006) Calcineurin is required for sclerotial development and pathogenicity of *Sclerotinia sclerotiorum* in an oxalic acid-independent manner. Mol Plant-Microbe Interact 19:682–693

Hemmati R, Javan-Nikkhah M, Linde CC (2009) Population genetic structure of *Sclerotinia sclerotiorum* on canola in Iran. Eur J Plant Pathol 125:617–628

Holst-Jensen A, Vaage M, Schumacher T (1998) An approximation to the phylogeny of *Sclerotinia* and related genera. Nordic J Bot 18:705–719

Jamaux I, Gelie B, Lamarque C (1995) Early stages of infection of rapeseed petals and leaves by *Sclerotinia sclerotiorum* revealed by scanning electron microscopy. Plant Pathol 44:22–30

Jones D (1976) Infection of plant tissue by *Sclerotinia sclerotiorum*: a scanning electron microscope study. Micron 7:275–279

Kabbage M, Williams B, Dickman MB (2013) Cell death control: the interplay of apoptosis and autophagy in the pathogenicity of *Sclerotinia sclerotiorum*. PLoS Path 9:e1003287

Karimi E, Safaie N, Shams-Bakhsh M (2012) Mycelial compatibility groupings and pathogenic diversity of *Sclerotinia sclerotiorum* (Lib.) de Bary populations on canola in Golestan Province of Iran. J Agric Sci Technol 14:421–434

Kim H-J, Chen C, Kabbage M, Dickman MB (2011) Identification and characterization of *Sclerotinia sclerotiorum* NADPH oxidases. Appl Environ Microbiol 77:7721–7729

Kim H-J, Min J-Y, Dickman MB (2008) Oxalic acid is an elicitor of plant programmed cell death during *Sclerotinia sclerotiorum* disease development. Mol Plant-Microb Interact 21:605–612

Kim YT, Prusky D, Rollins JA (2007) An activating mutation of the *Sclerotinia sclerotiorum pac1* gene increases oxalic acid production at low pH but decreases virulence. Mol. Plant Pathol. 8:611–622

Kohli Y, Morrall RAA, Anderson JB, Kohn LM (1992) Local and trans-Canadian clonal distribution of *Sclerotinia sclerotiorum* on canola. Phytopathology 82:875–880

Kohn LM (1995) The clonal dynamic in wild and agricultural plant-pathogen populations. Can J Bot 73:S1231–S1240

Kohn LM, Grenville DJ (1989) Anatomy and histochemistry of stromatal anamorphs in the Sclerotiniaceae. Can J Bot 67:371–393

Kohn LM, Carbone I, Anderson JB (1990) Mycelial interactions in *Sclerotinia sclerotiorum*. Exp Mycol 14:255–267

Kohn LM, Schaffer MR, Anderson JB, Grünwald NJ (2007) Marker stability throughout 400 days of in vitro hyphal growth in the filamentous ascomycete, *Sclerotinia sclerotiorum*. Fung Genet Biol 45:613–617

Kohn LM, Stasovski E, Carbone I, Royer J, Anderson JB (1991) Mycelial incompatibility and molecular markers identify genetic variability in field populations of *Sclerotinia sclerotiorum*. Phytopathology 81:480–485

Kull LS, Vuong TD, Powers KS, Eskridge KM, Steadman JR, Hartman GL (2003) Evaluation of resistance screening methods for *Sclerotinia* stem rot of soybean and dry bean. Plant Dis 87:1471–1476

Li M, Liang X, Rollins JA (2012) *Sclerotinia sclerotiorum* $\gamma$-glutamyl transpeptidase (Ss-Ggt1) is required for regulating glutathione accumulation and development of sclerotia and compound appressoria. Mol Plant-Microb Interact 25:412–420

Li M, Rollins JA (2009) The development-specific protein (Ssp1) from *Sclerotinia sclerotiorum* is encodedby a novel gene expressed exclusively in sclerotium tissues. Mycologia 101:34–43

Liang X, Rollins JA (2013) An oxalate decarboxylase gene functions in the early infection processes of *Sclerotinia sclerotiorum*. Phytopathology 103(S2):81

Liang X, Liberti D, Li M, Kim Y-T, Wilson R, Rollins J (2013) The biosynthesis of oxalate is entirely dependent on oxaloacetate acetylhydrolase in *Sclerotinia sclerotiorum*. Fungal Genet Rep 60S:277

Liang Y, Strelkov SE, Kav NNV (2010) The proteome of liquid sclerotial exudates from *Sclerotinia sclerotiorum*. J Proteome Res 9:3290–3298

Li Z, Wang Y, Chen Y, Zhang J, Fernando WGD (2009) Genetic diversity and differentiation of *Sclerotinia* populations in sunflower. Phytoparasitica 37:77–85

Litholdo Júnior CG, Gomes EV, Lobo Júnior M, Nasser LCB, Petrofeza S (2011) Genetic diversity and mycelial compatibility groups of the plant-pathogenic fungus *Sclerotinia sclerotiorum* in Brazil. 2011. Gen Mol Res 10:868–877

Lumsden RD (1979) Histology and physiology of pathogenesis in plant diseases caused by Sclerotinia species. Phytopathology 69:890–896

Lumsden RD, Dow RL (1973) Histopathology of *Sclerotinia sclerotiorum* infection of bean. Phytopathology 63:708–715

Lumsden RD, Wergin WP (1980) Scanning-electron microscopy of infection of bean by species of Sclerotinia. Mycologia 72:1200–1209

Madal AK, Dubey SC (2012) Genetic diversity analysis of *Sclerotinia sclerotiorum* causing stem stem rot in chickpea using RAPD, ITS-RFLP, ITS sequencing and mycelial compatibility grouping. World J Microb Biot 28:1849–1855

Malvárez M, Carbone I, Grünwald NJ, Subbarao KV, Schafer M, Kohn LM (2007) New populations of *Sclerotinia sclerotiorum* from lettuce in California and peas and lentils in Washington. Phytopathology 97:470–483

Marciano P, Di Lenna P, Magro P (1983) Oxalic acid, cell wall-degrading enzymes and pH in pathogenesis and their significance in the virulence of two *Sclerotinia sclerotiorum* isolates on sunflower. Physiol Plant Pathol 22:339–345

Maxwell DP, Lumsden RD (1970) Oxalic acid production by *Sclerotinia sclerotiorum* in infected bean and in culture. Phytopathology 60:1395

Patsoukis N, Georgiou CD (2007) Effect of sulfite-hydrosulfite and nitrite on thiol redox state, oxidative stress and sclerotial differentiation of filamentous phytopathogenic fungi. Pestic Biochem Physiol 88:226–235

Phillips DV, Carbone I, Gold SE, Kohn LM (2002) Phylogeography and genotype-symptom associations in early and late season infections of canola by *Sclerotinia sclerotiorum*. Phytopathology 92:785–793

Riou C, Freyssinet G, Feure M (1991) Production of cell wall degrading enzymes by the phytopathogenicfungus *Sclerotinia sclerotiorum*. Appl Env Microbiol 57:1478–148

Rollins JA (2003) The *Sclerotinia sclerotiorum pac1* gene is required for sclerotial development and virulence. Mol Plant-Microb Interact 16:785–795

Rollins JA, Dickman MB (1998) Inhibition of sclerotial development in *Sclerotinia sclerotiorum* by increasing endogenous and exogenous cAMP levels. Appl Environ Microbiol 64:2539–2544

Rollins JA. and Dickman MB. 2001. pH signaling in *Sclerotinia sclerotiorum*: Identification of a*pacC/RIMI* homolog. Appl Env Microbiol 67:75–81

Roper M, Seminara A, Bandi MM, Cobb A, Dillard HR, Pringle A (2010) Dispersal of fungal spores on acooperatively generated wind. PNAS 107:17474–17479

Sexton AC, Whitten AR, Howlett BJ (2006) Population structure of *Sclerotinia sclerotiorum* in an Australian canola field at flowering and stem-infection stages of the disease cycle. Genome 46:1408–1415

Sirjusingh C, and Kohn LM 2005. Characterization of microsatellites in the fungal plant pathogen, *Sclerotiniasclerotiorum*. Molecular Ecology Notes 1:267–269

Steadman JR (1979) Control of plant diseases caused by *Sclerotinia species*. Phytopathology 69:904–907

Steadman JR (1983) White mold - a serious yield-limiting disease of bean. Plant Disease 3:346–350

Stefanato FL, Abou-Mansour E, Van Kan J, Metraux JP, Schoonbeek HJ (2008) Oxaloacetate acetylhydrolase is responsible for oxalic acid production in *Botrytis cinerea* and required for lesion expansion on some, but not on most host plants. In: 3rd Botrytis Genome Workshop, Tenerife, Spain, Third Session, Pathogenicity, p 24

Sutton DC, Deverall BJ (1983) Studies on infection of bean (*Phaseolus vulgaris*) and soybean (*Glycine max*) by ascospores of *Sclerotinia sclerotiorum*. Plant Pathol 32:251–261

Tariq VN, Jeffries P (1984) Appressorium formation by *Sclerotinia sclerotiorum*: scanning electron microscopy. Trans Brit Mycol Soc 82:645–651

Tariq VN, Jeffries P (1986) Ultrastructure of penetration of *Phaseolus* spp. by *Sclerotinia sclerotiorum*. Botany 64:2909–2915

Van Kan, JAL (2006) Licensed to kill: the lifestyle of a necrotrophic plant pathogen. Trends Plant Sci 11:247—253

Veluchamy S, Williams B, Kim K, Dickman MB (2012) The CuZn superoxide dismutase from *Sclerotinia sclerotiorum* is involved with oxidative stress tolerance, virulence, and oxalate production. Physiol Mol Plant Pathol 78:14–23

Willetts HJ, Bullock S (1992) Developmental biology of sclerotia. Mycol Res 96:801–816

Williams B, Kabbage M, Kim H-J, Britt R, Dickman MB (2011) Tipping the balance: *Sclerotinia sclerotiorum* secreted oxalic acid suppresses host defenses by manipulating the host redox environment. PLoS Path 7:e1002107

Xu L, Chen W (2013) Random T-DNA Mutagenesis identifies a Cu/Zn Superoxide dismutase gene as a virulence factor of *Sclerotinia sclerotiorum*. Mol Plant-Microb Interact 26:431–441

Xu Z, Harrington TC, Gleason ML, Batzer JC (2010) Phylogenetic placement of plant pathogenic *Sclerotium* species among teleomorph genera. Mycologia 102:337–346

Yajima W, Kav NNV (2006) The proteome of the phytopathogenic fungus *Sclerotinia sclerotiorum*. Proteomics 6:5995–6007

Yanar Y, Onaran A (2011) Mycelial compatibility groups and pathogenicity of *Sclerotinia sclerotiorum* (Lib.) de Bary causal agent of white mold disease of greenhouse grown cucumber in Antalya-Turkey. Afr J Biotechnol 10:3739–3746

# The Genome of *Botrytis cinerea*, a Ubiquitous Broad Host Range Necrotroph

2

Matthias Hahn, Muriel Viaud, and Jan van Kan

**Abbreviations**

| | |
|---|---|
| SM | Secondary metabolism |
| TF | Transcription factor(s) |

## 2.1 Introduction

The genus *Botrytis*, together with the well-known genera *Sclerotinia* and *Monilinia*, belongs to the family Sclerotiniaceae within the class Ascomycetes. The family Sclerotiniaceae represents a rather recently evolved fungal lineage, and is assumed to consist mainly of necrotrophic plant pathogens. These include the two prominent wide host range species *Botrytis cinerea* and *Sclerotinia sclerotiorum*, many host specialists, but also some species with biotrophic and endophytic lifestyles (Andrew et al. 2012; Shipunov et al. 2008). The genus *Botrytis* currently comprises roughly 25–30 species, all of which are considered to be necrotrophic plant pathogens. Since the inference of a sequenced-based phylogeny of 22 recognized species (Staats et al. 2005), several new species have been reported: *B. pseudocinerea* (Walker et al. 2011), *B. caroliniana* (Li et al. 2012a, b), *B. fabiopsis* (Zhang et al. 2010a, b), *B. sinoallii* (Zhang et al. 2010a, b), *B. deweyae* (Grant-Downton et al. 2013). Additional *Botrytis* isolates have been described which are phylogenetically distinct from described species but remain unnamed (Shipunov et al. 2008). There is one hybrid, polyploid species, which arose from hybridization between *B. aclada* and *B. byssoidea* (Nielsen and Yohalam 2001; Staats et al. 2005). *Botrytis* species were grouped into two clearly distinct phylogenetic clades. The major clade, which is divided into five subclades, contains mainly species with monocot hosts, but also a few dicot pathogens. The other clade contains five species, including *B. cinerea*, that only infect dicot hosts (Staats et al. 2005; Walker et al. 2011).

*Botrytis cinerea* Pers. Fr. (teleomorph *Botryotinia fuckeliana* (de Bary) Whetzel) is by far the most thoroughly studied species, because of its broad host range (>200 species, Fig. 2.1a, b) and its huge economic impact, causing pre- and postharvest crop losses worldwide (Dean et al. 2012). Its ability to infect many different plant species and tissues under a wide range of environmental conditions, as well as the ability of sclerotia to survive in the soil, contribute to its

M. Hahn (✉)
Department of Biology, University of Kaiserslautern, Kaiserslautern, Germany
e-mail: hahn@biologie.uni-kl.de

M. Viaud
BIOGER, INRA, Grignon, France

J. van Kan
Laboratory of Phytopathology, Wageningen University, Wageningen, The Netherlands

R. A. Dean et al. (eds.), *Genomics of Plant-Associated Fungi and Oomycetes: Dicot Pathogens*,
DOI: 10.1007/978-3-662-44056-8_2, © Springer-Verlag Berlin Heidelberg 2014

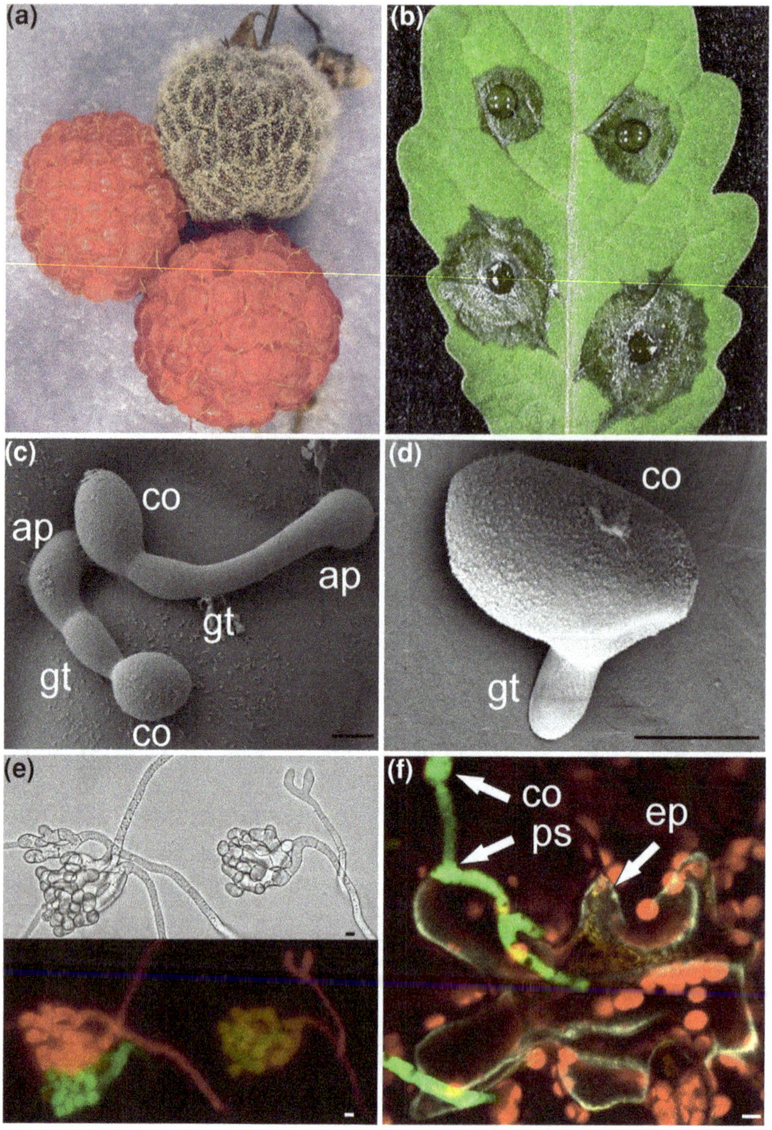


**Fig. 2.1** The gray mold fungus *Botrytis cinerea*. **a** Sporulation on raspberry. **b** Expanding lesions on a detached tomato leaf, 72 h after inoculation with 10 µl droplets containing $10^2$ (*top*) and $10^3$ (*bottom*) conidia. **c** Conidia (co) germinated in 10 mM fructose on a bean leaf surface, with germ tubes (gt) and appressoria (ap) (24 h). **d** Conidium germinated in water on polyethylene. A short germ tube is formed, with no apparent appressoria (48 h). **e** Complex appressoria (infection cushions), formed on a glass surface by a mixture of GFP and mCherry expressing strains. On the *right side*, hyphal fusion resulted in hyphae expressing both fluorescent proteins. **f** GFP-marked hyphae during early invasion of *Arabidopsis* leaf cells. From a germinated conidium (co), a hypha penetrated (*ps* penetration site) into an epidermis cell (ep), which collapsed and showed cell wall autofluorescence. Red autofluorescence is from mesophyll cell chloroplasts. **c**, **d** Kindly provided by K. Mendgen. **e**, **f** Modified from Leroch et al. 2013a, with permission

persistent and widespread nature. Dispersal of *B. cinerea* is via airborne conidia, and to lesser extent via ascospores. Although fruiting bodies (apothecia) are rarely observed, population genetics data suggest that sexual reproduction occurs frequently. The availability of molecular tools has considerably advanced our understanding of the infection strategies of *B. cinerea*, which we will summarize in the next paragraph. In the remaining paragraphs, we will discuss several topics with special '-omics' relevance.

## 2.2 Current Knowledge of *B. cinerea* Infection Strategies

Conidia, produced in large quantities in necrotized host tissue, are the main sources of infection by *B. cinerea*. In the presence of free water or high humidity (>93 % RH), germination is triggered by a combination of physical and chemical signals. While the mechanisms of signal perception are largely unknown yet, both a cAMP-dependent pathway and a MAP kinase cascade are involved in signal transduction leading to germination (Schumacher et al. 2008a; Schamber et al. 2010). On the host cuticle, the germlings form nonmelanized appressoria that are morphologically not always distinct, and not separated by a septum from the germ tube (Fig. 2.1c, d). Appressorium formation and function requires the interplay of many regulatory components, as exemplified by various kinds of penetration defects observed in mutants lacking heterotrimeric G protein components, adenylate cyclase, and either of the three MAP kinase cascades (Schumacher et al. 2008a, b, c; see Sect. 2.6.1). A specific role in appressorium-mediated penetration of *B. cinerea* and other plant pathogens could be assigned to the tetraspanin Pls1, a membrane-spanning regulatory protein (Gourgues et al. 2004, Siegmund et al. 2013). Besides appressoria, the fungus is able to form more complex penetration structures called infection cushions (Choquer et al. 2007). Because of their rather low degree of differentiation, the physical penetration ability of *B. cinerea* appressoria into intact host tissue seems to be limited, and probably needs to be supported by secreted lytic enzymes. Moreover, the need of a stable cell wall of the penetration structures may require the activity of specific chitin synthase isoenzymes, several of which were shown to be important for pathogenicity (Arbelet et al. 2010; Morcx et al. 2013).

Penetration is usually accompanied by rapid death of the host cells. *B. cinerea* secretes several toxic compounds that play a role in host killing. The toxic secondary metabolites (SM), botrydial and botcinic acid derivatives, will be described in Sect. 2.4.3. Hydrogen peroxide ($H_2O_2$) is released by the invading fungus into the host tissue (Tenberge 2004), presumably by an extracellular Cu–Zn superoxide dismutase which is important for virulence (Rolke et al. 2004). However, the situation is complicated by the production of defense-related $H_2O_2$ from the invaded host cell. Several secreted proteins have been studied for their role in pathogenesis. The role of necrosis-inducing proteins of the NLP family is discussed in Sect. 2.6.5. A xylanase, Xyn11A, was shown to be required for full virulence (Brito et al. 2006). This function was not related to its enzymatic activity, but rather to its necrosis-inducing activity which was retained in a catalytically-impaired Xyn11A variant (Noda et al. 2010). A cerato-platanin family protein, Spl1, was also found to contribute to full virulence. Purified Spl1 induces hypersensitive cell death and salicylic acid-mediated systemic acquired resistance (Frías et al. 2011, 2013). Plant wall degradation is another tool of necrotrophic fungi for efficient colonization of host tissue. Due to the redundancy of lytic enzymes, however, their role is experimentally difficult to assess. The two endopolygalacturonases, PG2 and PG1, have been shown to be involved in primary lesion formation and lesion outgrowth (ten Have et al. 1998; Kars et al. 2005a). In contrast, two pectin methylesterases (PME1 and PME2) were dispensable for pathogenesis, indicating that pectin demethylation is not a prerequisite for PG activity on the host cell wall, despite a previous report which indicated some importance for PME1 (Valette-Collet et al. 2003; Kars et al. 2005b). For cellulases, hemicellulases and proteases, no evidence for their role in the infection process has yet been obtained, although they contribute in decomposition of the killed host tissue.

While oxalic acid is an essential virulence factor for *S. sclerotiorum*, its role in *B. cinerea* is less prominent. A *B. cinerea* deletion mutant lacking oxaloacetate acetyl hydrolase was unable to produce oxalic acid (Han et al. 2007). On several host plants, this mutant had no effect on virulence, but it was reduced in lesion outgrowth on *Arabidopsis* and cucumber leaves (van Kan unpublished). Other *B. cinerea* mutants have been described that were defective in oxalic acid production and impaired in virulence, but a causal relationship has not been proven (Kunz et al. 2006; Schumacher et al. 2012). Recent studies with *S. sclerotiorum* indicate that oxalic acid, on one hand promotes the apoptotic oxidative burst in the host plant that supports necrotrophic development, and on the other hand suppresses an early, defense-associated oxidative burst during the initial stages of infection (Williams et al. 2011). *B. cinerea* can infect a variety of host tissues that greatly differ in their pH values, and the fungus itself modulates the pH during infection (Manteau et al. 2003; Billon-Grand et al. 2010). Infection of sunflower cotyledons by *B. cinerea* was accompanied by a first phase of acidification and accumulation of several organic acids, and a second phase of alkalinization and ammonium formation in the colonized tissue (Billon-Grand et al. 2010). Because expression and activity of pectinolytic and proteolytic enzymes is pH dependent, *B. cinerea* seems to be able to release different sets of virulence factors, which illustrates its remarkable pathogenic plasticity.

According to a long-standing hypothesis, *B. cinerea* and other necrotrophic fungi induce the plant hypersensitive response (HR) as part of their infection strategy (Govrin and Levine 2000; van Baarlen et al. 2007; Williamson et al. 2007). This is supported by the observation that cell death suppression in transgenic tobacco expressing animal anti-apoptotic genes results in resistance to necrotrophic fungi (Dickman et al. 2001).The role of necrotrophic effector proteins in triggering host plant HR will be discussed in Sect. 2.6.5. How *B. cinerea* overcomes the strong defense reactions that accompany the HR, such as the release of reactive oxygen species (ROS), secretion of PR proteins and production of phytoalexins, is poorly understood. Surprisingly, a TF with a central role in ROS detoxification, AP-1, was found to be not required for infection, which indicated that the fungus does not seem to suffer ROS stress during host invasion (Temme and Tudzynski, 2009). The situation is further complicated by the discovery that modulation of fungal apoptosis also plays a role in pathogenesis of *B. cinerea* (Shlezinger et al. 2011a).

Detoxification of host defense compounds has been demonstrated to be important for fungal pathogenesis. *B. cinerea* is capable of enzymatic conversion of the phytoalexin resveratrol into ε-viniferin by a substrate-inducible laccase, Lcc2. Surprisingly, loss of Lcc2 resulted in higher tolerance to resveratrol in vitro, because the mutant was unable to convert it to ε-viniferin which has a higher fungitoxicity than resveratrol (Schouten et al. 2002). Another detoxification mechanism is mediated by drug efflux transporters (De Waard et al. 2006). The *B. cinerea* ABC-type efflux transporter, AtrB, was found to lower the toxicity of camalexin, the major antifungal phytoalexin of Arabidopsis. *B. cinerea atrB* mutants are impaired in virulence on Arabidopsis wild type, but similarly aggressive as the wild type strain on the camalexin-deficient Arabidopsis *pad3* mutant (Stefanato et al. 2009). In addition to its role in infection, AtrB was found to be involved in a multidrug-resistance phenotype (MDR) in *B. cinerea*, which considerably reduces its sensitivity to several fungicides (see Sect. 2.5.3).

Being known as a devastating necrotroph, the discovery that *B. cinerea* is able to grow as an endophyte was surprising. In several host plants, such as *Lactuca* and *Primula* spp., the fungus has been found to grow in nonsymptomatic plant tissue (Sowley et al. 2010). A population genetic study indicated no differentiation between endophytic and necrotrophic forms of the fungus (Rajaguru and Shaw 2010). Another study recently demonstrated the occurrence of several new putative *Botrytis* spp. from a symptomless plant (Shipunov et al. 2008). It will be of great interest to understand the molecular mechanism of this 'hidden' life form, and transcriptomics would be a tool of choice to decipher the changes in regulatory pathways compared to the necrotrophic lifestyle. A prerequisite for further investigation, however, would be the establishment of a controlled infection leading to endophytic growth.

## 2.3 Technical Advances in Molecular Research on *Botrytis*

In the pregenomic era, the identification of putative virulence genes was tedious, requiring their cloning from genomic or cDNA libraries usually based on heterologous hybridization, or by PCR using gene-specific, degenerate primers. This changed rapidly with the availability of the genome sequence, which allowed easy access to genes encoding protein families, such as aspartic peptidases (ten Have et al. 2010), chitin synthases (Morcx et al. 2013), or to proteins involved in a common regulatory process, such as the BMP1 MAP kinase cascade (see below) or regulators of fungal apoptosis (Shlezinger et al. 2011b). Transformation of *B. cinerea*, either via protoplasts or by using *Agrobacterium tumefaciens*, is a standard tool, but methods for functional analysis of multiple genes remain to be optimized. At present, only double or, with some effort, triple mutants can be generated in *B. cinerea*, due to the limitation of commonly available selection markers (for hygromycin, nourseothricin, and phleomycin resistance). For example, a double knock-out mutant lacking cutinase A and the cutinolytic lipase 1 has been constructed, and was found to be fully virulent on several hosts (Reis et al. 2005). This result, however, did not exclude a role of cutinases in penetration because the *B. cinerea* genome contains several additional genes encoding cutin degrading enzymes. In another study, a mutant lacking all three hydrophobin encoding genes was constructed, to evaluate their role in the formation of hydrophobic surface layers in conidia and hyphae. The triple mutant did not show any differences in the hydrophobicity of conidia and aerial hyphae, which led to the conclusion that *in B. cinerea*, hydrophobins do not contribute to surface coating of these structures (Mosbach et al. 2011). For simultaneous mutational studies on multiple genes, serial gene replacement strategies by iterative use of a single resistance marker are now available for some filamentous fungi such as *Aspergillus fumigatus* (Krappmann et al. 2005), but remain to be developed for *B. cinerea*. Another strategy for targeting multiple genes, simultaneous RNA silencing, has been successfully applied in *Magnaporthe oryzae*. Simultaneous silencing of ten xylanase genes (Nguyen et al. 2011) and nine cellulase genes (van Vu et al. 2012) resulted in significant reduction of lesion formation, supporting a role of xylan and cellulose degradation in infection of the rice blast fungus. Although this approach suffers from the fact that complete suppression of the target genes is not achieved, it could be used to analyze the function of similar gene families in *B. cinerea*. A comprehensive, unbiased and genome-wide approach for analyzing gene function is random insertion mutagenesis. In a recent study, 2367 *A. tumefaciens*-induced *B. cinerea* insertion mutants were generated, of which 68 mutants showed a significant reduction in virulence (Giesbert et al. 2012). This mutant collection, although as yet not saturated, provides a valuable source for identification of novel genes involved in *B. cinerea* pathogenicity.

To improve the efficiency and versatility of constructs used for manipulation of *B. cinerea*, a set of vectors has been generated (Schumacher 2012). These vectors allow targeted integration of constructs into two nonessential gene loci (*niaD* encoding nitrate reductase, *niiA* encoding nitrite reductase), using three selection markers, and the integration of coding sequences into suitable expression cassettes. They simplify the fusion of proteins with codon-optimized reporters such as GFP for analysis of subcellular localization and in vivo protein-protein interactions. Using one of these vectors, a redox-sensitive GFP has been demonstrated to be functional in *B. cinerea*, which is useful for investigating the role of infection-related redox processes (Heller et al. 2012a). A valuable addition for this toolbox would be a chemically inducible promoter system, such as a tetracyclin-inducible promoter that has been developed for *Aspergillus fumigatus* (Vogt et al. 2005).

## 2.4 Lessons Learned from the *B. Cinerea* Genome

### 2.4.1 Genome Organization

Two *B. cinerea* strains were sequenced, with approximately five-year intervals. Strain B05.10 was among the first filamentous fungi to be sequenced, in the mid 1990s, whereas strain T4 was sequenced in the early years of the twenty-first century, when methods were more established, affordable and reliable, and more fungal genomes were available for comparison. This time lapse between the two genome projects had implications for the sequence strategy, methodology, and data quality, eventually leading to sequence assemblies of overall distinct quality (Amselem et al. 2011). While the assembly of B05.10 had low (4.5x) genome coverage and consisted of >4500 contigs that were merged into 588 scaffolds, the assembly of T4 (10x coverage) had 679 contigs merging into 118 scaffolds. In the case of T4, a genetic map allowed to further group the 118 scaffolds in 36 superscaffolds. A recent effort to resequence and reassemble the genomes of both these strains using Illumina technology (Staats and van Kan 2012) led to a much higher coverage (>100x), lower numbers of scaffolds (82 and 56 in B05.10 and T4, respectively) and concomitantly higher N50 (970 and 1710 kb, respectively). More importantly, the resequencing corrected many tens of thousands of wrong base calls and indels, closed hundreds of gaps and reduced the number of Ns from 3.9 to 1.6 million (version 1) to <0.5 million for both genomes (version 2; Staats and van Kan 2012; Table 2.1). The numbers of predicted protein-coding genes in version 1 genome assemblies were in the order of 16,400, of which some 2,100–2,800 were considered dubious (Amselem et al. 2011) and many hundreds were split gene models. The version 2 genomes resulting from the resequencing contain an estimate of 10,400 predicted protein-coding genes (Staats and van Kan 2012), however, it was noted that the gene prediction tool had failed to predict roughly 1,100 bonafide gene models for various technical (bioinformatics) reasons (van Kan, unpublished). The number of protein-coding genes in *B. cinerea* is thus estimated to mount to approximately 11,500 (Table 2.1). Further optimization of the genome assembly of strain B05.10 and the gene models and annotations is ongoing (Stassen and van Kan, unpublished).

The average GC contents of the *B. cinerea* and *S. sclerotiorum* genomes (~42 %), are remarkably low compared to the GC contents from other filamentous ascomycetes, which are above 50–52 %. Interestingly, they are similar to the GC content of another Leotiomycete, the obligate biotroph *Blumeria graminis*, indicating that this is a general character of this class within the ascomycetes. The evenly distributed, low GC content might have consequences for chromosome organization and gene expression. For example, poor expression of the commonly used eGFP gene in *B. cinerea* was found to be due to its high GC content. Synthesis of a gene with lower GC content and a codon usage adapted to that of *B. cinerea* resulted in strongly increased expression and fluorescence of eGFP (Leroch et al. 2011; Fig. 2.1e, f).

Despite their close relatedness, the *B. cinerea* and *S. sclerotiorum* genomes differ remarkably in their content of transposable elements. While 7 % of the genomic DNA in *S. sclerotiorum* is made up of transposable elements, *B. cinerea* strains T4 and B05.10 contain <1 % transposable elements. Phylogenetic analysis indicates that this difference was caused by recent bursts of transposition and amplification events in *S. sclerotiorum*. Despite the similar total number of predicted genes in the two species, *B. cinerea* tends to have higher numbers of genes than *S. sclerotiorum* for several classes of proteins

**Table 2.1** Assembly and gene statistics of *B. cinerea* and *S. sclerotiorum* genomes

| | *B. cinerea* | | *S. sclerotiorum* |
|---|---|---|---|
| | Strain T4 | Strain B05.10 | Strain 1980 |
| Coverage | 10x (**>100x**) | 4.5x (**>100x**) | 9.1x |
| Assembly size (Mb) | 39.5 (**41.6**) | 42.3 (**41.2**) | 38.3 |
| Scaffolds | 118 (**56**) | 588 (**82**) | 36 |
| Scaffold N50 (kb) | 562 (**1,710**) | 257 (**970**) | 1,630 |
| GC% | 43.2 (**42.4**) | 43.1 (**42.8**) | 41.8 |
| GC exons (%) | 46.2 | 46.2 | 45.7 |
| GC introns (%) | 40.0 | 40.9 | 39.9 |
| Transposable elements (% of total genome size) | 0.7 | 0.9 | 7.0 |
| Predicted protein coding regions | 13,664 | 14,270 | 11,860 |
| Median exon length (nt) | 208 | 190 | 182 |
| Median intron length (nt) | 62 | 74 | 78 |
| Carbohydrate Active Enzymes (CAZymes) | 367 | | 346 |
| Plant cell wall (PCW) degrading CAZymes | 118 | | 106 |
| Pectin degrading enzymes (percent of PCW CAZymes) | 44 (37 %) | | 33 (31 %) |
| Total secondary metabolism key enzymes | 43 | | 28 |
| PKS | 16 | | 15 |
| NRPS | 9 | | 6 |
| PKS-NRPS | 5 | | 3 |
| Terpene synthases | 11 | | 2 |
| Total transporters | 432 | | 377 |
| ABC transporters | 40 | | 33 |
| MFS transporters | 286 | | 218 |
| P-type transporters | 20 | | 17 |
| Total transcription factors | 419 | | 335 |
| $Zn(II)_2Cys_6$ TF | 222 | | 155 |
| $C_2H_2$ TF | 116 | | 98 |
| Secreted proteins (minus CAZymes and peptidases) | 879 | | 603 |
| Secreted proteins >300 aa | 521 | | 363 |

Data from Amselem et al. (2011), bold numbers in parenthesis are data from resequencing by Staats and van Kan (2012). Data for *B. cinerea* strain T4 (version 1, as published by Amselem et al. 2011) are accessible via http://urgi.versailles.inra.fr/Species/Botrytis/Sequences-Databases. Data for strain T4 (version 2, as published by Staats and van Kan 2012) and both versions of B05.10 (as published by Amselem et al. 2011; Staats and van Kan 2012) are accessible via http://www.broadinstitute.org/annotation/genome/botrytis_cinerea/MultiHome.html

(Table 2.1). The content of carbohydrate-active enzymes (CAZymes; Cantarel et al. 2009) encoding genes is of special interest because of their potential role in host cell wall penetration and plant tissue breakdown. Overall, the CA-Zome of *B. cinerea* and *S. sclerotiorum* is not exceptionally large as compared to other necrotrophic and hemibiotrophic ascomycetes. However, the proportion of pectin-degrading enzymes is high (37 and 31 %, respectively, of *B. cinerea* and *S. sclerotiorum* plant cell wall CAZymes), being exceeded currently only by *Aspergillus niger* (38 %), a saprophyte that also develops well on fruits.

Analyses of fungal genomes for the content of genes encoding membrane transporters did not reveal major differences between saprotrophs and pathogens. Both, ABC transporters (40 in *B. cinerea*, 33 in *S. sclerotiorum*) and MFS transporters (286 in *B. cinerea*, 218 in *S.*

*sclerotiorum*) occur in higher numbers in *B. cinerea* than in *S. sclerotiorum*. In contrast, no pronounced differences were observed for P-type ATPases, amino acid transporters, mitochondrial carriers, and aquaporins (Amselem et al. 2011). Since the majority of *B. cinerea* transporter genes that are absent in *S. sclerotiorum* do not have orthologs in other fungi, gene expansion in *B. cinerea* rather than gene loss in *S. sclerotiorum* has probably led to the observed differences.

Another striking difference between both species is in their content of transcription factors (TFs). A total of (approximately) 419 predicted TF-encoding genes were identified in *B. cinerea*, but only 330 in *S. sclerotiorum*. This difference is mostly due to the number of genes of the $Zn(II)_2Cys_6$ binuclear cluster family (*B. cinerea* 222; *S. sclerotiorum* 155) and the Zn(II)-coordinating $C_2H_2$ family (*B. cinerea* 116; *S. sclerotiorum* 98). *B. cinerea* has 96 TFs that are absent in *S. sclerotiorum*, while only 12 TFs are unique to *S. sclerotiorum*. Among the 96 *B. cinerea* TFs that have no counterpart in *S. sclerotiorum*, 57 have orthologs in other fungal genomes. This suggests that these 57 TFs have been lost in *S. sclerotiorum*, while the remaining 39 TFs resulted from gene duplication or horizontal gene transfer in *B. cinerea* after separation between the two species. Nine of these *B. cinerea*-specific TF genes are located in SM gene clusters that are not present in *S. sclerotiorum* (Sect. 2.4.3).

Effector proteins that are translocated into plant host cells are typically major determinants of pathogenesis in biotrophic and hemibiotrophic fungi. The potential role of necrotrophic effectors for *Botrytis* spp. is discussed in Sect. 2.6.5. The size of the secretome (secreted proteins without CAZymes and peptidases) of *B. cinerea* is considerably larger (879) than that of *S. sclerotiorum* (603). So far, the role of most of these secreted proteins in *B. cinerea* remains unclear.

Overall, the genome sequences and their gene contents of *B. cinerea* and *S. sclerotiorum* did not reveal unique features that could explain their lifestyles as wide host range necrotrophs, and set them apart from other pathogenic of saprotrophic fungi.

### 2.4.2 Mating Type Genes

*B. cinerea* is considered to be a heterothallic (obligate outcrossing) species with two distinct mating types, designated MAT1-1 and MAT1-2. The two sequenced *B. cinerea* strains are of opposite mating type (B05.10 is of MAT1-1 identity and T4 is MAT1-2), and have the typical *MAT* organization of heterothallic Pezizomycotina, with the presence of dissimilar idiomorph sequences at a single locus. The *MAT1-1* idiomorph of B05.10 contains a *MAT1-1-1* alpha-domain encoding gene, whereas the *MAT1-2* idiomorph of T4 contains a *MAT1-2-1* HMG-domain encoding gene. Besides these archetypal genes, two novel open reading frames (ORFs) were detected that have no known PFAM domain and were not previously reported from other fungi. The first novel ORF is located in the *MAT1-1* idiomorph of B05.10 and was designated *MAT1-1-5*. The second novel ORF is located in the *MAT1-2* idiomorph of T4 and named *MAT1-2-4*. Gene knock-out mutants in the *MAT1-1-5* and *MAT1-2-4* genes showed identical phenotypes, i.e., they produced apothecial stipes of normal appearance, but the sti pes were arrested in the transition to apothecial disk development (Terhem and van Kan, unpublished). These observations suggest that *MAT1-1-5* and the *MAT1-2-4* gene might be transcriptional regulators that jointly control differentiation of the apothecial disk. Transcriptome analysis of *B. cinerea* apothecia in different developmental stages is ongoing (van Kan, unpublished).

A model was proposed for the evolution of the *B. cinerea* mating system from a homothallic ancestral locus, much like the locus present in contemporary *Sclerotinia sclerotiorum*, following (hypothetical) occurrence of two simple inversions and two distinct deletion events, yielding the two separate (MAT1-1 and MAT1-2) heterothallic configurations of the locus

(Amselem et al. 2011). This model was recently challenged by Chitrampalam et al. (2013), who reported the occurrence of an inversion of a 3.6 kb segment of the *S. sclerotiorum* MAT locus in every meiotic generation in half the progeny, due to a recombination event between two short inverted repeats within the MAT locus. Based on these observations, Chitrampalam et al. (2013) postulated that the homothallic behavior of *S. sclerotiorum* evolved from a heterothallic ancestor by a recombination event that brought both idiomorphs together in a single genome.

### 2.4.3 Secondary Metabolism

Before sequencing of its genome, *B. cinerea* was known to produce three SM with potential or confirmed role in the interaction with host plants: two families of sesquiterpenes, i.e., botrydial (Colmenares et al. 2002) and abscisic acid (ABA; Siewers et al. 2006) and one family of polyketides, i.e., botcinic acid (Tani et al. 2006). While ABA is a plant hormone, botrydial and botcinic acid are nonspecific toxins that have been only isolated in *Botrytis* so far. Identification of the genes involved in the biosynthesis of these particular compounds was initiated by pregenomics approaches such as the use of ESTs for library screening (*bot1*; Siewers et al. 2005) but the sequence of the whole genome revealed that *B. cinerea* has a significantly higher potential for SM biosynthesis. In contrast to genes involved in primary metabolism, genes that contribute to the biosynthesis of a single SM or multiple structurally closely-related SMs are usually clustered at one genomic locus (Hoffmeister and Keller 2007; Brakhage 2013) which helps to identify them on the genome sequence. The enzyme responsible for the committed biosynthetic step, often referred to as the ''key'' enzyme, can be a PolyKetide Synthase (PKS), a Non-Ribosomal Peptide Synthetase (NRPS), a hybrid PKS/NRPS, a DiMethylAllylTryptophan Synthase (DMATS), or a terpene synthase. The *B. cinerea* genome contains 43 genes that were predicted to encode key enzymes: 16 type I iterative PKS, nine NRPS, five PKS/NRPS, one DMATS, one type III PKS (chalcone synthase), and 11 terpene synthases. Most of these genes (37) are clustered with other genes encoding biosynthesis enzymes such as P450 monooxygenases. About one-third of the clusters also encodes an ABC or MFS transporter and a similar proportion encodes a $Zn(II)_2Cys_6$ zinc finger TF.

The total number of SM clusters is similar to the average in ascomycetes but strikingly, *B. cinerea* has the highest number (i.e., six) of terpene synthases from the sesquiterpene cyclase (STC) subfamily, although similar numbers of STC genes were identified in basidiomycetes (Agger et al. 2009). One of the six *B. cinerea* STC, BOT2, is the first presilphiperfolan-8$\beta$-ol synthase identified in fungi and the key enzyme for botrydial synthesis (Pinedo et al. 2008). The gene encoding BOT2 is clustered with four other genes (Fig. 2.2), all required for the successive steps of botrydial synthesis (Siewers et al. 2005; I. G. Collado, *unpublished*). Due to AT-rich regions upstream and downstream of these five *bot* genes, the surrounding genomic regions and genes could not be investigated through the different sequencing projects or PCR approaches. Although the five *bot* genes may represent the complete toolbox for botrydial synthesis, additional (unidentified) genes are likely required for regulation of the cluster and toxin export. The terpenes produced by the five other STC and the three DiTerpene Cyclases (DTC) are unknown, although one of the STC is probably the key enzyme for ABA biosynthesis. Finally, two other clusters involved in the biosynthesis of terpenes were predicted. The first one is a carotenoid gene cluster with genes encoding a phytoene synthase, a putative oxidase, and an opsin photoreceptor with a conserved retinal-binding domain. This cluster could be involved in retinal biosynthesis as described in *Fusarium fujikuroi* (Prado-Cabrero et al. 2007). The second cluster contains a terpene synthase encoding gene orthologous to the *Penicillium paxilli PaxC* gene, involved in biosynthesis of the indole-diterpene paxillin (Saikia et al. 2008).

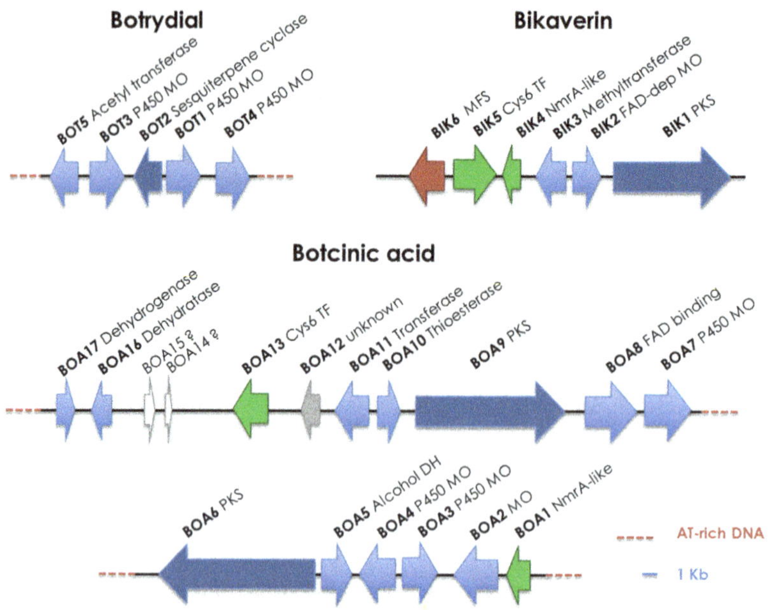


**Fig. 2.2** Examples of secondary metabolism gene clusters in *B. cinerea*. Genes encoding key enzymes (sesquiterpene cyclase and polyketide synthases) are in *dark blue* while genes encoding other biosynthetic enzymes (P450 monooxygenases, dehydrogenases, transferases…) are in *light blue*. Genes encoding regulators (Zn(II)$_2$Cys$_6$ transcription factors, NmrA-like regulators) are in *green*, and a Major Facilitator Superfamily transporter gene is in *red*. *Dotted lines* indicate AT-rich regions exhibiting AT contents higher than 70 % that could not be sequenced. The cluster responsible for the synthesis of the pigment bikaverin is from the *pinkish strain* 1750. Note that in *gray strains* such as B05.10 and T4, *bik1* is missing

The repertoire of *B. cinerea* PKS genes predicted from the genome of B05.10 (Kroken et al. 2003) enabled identification through mutagenesis of the genes responsible for the production of the reduced polyketide botcinic acid. This biosynthesis requires two PKSs, BOA6, and BOA9 (Fig. 2.2; Dalmais et al. 2011). The iterative nature of fungal PKSs means that, in the majority of cases, only a single PKS is involved in the synthesis of one particular polyketide. However, a few fungal polyketides such as lovastatin in *Aspergillus terreus* (Sutherland et al. 2001) are known to be assembled by the action of two PKSs. Isotopic labeling experiments showed that BOA6 produces the pentaketide core of botcinic acid, while BOA9 was responsible for the tetraketide side chain (Massaroli et al. 2013). In the successive T4 and B05.10 genome assemblies (Amselem et al. 2011), the genes encoding BOA6 and BOA9 were on different small contigs, each being adjacent to other putative botcinic acid biosynthetic genes (*boa1*-17). The compounds produced by the other PKS genes in the genomes of B05.10 and T4 remain to be identified. Among the nonreducing PKSs, phylogenetic analysis suggested that PKS12 and PKS13, however, are candidates for melanin biosynthesis (Kroken et al. 2003). Two other genes encoding enzymes involved in DHN melanin synthesis (scytalone dehydratase and hydroxynaphthalene reductase; Pihet et al. 2009) are adjacent to the *pks13* gene.

The *B. cinerea* genome contains nine mono- or multimodular NRPS encoding genes. Recent phylogenetic analysis of fungal NRPS by Bushley and Turgeon (2010) indicates that NRPS6 is involved in the biosynthesis of coprogen siderophore, while NRPS2 and NRPS3 belong to the intracellular siderophore synthetases subfamily.

In conclusion, sequencing the *B. cinerea* genome revealed that, in addition to the two previously known phytotoxins i.e. botrydial and botcinic acid, the fungus has the capacity to produce about 38 other SM. Several of them are shared by many ascomycetes i.e. melanin and siderophores, while the majority represent compounds with unknown structures and functions.

The role of the SM in pathogenicity of *B. cinerea* was investigated by inactivation of genes encoding SM biosynthesis enzymes. This approach allowed the identification of the botrydial (Siewers et al. 2005; Pinedo et al. 2008) and the botcinic acid gene clusters (Dalmais et al. 2011) described in Fig. 2.2. Deletion of the key enzyme BOA6 or BOT2 encoding genes did not modify significantly the virulence of B05.10, while *boa6 bot2* double mutants that do not produce either of the two toxins were severely impaired in virulence on several hosts suggesting that the compounds have a redundant function in plant tissue colonization (Dalmais et al. 2011).

Transcriptomic studies based on microarrays (see Sect. 2.5) indicate that the SM biosynthesis clusters have different expression patterns. Some, like the botrydial and botcinic acid clusters, are expressed on several media as well as during plant infection, while other clusters are more specifically expressed at one specific stage of the life cycle or under specific nutritional conditions (e.g., nitrogen starvation). Of particular interest were the clusters that are upregulated during appressorium formation and/or host colonization. Out of seven PKS- and STC-encoding genes inactivated so far, three resulted in mutants that were altered in virulence and sporulation (Viaud unpublished). The polyketide and sesquiterpenes produced by the three corresponding clusters remain to be identified as well as their mode of action. In conclusion, although botrydial and botcinic acid toxins play together a significant role in virulence, other SM seems to be crucial in *B. cinerea* lifestyle. The current hypothesis is that some compounds have both a role in necrotrophy and in fungal development.

## 2.5 Genome-Enabled Approaches/Applications of the Genome

### 2.5.1 Transcriptomics

Before the advent of high-throughput sequencing, approaches to identify genes expressed in a stage-specific manner were tedious. For example, suppression subtractive hybridization was used to identify genes controlled by the G$\alpha$1 subunit of the heterotrimeric G protein, BCG1 (Schulze Gronover et al. 2004). Transcriptomic approaches started before the availability of the *B. cinerea* genome sequence, using EST sequences bound to Nylon filters as macroarrays. They were used for the discovery of genes that were upregulated during infection of *Arabidopsis* leaves (Gioti et al. 2006), genes that were regulated by BCG1 and the calcineurin phosphatase (Viaud et al. 2003; Schumacher et al. 2008a), and genes controlled by the ROS regulator AP-1 (Temme and Tudzynski 2009).

After the annotated *B. cinerea* genome sequences became available, Nimblegen microarrays were designed, containing 21,200 gene models identified originally in the B05.10 and T4 genomes (Amselem et al. 2011). RNA isolated either from mycelium grown on agar or from inoculated sunflower cotyledons (48 hpi) was used as labeled hybridization probes for a comparative transcriptome analysis of saprotrophic and necrotrophic growth. A total of 253 genes were upregulated, and 247 genes were downregulated in leaves. The upregulated genes were enriched in genes encoding CAZymes, P-type ATPases, MFS-type sugar transporters and peptidases. When compared to the results of a parallel experiment performed with *S. sclerotiorum*, an overlap in patterns of *in planta* upregulated genes was observed only for the CAZyme-encoding genes. The transcriptomic data provided no evidence for a major role of small secreted proteins during infection. Among the 521 genes encoding secreted proteins smaller than 300 amino acids in *B. cinerea*, only 20 (4 %) were significantly upregulated *in planta*

(Amselem et al. 2011). This is in contrast to the related biotroph, *Blumeria graminis*, which was found to encode 195 candidate secreted effector proteins (with no homologs to nonpowdery mildew fungi) that were strongly upregulated in haustoria (Spanu et al. 2010).

In another study, microarrays were used for transcriptome profiling during germination of *B. cinerea* conidia (Leroch et al. 2013a). On apple wax-coated surfaces, which mimicked the hydrophobic plant cuticle and allowed rapid germination in a minimal medium containing fructose, massive changes in gene expression were observed within the first hour of incubation, before germ tube emergence, with 2,864 genes (23.7 % of the expressed genes) being up- or downregulated. Smaller changes were observed in the time window between 1 h and 4 h, when the germ tubes grew out and appressoria were formed. Among the genes that were induced during germination, a significant enrichment was observed of genes encoding secreted proteins, including many CAZymes. A comparison between the wild type and a mutant deficient in the MAP kinase BMP1 revealed that a high proportion of the germination-induced genes encoding secreted proteins were not upregulated in the *bmp1* mutant. These data indicated that germination is accompanied by the production of many secreted proteins, and the BMP1 MAP kinase cascade plays an important role in this process (Leroch et al. 2013a; Sect. 2.6.1).

### 2.5.2 Proteomics

Proteomics approaches were launched by several groups soon after the annotated *B. cinerea* genome sequences were available (Fernandez–Acero et al. 2009; Shah et al. 2009; Espino et al. 2010; Fernandez–Acero et al. 2010). They primarily focused on the fungal secretome, as secreted proteins are often considered to be crucial determinants in host plant invasion. In some of these studies, a host environment was simulated by culturing *B. cinerea* in the presence of plant extracts (Shah et al. 2009; Espino et al. 2010; Shah et al. 2012). The observed protein profiles demonstrated that the *B. cinerea* secretome consists of a common set of proteins, but varies in response to different nutrients. In several studies, proteins which have been previously characterized by mutant analysis as virulence factors, including the polygalacturonases PG1 and PG2, the xylanase Xyn11, and the cerato-platanin family protein Spl1 were detected. A recent study investigated the response of the *B. cinerea* secretome to cultivation at different pH values, and revealed distinct profiles between pH 4 and pH 6. Among the 21 proteins that were uniquely detected at only one pH value, proteins involved in proteolysis were found to be induced at pH 4, while most of the proteins produced at pH 6 were cell wall degrading enzymes (Li et al. 2012a). Most of these proteome investigations were, however, performed when only the first version of the B05.10 genome was publicly accessible. Several proteins may have remained undetected of which the peptide mass peaks could not be assigned because of incorrect or incomplete gene models resulting from the low genome coverage. Future studies employing new versions of the genome assemblies will increase sensitivity of proteomics approaches.

An in silico approach was followed to predict secreted proteins with O-glycosylation based on their content of serine/threonine-rich regions. According to prediction by NetOGlyc, most secretory proteins in filamentous fungi are O-glycosylated (González et al. 2012). The importance of O-glycosylation in fungi was established by the analysis of mutants defective in genes encoding protein O-mannosyltransferases (PMT), which showed reduced ability to add mannoses to Ser/Thr residues during the secretion pathway. O-glycosylation of secreted proteins seems to increase their stability and solubility. In *Ustilago maydis*, deletion of a singly PMT (Pmt4) resulted in complete loss of pathogenicity, without affecting growth rate and morphology (Fernández-Álvarez et al. 2009). A substrate of *U. maydis* Pmt4 that was partly responsible for the defect in penetration, was identified as the signaling mucin Msb2 (Fernández-Álvarez et al. 2012; Sect. 2.6.1). Individual inactivation of the three *pmt* genes in *B. cinerea* revealed that they are important for cell wall stability, extracellular matrix formation, adherence to plant surface, and host invasion. These data support an important role of O-glycosylated secreted proteins in the infection process (González et al. 2013).

Protein phosphorylation is another important posttranslational modification. In particular, signal transduction cascades modulate cellular responses to environmental stimuli through phosphorylation and dephosphorylation. Tracing protein phosphorylation profiles through proteomics has been developed for model yeasts (Amoutzias et al. 2012), but only few data are available for filamentous fungi (Rampitsch et al. 2012). Off-gel phosphoproteomics, which is more sensitive than 2D-gel phosphoprotein detection (Villen and Gygi 2008), were applied to *B. cinerea* proteins extracted from exponentially grown mycelia and resulted in a set of 1361 phosphopeptides representing 818 phosphoproteins (82.7 % phosphoserine, 15.8 % phosphothreonine, and 1.5 % phosphotyrosine) (S. Fillinger, pers. comm.). Comparative phosphoproteomics of different growth conditions coupled with signal-transduction mutants will reveal further insight into the precise signaling pathways and their interactions.

### 2.5.3 Map-Based Cloning

The availability of genome sequences and the possibility to perform sexual crosses with *B. cinerea* under controlled laboratory conditions has allowed the application of map-based-cloning approaches for interesting genes. Two examples are presented here. The first example concerns the TF Mrr1, the determinant of a multidrug-resistance phenotype called MDR1. *B. cinerea* strains showing low-to-medium levels of resistance to several fungicides, were isolated for the first time in Champagne vineyards in the mid 1990s, and are now widely distributed in French and German wine producing regions. In contrast to the common type of fungicide resistance caused by mutations that alter the fungicide target protein, MDR strains show increased fungicide efflux (Chapeland et al. 1999). This was confirmed by the observation that *B. cinerea* strains with MDR1 showed a massive constitutive overexpression of the gene encoding the ABC-type efflux transporter AtrB. Crosses between MDR1 and sensitive strains revealed that MDR1 is controlled by a single dominant gene. Microsatellite-assisted mapping was able to pinpoint the gene down to a region of more than 100 kb. Based on the knowledge from *Candida albicans* MDR phenotypes, previously described to be caused by gain-of-function mutations in TF-encoding genes (Morschhäuser 2010), it was hypothesized that similar mutations could be responsible for *B. cinerea* MDR1. Among the few candidate genes in the mapped region, one gene encoding a $Zn(II)_2Cys_6$ zinc cluster TF, called Mrr1 (MDR1-related regulator), was found to carry MDR1-related point mutations. Indeed, all *B. cinerea* field isolates with MDR1 phenotypes contained at least one mutation that resulted in an amino acid substitution in Mrr1, compared to the Mrr1 sequences of sensitive strains T4 and B05.10. Knock-out mutants in *mrr1* failed to activate *atrB*, even in the presence of inducing fungicides, and confirmed the regulatory role of Mrr1 in efflux-based fungicide tolerance (Kretschmer et al. 2009).

Another example of map-based cloning assisted by genome sequence data pertains to the phenotypic differences between the two sequenced *B. cinerea* strains, B05.10 and T4. While B05.10 is highly virulent and conidiates in a light-dependent manner, T4 shows weak virulence, lack of sclerotia formation and oxalic acid production, and light-independent conidia formation. All these phenotypes co-segregated in the progeny of a cross, in a 1:1 ratio, indicating that they are all collectively controlled by a single gene locus. Polymorphic microsatellite markers were generated that co-segregated with the locus, and allowed the locus to be narrowed down to a 115 kb genomic region which included 44 predicted genes in the T4 genome sequence. A single nucleotide polymorphism (SNP) between the two strains was found in a gene encoding the velvet protein, VEL1. This SNP resulted in a truncation of the protein (184 amino acids) in T4 instead of the 575 amino acids of the full-length protein (Schumacher et al. 2012). As described below, VEL1 is also involved in the regulation of SM synthesis. With the decreasing costs of generating genome sequences, we expect that map-based cloning, or similar approaches that allow the rapid correlation of point mutations with phenotype alterations, will be used more and more frequently as a tool in fungal biology.

## 2.6 The Impact of '-Omics' Approaches in Our Understanding of *Botrytis* Biology

### 2.6.1 MAP Kinase Signal Transduction Pathways

Fungi contain highly conserved mitogen-activated protein kinase (MAPK) cascades that serve as key signal transducers in a variety of cellular processes. Because of their conservation, most MAPK components in filamentous fungi have been identified by comparative genomics, using yeast as a reference. MAPK cascades consist of three interlinked protein kinases that are sequentially activated by phosphorylation, namely MAPK kinase kinases (MAP3 K), MAPK kinase (MAP2 K), and MAPK. In filamentous fungi, three major MAPK pathways exist. In addition to their crucial roles in the regulation of basic developmental processes, they control various aspects of infectious growth in plant pathogens. The MAPK cascade controlling filamentous and invasive growth in yeast, Sc-Ste11 (MAP3 K)- Sc-Ste7 (MAP2 K)- Sc-Kss1 (MAPK), is mirrored in *B. cinerea* by the orthologous Ste11-Ste7-BMP1 cascade. Mutants in each of these components, and in the associated adapter protein Ste50, are unable to differentiate appressoria, do not penetrate into host cells, and do not even colonize wounded host tissue. In addition, such mutants show reduced growth, impaired germination on hydrophobic surfaces, lack of sclerotium formation and sexual development, and inability to form hyphal fusions (Doehlemann et al. 2006; Schamber et al. 2010; U. Siegmund, personal communication). The similarity of phenotypes of the *ste11*, *ste7*, *bmp1,* and *ste50* mutants indicates that there are no branch points in the cascade upstream of BMP1, and that, in contrast to yeast, Ste11 displays no cross talk with the high osmolarity MAP kinase pathway (see below). The TF Ste12, one of the major targets of yeast Kss1, also seems to be under MAPK control in filamentous fungi. In *B. cinerea*, *ste12* mutants have defects in host penetration, and show reduced virulence (Wong Sak Hoi and Dumas 2010; Schamber et al. 2010). Recent data obtained with *Aspergillus nidulans* impressively demonstrate in vivo interaction of all four MAPK components at the plasma membrane during initiation of sexual development, and the interaction of AnFus3 with AnSte12 and VeA (the homolog of VEL1, see Sects. 2.5.3 and 2.6.2) in the nucleus (Bayram et al. 2012). It can be expected that similar dynamic interactions involving MAPK modules occur in *B. cinerea* during pathogenic differentiation.

To identify target genes of the BMP1 MAPK pathway that play a role in the infection process, a microarray-based transcriptomics approach was followed (see Sect. 2.5.1). The sensory mechanisms and the upstream components that activate the BMP1 MAPK pathway are still poorly understood. In yeast, *U. maydis* and *M. oryzae*, the sensory mucin Msb2 has been shown to be involved in surface sensing which is transduced via the filamentous and invasive growth MAPK pathway (Lanver et al. 2010; Liu et al. 2011a). In *B. cinerea*, a similar function for Msb2 was indicated by the phenotype of an *msb2* mutant, which failed to form appressoria on artificial or plant surfaces, and showed reduced phosphorylation of BMP1 during germination on hard surfaces (Leroch and Hahn, unpublished).

The fungal high osmolarity MAPK pathway of fungi has attracted particular attention because of its involvement in the sensitivity and resistance to phenylpyrrole (e.g., fludioxonil) and dicarboximide (e.g., iprodione) fungicides. The homolog of the high osmolarity pathway MAPK in yeast, Sc-Hog1, was called Sak1 ('stress-activated kinase'). Mutants in Sak1 are hypersensitive to hyperosmotic and oxidative stress, they do not form conidiospores, are unable to differentiate appressoria, and fail to invade intact host tissue. Nevertheless, they retain the ability to form primary and secondary lesions if inoculated on wounded tissue (Segmüller et al. 2007). Similar phenotypes were observed for mutants defective in the kinases upstream of Sak1, BOS5 (MAP2 K) and Os4 (MAP3 K) (Yan et al. 2010; Yang et al. 2012). The *bos5* and *os4* mutants were unable to phosphorylate Sak1 in response to osmotic stress, which confirmed their upstream position in the pathway. The Bos1 histidine kinase, which is believed to be involved in osmosensing, has also been shown to be required for full virulence in *B. cinerea* (Viaud et al. 2006; Liu et al. 2008). In contrast to the situation in other filamentous fungi, Bos1 seems to negatively regulate the activity of Sak1, because Sak1 phosphorylation levels were increased in the *bos1* mutant (Liu et al. 2008). In addition to its role in osmoregulation, Bos1 is also involved in the sensitivity to fludioxonil and iprodione. Several mutations in *bos1* were mapped in mutants that were selected for resistance to fludioxonil and iprodione. Site-directed mutagenesis confirmed that these mutations lead to resistance to the fungicides and to hypersensitivity to osmotic stress (Fillinger et al. 2012). Another component of this pathway is the putative response regulator Brrg-1 (Yan et al. 2011). Transcriptomic studies with macroarrays with the *sak1* mutant revealed that the majority of Sak1-regulated genes are unrelated to stress responses, but rather involved in general metabolic functions (Heller et al. 2012b).

The 'cell wall integrity' pathway, represented by the (Stl2-like) Bmp3 MAPK, is poorly analyzed in *B. cinerea*. Mutant analysis revealed an involvement of Bmp3 in vegetative growth, penetration, and lesion development, but surprisingly, no role in cell wall integrity (Rui and Hahn 2007). Instead of Bmp3, Sak1 was identified as the MAPK mainly involved in cell wall integrity maintenance, as indicated by aberrations in the cell wall structure in *sak1* mutants, and the observation that Sak1 was found to be phosphorylated in response to cell wall stress (Liu et al. 2011b). The three MAPK cascades show several interconnections. All are involved in surface sensing by germ tubes or hyphae leading to the appressorium formation, and in the regulation of melanin biosynthesis genes. Sak1 was found to interfere with the phosphorylation status of BMP3 during cell wall stress, hinting to cross talk between both MAPK pathways (Liu et al. 2008; Liu et al.,2011b). Both Sak1 and Bmp3 are required for normal expression of Reg1, a transcriptional regulator involved in secondary metabolism (SM) and pathogenicity (Michielse et al. 2011; see Sect. 2.6.1).

### 2.6.2 Regulation of Secondary Metabolism

In recent years, important progress has been made in the identification of signal transduction pathways that regulate the production of the two identified phytotoxins botrydial and botcinic acid. Expression of the *bcbot* and *bcboa* genes has been shown to depend on two pathways. The first involves the $\alpha$ subunit BCG1 of a heterotrimeric G protein, phospholipase C (PLC1), calcineurin phosphatase (Viaud et al. 2003; Pinedo et al. 2008; Schumacher et al. 2008a; Dalmais et al. 2011) and the downstream Cys2His2 zinc finger TF CRZ1 (Schumacher et al. 2008c). The second transduction pathway is the stress-activated MAPK cascade (Sect. 2.6.1) which controls the basic region leucine zipper (bZIP) TF Atf1 (Temme et al. 2012) and the transcriptional regulator Reg1 (Michielse et al. 2011). The pleiotropic phenotypes of the *crz1*Δ, *atf1*Δ and *reg1*Δ mutants and analyses of the CRZ1- and Atf1-dependent genes has confirmed that these three proteins regulate broad sets of genes involved in different cellular processes (Schumacher et al. 2008b; Temme et al. 2012; Michielse et al. 2011), suggesting that they play only an indirect role in *bcbot* and *bcboa* genes expression. Indeed, in fungi, pathway-specific TF (often of the $Zn(II)_2Cys_6$ family) and broad-domain TF responsive to general environmental factors seem to act jointly to regulate SM biosynthesis (Yin and Keller 2011). In *B. cinerea*, 11 SM clusters, including the botcinic acid cluster (Fig. 2.2), contain a gene encoding a putative pathway-specific $Zn(II)_2Cys_6$ TF, whereas the remaining 29 SM clusters contain no identifiable TF-encoding gene. Gene inactivation of *boa13* encoding a $Zn(II)_2Cys_6$ zinc finger TF showed that it indeed regulates all the *boa* genes (Viaud, unpublished). Unlike the botcinic acid cluster, the botrydial biosynthesis gene cluster contains no TF gene that could specifically co-regulate the expression of *bcbot* genes. To identify which TF(s) interact with the promoters of this cluster, a postgenomic strategy based on the Yeast One-Hybrid (Y1H) method was developed (Simon et al. 2013). A Y1H library consisting of 393 TF DNA-binding sites out of the 406 predicted from the *B. cinerea* genome was screened with the promoter of *bot2* which allowed to identify a Cys2His2 TF called YOH1. Inactivation of *yoh1* confirmed that this TF positively regulates *bot* genes but also revealed that it is not a pathway-specific TF. Indeed, microarray data showed that YOH1 regulates the expression of genes from more than half (22 of 40) the SM clusters of *B. cinerea* including the botrydial and botcinic acid clusters, suggesting that this TF is a global regulator of SM. In addition, the list of genes regulated by YOH1 is enriched in two other main categories of genes encoding proteins thought to be associated with the necrotrophic processes of *B. cinerea*: CAZymes and MFS transporters, involved in carbohydrate uptake or in efflux of SM from the hyphae into the host.

The global regulator *velvet* also takes part in regulation of SM production in *B. cinerea.* This protein complex is conserved in filamentous fungi and coordinates SM and light-dependent development (Bayram and Braus 2012). The *velvet* complex of *A. nidulans* consists of at least three proteins: LaeA, VeA, and VelB; LaeA links light regulation to chromatin modification (Brakhage 2013). The VeA ortholog in *B. cinerea* (*vel1*) was identified by a map-based cloning approach (Sect. 2.5.3). The T4 strain harbors a truncated BcVEL1 protein (184 aa instead of 575 aa), which accounts for the lack of sclerotia development and oxalic acid formation, as well as the reduced virulence (Schumacher et al. 2012). Strikingly, a *B. cinerea* strain producing a red pigment, bikaverin, which has been previously characterized in *Fusarium* spp., was recently shown to be another natural *velvet* mutant in which the VEL1 protein is even shorter (100 aa). By comparing the phenotypes due to the different truncated BcVEL1 proteins or to the complete KO, it was concluded that the presence of the C-terminal part (aa 185–575) of BcVEL1 is a prerequisite for nuclear localization, sclerotia formation, oxalic acid formation, and full virulence while only the first 100 aa of BcVEL1 are required for regulation of pigment formation, i.e., repression of melanin formation and stimulation of bikaverin formation (Schumacher et al. 2013). Further knowledge on the transcriptional and epigenetic regulation mechanisms that control SM biosynthesis is crucial to understand the development and virulence of the gray mold fungus. This knowledge could also provide approaches to "wake up" SM clusters during in vitro growth, so that the compounds could be characterized by chemists. Possible approaches include genetic engineering (e.g., overexpression of key enzymes encoding genes or pathway-specific TFs), modification of the chromatin structure or stimulation by coculture with other microorganisms (Brakhage 2013).

### 2.6.3 Intra- and Interspecific Diversity of *B. Cinerea* and Related Species

Rowe and Kliebenstein (2010) illustrated that intraspecific diversity is often ignored in studies with necrotrophic pathogens. In a survey of literature from 2000–2010 about plant defense against *B. cinerea*, only 12 % of the studies reported experiments with more than one isolate. Studies on biotrophic pathogens are more often performed with multiple isolates to study gene-for-gene interactions of pathogen isolates with host genotypes. However, this is usually not the case in studies with necrotrophs such as *B. cinerea*, for which the genetic basis of (low) host specificity is unknown. Indeed, it remains to be clarified whether the reported host range of more than 200 plant species could be attributed to a single isolate of *B. cinerea*. Isolate-specific interactions between *Arabidopsis thaliana* and *B. cinerea* were observed upon infection of *Arabidopsis* mutants in jasmonate-dependent defense with diverse isolates of *B. cinerea*. Pathogen isolates were inhibited to variable extent by jasmonate-mediated defense responses. Transcriptome analysis of plants infected by two distinct *B. cinerea* isolates showed only minor differences in transcriptional responses of wild-type *Arabidopsis*, but notable isolate-specific transcript differences in jasmonate-insensitive mutants (Rowe et al. 2010).

Field isolates of *B. cinerea* have been shown to exhibit considerable phenotypic variability in vegetative growth, conidiation, sclerotium formation, SM, and virulence (Kerssies and Bosker-van Zessen 1997; Martinez et al. 2003; Schumacher et al. 2013). This is reflected by a similar variability in genotypic characters, which have been analyzed by PCR-assisted techniques such as random amplification of polymorphic DNA, amplified restriction length polymorphism, detection of transposable elements and microsatellite heterogeneity (Kerssies and Bosker-van Zessen 1997; Giraud et al. 1997; Fournier et al. 2002; Moyano et al. 2003). In some of these studies, the majority of field isolates could be assigned to different haplotypes (Giraud et al. 1999; Kretschmer and Hahn 2008). However, a subdivision of *B. cinerea* into distinct subgroups turned out to be difficult. Studies performed with isolates from French vineyards, initially indicated the occurrence of genetically distinct groups, which could be distinguished by the presence (*transposa* strains) or absence (*vacuma* strains) of two transposable elements called Boty and Flipper (Giraud et al. 1997, 1999). This distinction between *transposa* and *vacuma* isolates, although widely used, turned out to be inadequate for subdivision of *B. cinerea* populations into genetically distinct groups. Indeed, gene sequencing demonstrated that only a subgroup (called 'group I') mostly of the *vacuma* strains represented a species distinct from *B. cinerea*, which has recently been named *B. pseudocinerea* (Fournier et al. 2005; Walker et al. 2011). *B. pseudocinerea* has been found to occur as a minor gray mold species in French vineyards, and on several other plants. Seasonal variations in the frequency of *B. pseudocinerea* in vineyards, and its rare occurrence on molded grape berries suggests that it has a lower pathogenic performance than *B. cinerea*, but might survive better as a saprophyte (Martinez et al. 2005). *B. pseudocinerea* is morphologically indistinguishable from *B. cinerea*, but shows intrinsic resistance to the fungicide fenhexamid. This resistance is correlated with a reduced affinity of the fenhexamide target enzyme, 3-ketoreductase, compared to the target enzyme in *B. cinerea* (Debieu et al. 2013). Furthermore, a fenhexamid-induced cytochrome P450 monooxygenase, presumably is involved in detoxification, has been found to contribute to fenhexamid resistance in *B. pseudocinerea* (S. Azeddine and S. Fillinger, personal communication). Together with the host-specific *B. pelargonii*, *B. fabae* and *B. calthae*, *B. cinerea* and *B. pseudocinerea* represent the so-called dicot clade of *Botrytis* spp., comprising species that only infect dicot plants (Staats et al. 2005; Walker et al. 2011).

Although the genetic variability of *B. cinerea* is well documented, correlations between biological or pathogenic traits and genetic patterns have rarely been identified. Nevertheless, evidence of genetic differentiation within populations has been obtained. Populations of *B. cinerea*, sampled from grapes and blackberries, growing side-by-side in six regions of France, were shown by microsatellite-based population genetic studies to be significantly differentiated, indicating restricted gene flow between strains that infect these different hosts (Fournier and Giraudn 2008). A recent study on gray mold isolates from intensively fungicide-treated strawberry fields revealed a hitherto unknown genotype that is closely related but distinct from *B. cinerea*. It was detected during analysis of the molecular basis of a stronger variant of the multidrug-resistance phenotype 1 (MDR1), called MDR1h. Sequencing of the *mrr1* gene of MDR1 h-isolates, and of some non-MDR1h isolates from strawberries, revealed considerable sequence divergence from the *mrr1* genes from *B. cinerea*. A phylogenetic tree based on the combined sequence analysis of five genes placed *Botrytis* group S between *B. cinerea* and the host-specific *B. fabae* (Leroch et al. 2013b). Isolates with sequence signatures of *Botrytis* group S were found in different European countries, which indicated that little genetic exchange occurs between *B. cinerea* and *Botrytis* group S. To clarify the taxonomic status of *Botrytis* group S, genome sequencing and sexual crosses are underway. A population study revealed distinct patterns in the prevalence of *Botrytis* genotypes: while *Botrytis* group S was more abundant than *B. cinerea* in fungicide-treated strawberry fields, wine grapes were infected almost exclusively by *B. cinerea*. These data, together with those of putative new endophytic *Botrytis* species that were grouped close to *B. cinerea* (Shipunov et al. 2008), strongly indicate that the gray mold fungus indeed is not represented by a single species but rather a species complex, comprising genetically distinct taxa that are differentiated by host preference, virulence, accumulation of fungicide resistance mutations and other, as yet unknown physiological properties.

### 2.6.4 Genes under Positive Selection

It has often been observed that genes which are involved in interactions with an organism's (biotic) environment display an increased evolutionary rate, and are under positive (also called 'diversifying') selection. When coding sequences of such genes are compared between individuals within a population of a species, or between closely related species, positive selection is reflected in a high overall ratio of nonsynonymous nucleotide substitutions versus synonymous substitutions (dN/dS), often at positions within the protein that are potentially important for its interactions with receptors or ligands. In *B. cinerea*, two endopolygalacturonase genes, *Bcpg1* and *Bcpg2*, both of which contribute to full virulence (ten Have et al. 1998; Kars et al. 2005a, b) are under positive selection while other endopolygalacturonase genes are not (Rowe and Kliebenstein 2007; Cettul et al. 2008). Similarly, a study on NLP genes in the entire genus *Botrytis*, showed that the phytotoxic protein-encoding genes NEP1 and NEP2 are under positive selection (Staats et al. 2007a), however, functional analyses using mutants in *B cinerea* and *B. elliptica* showed that these NLP genes do not make any detectable contribution to virulence (Staats et al. 2007b; Cuesta Arenas et al. 2010).

In other necrotrophic fungi, it was observed that genes that encode important virulence determinants are under positive selection. Building on such concepts, the availability of *Botytis* genome or transcriptome sequences enables genome-wide studies to identify genes under positive selection, and to mark those genes as candidate virulence determinants. Using such an approach, Aguileta et al. (2012) compared the sequences of ESTs from *B. tulipae* and *B. ficariarum* with gene models of *B. cinerea* and *S. sclerotiorum*, and detected 21 genes under positive selection. For nine of these 21 genes, additional sequencing was done in 12 more *Botrytis* species and data confirmed the conclusion that these genes are under positive selection. Functional analysis of four of these genes by targeted knockout in *B. cinerea*, indicated that mutations in these genes did not significantly affect virulence on bean and tomato leaves (Aguileta et al. 2012). One of these genes, however, encoding the cerato-platanin family protein Spl1 was also analyzed by Frías et al. (2011) and found to be involved in virulence and induction of necrosis on tobacco leaves (Sect. 2.2). Interestingly the two sites found to be under positive selection in the Spl1 protein were situated precisely in the region with necrosis-inducing activity on tobacco leaves.

### 2.6.5 Evidence for Necrotrophic Effector Proteins

There are several lines of evidence for the ability of necrotrophs to subvert plants by co-opting the programmed cell death machinery of the host (Dickman et al. 2001; van Kan 2006; Oliver and Solomon 2010). The induction of host cell death provides an entry point for infection, and is essential for a necrotrophic fungus to successfully infect a host. Host cell death induction is achieved by releasing metabolites or proteins with phytotoxic action into a host plant (Stergiopoulos et al. 2013). Phytotoxic proteins of necrotrophic fungi are recently referred to as 'necrotrophic effectors,' by analogy to molecules that biotrophic pathogens secrete to promote virulence (Tan et al. 2010; Ciuffetti et al. 2010). There are several well-studied, host-specific necrotrophic effectors, e.g., the ToxA protein of *Pyrenophora tritici-repentis* (Ciuffetti et al. 2010). The wheat gene *Tsn*1, encoding an NBS-LRR gene (similar to resistance genes effective against biotrophs), governs sensitivity to ToxA, and thereby confers susceptibility to *P. tritici-repentis* as a dominant trait (Faris et al. 2010). Thus, recognition of necrotrophic effectors often relies on similar mechanisms as the recognition of effectors from biotrophs ('avirulence proteins') by plant receptor proteins, leading to disease resistance. However, the outcome of necrotrophic effector recognition is inverse to classical gene-for-gene-based pathosystems, i.e., effector recognition triggers susceptibility (Tan et al. 2010, 2012).

While *B. cinerea* has a wide host range, all other *Botrytis* species are considered to be much more restricted to rather few related host species, or to a single host species. One working model for their host specificity is that each species secretes effector proteins that are toxic to its host plant, but not to nonhosts. Several independent host jumps occurred during speciation in the genus *Botrytis* (Staats et al. 2005), which may have been caused by the acquisition or evolution of effector genes. An important aspect of genome projects in related necrotrophic fungi could be to identify unique effector genes, encoding proteins which only kill a specific host species. Such effector genes may be identified by a combination of comparative genomics and proteomics approaches. A limitation of this working model is the difficulty to explain the wide host range of *B. cinerea*, which includes hosts infected by the host-specific *Botrytis* spp.. For example, *B. cinerea* also infects *Vicia faba* (broad bean) and *Caltha palustris*, the preferred or exclusive host plants of *B. fabae* and *B. calthae*, respectively (Hahn, unpublished). *B. cinerea* is closely related to *B. fabae*, which raises the question as to the genomic differences that are responsible for their strikingly different host ranges.

*B. cinerea* produces two known phytotoxic SM, and several phytotoxic proteins, all of which can induce cell death in a range of dicot plants. The multiplicity of cell death-inducing compounds implies that they perform redundant functions in pathogenesis (Cuesta Arenas et al. 2010; Dalmais et al. 2011; Frías et al. 2011). Transcriptome studies did not provide strong evidence for the existence of effector proteins similar to those discovered in other necrotrophic fungi (Sect. 2.5.1). Some of the host-specific *Botrytis* species are known to produce phytotoxic botrydial and/or botcinic acid related compounds (Collado et al. 2000), and there are also some reports of phytotoxic effector proteins. Functional analysis of proteins of the NLP family from the lily pathogen *B. elliptica* (BeNEP1 and BeNEP2) showed that remarkably, these proteins were phytotoxic to all dicots tested, but not to lily, and the *BeNEP*1 or *BeNEP*2 genes were not required for pathogenicity on lily (Staats et al. 2007a, b). Instead, there is evidence for the production of proteinaceous host-specific effectors by *B. elliptica* (van Baarlen et al. 2004) and by *B. fabae* (Harrison 1980). Proteins secreted by *B. elliptica* specifically caused programmed cell death in lily (van Baarlen et al. 2004), but not in other plants (a.o. onion, tulip, rice, maize, tobacco, Arabidopsis; van Kan, unpublished). Furthermore, prior infiltration of the *B. elliptica* protein sample into lily leaves enabled the otherwise avirulent species *B. cinerea* and *B. tulipae* to infect lily, thereby proving the importance of host-specific effector protein activity for infecting lily (van Kan, unpublished). The *B. elliptica* protein sample was analyzed by LC-MS/MS and mass peaks were matched to the proteins predicted to be encoded in the *B. elliptica* genome. This resulted in identification of 10 candidate effector proteins, which remain to be further analyzed (van Kan, unpublished).

An alternative, or complementary, mechanism that could determine host specificity in *Botrytis* ssp. is their differential sensitivity to plant defense metabolites. *Vicia faba* leaves respond to infection by *Botrytis* ssp. with the synthesis of wyerone and related phytoalexins (Hargreaves et al. 1977). *B. fabae*, which is highly aggressive on *V. fabae*, was found to be significantly more tolerant to wyerone than *B. cinerea* (Rossall et al. 1980). In another study, the ability to detoxify α-tomatine was found in most natural isolates of *B. cinerea*. However, one isolate from grapevine was found to lack tomatinase activity, and this isolate was highly sensitive to α-tomatine and strongly impaired in virulence on tomato leaves, but normally virulent on bean leaves (Quidde et al. 1998).

## 2.7 Future Perspectives

Research in the last few years has revealed that the pathogenic mechanisms of *B. cinerea* are more complex than previously imagined, and still far from being understood. Recent studies provide evidence for the ability of *Botrytis* to suppress host defense by several mechanisms. These include the manipulation of plant hormone pathways, in particular those that are involved in defense responses, namely salicylic acid (which generally promotes infection by necrotrophs), jasmonic acid and ethylene (which both impair infection) (reviewed by Mengiste 2012). Consequently, '-omics' approaches that include the host plant will be increasingly applied to decipher the pathogen-induced changes in host metabolism. Examples are a recently performed, high resolution expression temporal transcriptomic analysis with Arabidopsis leaves during infection with *B. cinerea* (Windram et al. 2012), and a combined proteomic analysis of pathogen and host proteins in *B. cinerea*-infected ripening tomato fruits (Shah et al. 2012). An additional layer of complexity in our understanding of the infection mechanisms of *B. cinerea* has been recently uncovered by the discovery of small RNAs that suppress plant immunity by means of hijacking the host RNA interference pathway (Weiberg et al. 2013).

It is only a matter of time until the genome sequences of all known *Botrytis* species will be determined. Besides *B. cinerea*, genome assemblies have been generated for nine other *Botrytis* species (Staats and van Kan, unpublished), but remain to be annotated and analyzed in detail. For species of greater economic interest, multiple strains will be sequenced to explore intraspecific genetic diversity. An effort is in progress to sequence a large collection of *B. cinerea* field isolates and perform a genome-wide association analysis of phenotypic traits (Kliebenstein, pers. comm.). The genome sequences, together with transcriptome and proteome data will provide valuable resources for further studies, aimed at unraveling the biology of *Botrytis* spp..

**Acknowledgments** We thank our colleagues that provided their data prior to publication. We are grateful to S. Fillinger for helpful comments, and K. Mendgen for providing pictures.

## References

Agger S, Lopez-Gallego F, Schmidt-Dannert C (2009) Diversity of sesquiterpene synthases in the basidiomycete *Coprinus cinereus*. Mol Microbiol 72:1181–1195

Aguileta G, Lengelle J, Chiapello H, Giraud T, Viaud M, Fournier E, Rodolphe F, Marthey S, Ducasse A, Gendrault A, Poulain J, Wincker P, Gout L (2012) Genes under positive selection in a model plant pathogenic fungus, *Botrytis*. Infect Genet Evol 12:987–996

Amoutzias GD, He Y, Lilley KS, van de Peer Y, Oliver SG (2012) Evaluation and properties of the budding yeast phosphoproteome. Mol Cell Proteom 11(M111):009555

Amselem J, Cuomo CA, van Kan JA, Viaud M, Benito EP, Couloux A, Coutinho PM, de Vries RP, Dyer PS, Fillinger S, Fournier E, Gout L, Hahn M, Kohn L, Lapalu N, Plummer KM, Pradier JM, Quévillon E,

Sharon A, Simon A, ten Have A, Tudzynski B, Tudzynski P, Wincker P, Andrew M, Anthouard V, Beever RE, Beffa R, Benoit I, Bouzid O, Brault B, Chen Z, Choquer M, Collémare J, Cotton P, Danchin EG, Da Silva C, Gautier A, Giraud C, Giraud T, Gonzalez C, Grossetete S, Güldener U, Henrissat B, Howlett BJ, Kodira C, Kretschmer M, Lappartient A, Leroch M, Levis C, Mauceli E, Neuvéglise C, Oeser B, Pearson M, Poulain J, Poussereau N, Quesneville H, Rascle C, Schumacher J, Ségurens B, Sexton A, Silva E, Sirven C, Soanes DM, Talbot NJ, Templeton M, Yandava C, Yarden O, Zeng Q, Rollins JA, Lebrun MH, Dickman M (2011) Genomic analysis of the necrotrophic fungal pathogens *Sclerotinia sclerotiorum* and *Botrytis cinerea*. PLoS Genet 7:e1002230

Andrew M, Barua R, Short SM, Kohn LM (2012) Evidence for a common toolbox based on necrotrophy in a fungal lineage spanning necrotrophs, biotrophs, endophytes, host generalists and specialists. PLoS ONE 7:e29943

Arbelet D, Malfatti P, Simond-Côte E, Fontaine T, Desquilbet L, Expert D, Kunz C, Soulié MC (2010) Disruption of the Bcchs3a chitin synthase gene in *Botrytis cinerea* is responsible for altered adhesion and overstimulation of host plant immunity. Mol Plant Microbe Interact 23:1324–1334

Bayram O, Braus GH (2012) Coordination of secondary metabolism and development in fungi: the velvet family of regulatory proteins. FEMS Microbiol Rev 36:1–24

Bayram Ö, Bayram ÖS, Ahmed YL, Maruyama J, Valerius O, Rizzoli SO, Ficner R, Irniger S, Braus GH (2012) The *Aspergillus nidulans* MAPK module AnSte11-Ste50-Ste7-Fus3 controls development and secondary metabolism. PLoS Genet 8:e1002816

Billon-Grand G, Rascle C, Droux M, Rollins JA, Poussereau N (2010) pH modulation differs during sunflower cotyledon colonization by the two closely related necrotrophic fungi *Botrytis cinerea* and *Sclerotinia sclerotiorum*. Mol Plant Pathol 13:568–578

Brakhage AA (2013) Regulation of fungal secondary metabolism. Nat Rev Microbiol 11:21–32

Brito N, Espino JJ, González C (2006) The endo-beta-1,4-xylanase Xyn11A is required for virulence in *Botrytis cinerea*. Mol Plant-Microbe Interact 19:25–32

Bushley KE, Turgeon BG (2010) Phylogenomics reveals subfamilies of fungal nonribosomal peptide synthetases and their evolutionary relationships. BMC Evol Biol 10:26

Cantarel BL, Coutinho PM, Rancurel C, Bernard T, Lombard V, Henrissat B (2009) The Carbohydrate-active EnZymes database (CAZy): an expert resource for glycogenomics. Nucl Acids Res 37:D233–D238

Cettul E, Rekab D, Locci R, Firrao G (2008) Evolutionary analysis of endopolygalacturonase-encoding genes of *Botrytis cinerea*. Mol Plant Pathol 9:675–685

Chapeland F, Fritz R, Lanen C, Gredt M, Leroux P (1999) Inheritance and mechanisms of resistance to anilinopyrimidine fungicides in *Botrytis cinerea* (*Botryotinia fuckeliana*). Pestic Biochem Physiol 64:85–100

Chitrampalam P, Inderbitzin P, Maruthachalam K, Wu BM, Subbarao KV (2013) The *Sclerotinia sclerotiorum* mating type locus (*MAT*) contains a 3.6-kb region that is inverted in every meiotic generation. PLoS ONE 8:e56895

Choquer M, Fournier E, Kunz C, Levis C, Pradier JM, Simon A, Viaud M (2007) *Botrytis cinerea* virulence factors: new insights into a necrotrophic and polyphageous pathogen. FEMS Microbiol Lett 277:1–10

Ciuffetti LM, Manning VA, Pandelova I, Betts MF, Martinez JP (2010) Host-selective toxins, Ptr ToxA and Ptr ToxB, as necrotrophic effectors in the *Pyrenophora tritici-repentis*-wheat interaction. New Phytol 187:911–919

Collado G, Aleu J, Hernandez-Galan R, Duran-Patron R (2000) *Botrytis* species: an intriguing source of metabolites with a wide range of biological activities. Structure, chemistry and bioactivity of metabolites isolated from *Botrytis* species. Curr Organ Chem 4:1261–1286

Colmenares AJ, Aleu J, Duran-Patron R, Collado IG, Hernandez-Galan R (2002) The putative role of botrydial and related metabolites in the infection mechanism of *Botrytis cinerea*. J Chem Ecol 28:997–1005

Cuesta Arenas Y, Kalkman E, Schouten A, Dieho M, Vredenbregt P, Uwumukiza B, Osés Ruiz M, van Kan JAL (2010) Functional analysis and mode of action of phytotoxic Nep1-like proteins of *Botrytis cinerea*. Physiol Mol Plant Pathol 74:376–386

Dalmais B, Schumacher J, Moraga J, Le Pêcheur P, Tudzynski B, Collado IG, Viaud M (2011) The *Botrytis cinerea* phytotoxin botcinic acid requires two polyketide synthases for production and has a redundant role in virulence. Mol Plant Pathol 12:564–579

Dean R, Van Kan JA, Pretorius ZA, Hammond-Kosack KE, Di Pietro A, Spanu PD, Rudd JJ, Dickman M, Kahmann R, Ellis J, Foster GD (2012) The top 10 fungal pathogens in molecular plant pathology. Mol Plant Pathol 13:414–430

Debieu D, Bach J, Montesinos E, Fillinger S, Leroux P (2013) Role of sterol 3-ketoreductase sensitivity in susceptibility to the fungicide fenhexamid in *Botrytis cinerea* and other phytopathogenic fungi. Pest Manag Sci 69:642–651

Dickman MB, Park YK, Oltersdorf T, Li W, Clemente T, French R (2001) Abrogation of disease development in plants expressing animal antiapoptotic genes. Proc Natl Acad Sci USA 98:6957–6962

Doehlemann G, Berndt P, Hahn M (2006) Different signalling pathways involving a Galpha protein, cAMP and a MAP kinase control germination of *Botrytis cinerea* conidia. Mol Microbiol 59:821–835

Espino JJ, Gutierrez-Sanchez G, Brito N, Shah P, Orlando R, Gonzalez C (2010) The *Botrytis cinerea* early secretome. Proteomics 10:3020–3034

Faris JD, Zhang Z, Lu H, Lu SW, Reddy L, Cloutier S, Fellers JP, Meinhardt SW, Rasmussen JB, Xu SS, Oliver RP, Simons KJ, Friesen TL (2010) A unique wheat disease resistance-like gene governs effector-triggered susceptibility to necrotrophic pathogens. Proc Natl Acad Sci USA 107:13544–13549

Fernandez-Acero JF, Colby T, Harzen A, Cantoral JM, Schmidt J (2009) Proteomic analysis of the phytopathogenic fungus *Botrytis cinerea* during cellulose degradation. Proteomics 9:2892–2902

Fernandez-Acero JF, Colby T, Harzen A, Carbu M, Wieneke U, Cantoral JM, Schmidt J (2010) 2-DE proteomic approach to the *Botrytis cinerea* secretome induced with different carbon sources and plant-based elicitors. Proteomics 10:2270–2280

Fernández-Álvarez A, Elías-Villalobos A, Ibeas JI (2009) The O-mannosyltransferase PMT4 Is essential for normal appressorium formation and penetration in *Ustilago maydis*. Plant Cell 21:3397–3412

Fernández-Álvarez A, Marín-Menguiano M, Lanver D, Jiménez-Martín A, Elías-Villalobos A, Pérez-Pulido AJ, Kahmann R, Ibeas JI (2012) Identification of O-mannosylated virulence factors in *Ustilago maydis*. PLoS Pathog 8:e1002563

Fillinger S, Ajouz S, Nicot PC, Leroux P, Bardin M (2012) Functional and structural comparison of pyrrolnitrin- and iprodione-induced modifications in the class III histidine-kinase Bos1 of *Botrytis cinerea*. PLoS One 7(8):e42520

Fournier E, Giraud T, Loiseau A, Vautrin D, Estoup A, Solignac M, Cornuet JM, Brygoo Y (2002) Characterization of nine polymorphic microsatellite loci in the fungus *Botrytis cinerea* (Ascomycota). Mol Ecol Notes 2:253–255

Fournier E, Giraud T, Albertini C, Brygoo Y (2005) Partition of the *Botrytis cinerea* complex in France using multiple gene genealogies. Mycologia 97:1251–1267

Fournier E, Giraudn T (2008) Sympatric genetic differentiation of a generalist pathogenic fungus, *Botrytis cinerea*, on two different host plants, grapevine and bramble. J Evol Biol 21:122–132

Frías M, González C, Brito N (2011) BcSpl1, a cerato-platanin family protein, contributes to *Botrytis cinerea* virulence and elicits the hypersensitive response in the host. New Phytol 192:483–495

Frías M, Brito N, González C (2013) The *Botrytis cinerea* cerato-platanin BcSpl1 is a potent inducer of systemic acquired resistance (SAR) in tobacco and generates a wave of salicylic acid expanding from the site of application. Mol Plant Pathol 14:191–196

Giesbert S, Schumacher J, Kupas V, Espino J, Segmüller N, Haeuser-Hahn I, Schreier PH, Tudzynski P (2012) Identification of pathogenesis-associated genes by T-DNA-mediated insertional mutagenesis in *Botrytis cinerea*: a type 2A phosphoprotein phosphatase and an SPT3 transcription factor have significant impact on virulence. Mol Plant-Microbe Interact 25:481–495

Gioti A, Simon A, Le Pêcheur P, Giraud C, Pradier JM, Viaud M, Levis C (2006) Expression profiling of *Botrytis cinerea* genes identifies three patterns of up-regulation in planta and an FKBP12 protein affecting pathogenicity. J Mol Biol 358:372–386

Giraud T, Fortini D, Levis C, Leroux P, Brygoo Y (1997) RFLP markers show genetic recombination in *Botryotinia fuckeliana* (*Botrytis cinerea*) and transposable elements reveal two sympatric species. Mol Biol Evol 14:1177–1185

Giraud T, Fortini D, Levis C, Lamarque C, Leroux P, LoBuglio K, Brygoo Y (1999) Two sibling species of the *Botrytis cinerea* complex, *transposa* and *vacuma*, are found in sympatry on numerous host plants. Phytopathology 89:967–973

González M, Brito N, González C (2012) High abundance of Serine/Threonine-rich regions predicted to be hyper-O-glycosylated in the secretory proteins coded by eight fungal genomes. BMC Microbiol 12:213

Govrin EM, Levine A (2000) The hypersensitive response facilitates plant infection by the necrotrophic pathogen *Botrytis cinerea*. Curr Biol 10:751–757

González M, Brito N, Frias M, Gonzalez C (2013) *Botrytis cinerea* protein O-mannosyltransferases play critical roles in morphogenesis, growth, and virulence. PLos One 8(6):e65924

Gourgues M, Brunet-Simon A, Lebrun MH, Levis C (2004) The tetraspanin BcPls1 is required for appressorium-mediated penetration of Botrytis cinerea into host plant leaves. Mol Microbiol 51:619–629

Grant-Downton R, Terhem RA, Kapralov MV, Mehdi S, Rodriguez-Enriquez MJ, Gurr SJ, van Kan JAL, Dewey FM (2013) A novel *Botrytis* species is associated with a newly emergent foliar disease in cultivated *Hemerocallis* (Submitted for publication)

Han Y, Joosten HJ, Niu W, Zhao Z, Mariano PS, McCalman M, van Kan J, Schaap PJ, Dunaway-Mariano D (2007) Oxaloacetate hydrolase, the C-C bond lyase of oxalate secreting fungi. J Biol Chem 282:9581–9590

Hargreaves JA, Mansfield JW, Rossall S (1977) Changes in phytoalexin concentrations in tissues of the broad bean plant (*Vicia faba* L.) following inoculation with species of *Botrytis*. Physiol Plant Pathol 11:227–242

Harrison JG (1980) The production of toxins by *Botrytis fabae* in relation to growth of lesions on bean leaves at different humidities. Ann Appl Biol 95:63–72

ten Have A, Mulder W, Visser J, van Kan JAL (1998) The endopolygalacturonase gene Bcpg1 is required for full virulence of *Botrytis cinerea*. Mol Plant-Microbe Interact 11:1009–1016

ten Have A, Espino JJ, Dekkers E, van Sluyter SC, Brito N, Kay J, González C, van Kan JA (2010) The *Botrytis cinerea* aspartic proteinase family. Fungal Genet Biol 47:53–65

Heller J, Meyer AJ, Tudzynski P (2012a) Redox-sensitive GFP2: use of the genetically encoded biosensor of the redox status in the filamentous fungus *Botrytis cinerea*. Mol Plant Pathol 13:935–947

Heller J, Ruhnke N, Espino JJ, Massaroli M, Collado IG, Tudzynski P (2012b) The mitogen-activated protein kinase BcSak1 of *Botrytis cinerea* is required for pathogenic development and has broad regulatory functions beyond stress response. Mol Plant Microbe Interact 25:802–816

Hoffmeister D, Keller NP (2007) Natural products of filamentous fungi: enzymes, genes, and their regulation. Nat Prod Rep 24:393–416

Kars I, McCalman M, Wagemakers L, van Kan JA (2005a) Functional analysis of *Botrytis cinerea* pectin methylesterase genes by PCR-based targeted mutagenesis: Bcpme1 and Bcpme2 are dispensable for virulence of strain B05.10. Mol Plant Pathol 6:641–652

Kars I, Krooshof G, Wagemakers CAM, Joosten R, Benen JAE, van Kan JAL (2005b) Necrotising activity of five *Botrytis cinerea* endopolygalacturonases produced in *Pichia pastoris*. Plant J 43:213–225

Kerssies A, Bosker-van Zessen AI (1997) Variation in pathogenicity and DNA polymorphism among *Botrytis cinerea* isolates sampled inside and outside a glasshouse. Plant Dis 81:781–786

Krappmann S, Bayram Ö, Braus GH (2005) Deletion and allelic exchange of the *Aspergillus fumigatus veA* locus via a novel recyclable marker module. Eukaryot Cell 4:1298–1307

Kretschmer M, Hahn M (2008) Fungicide resistance and genetic diversity of *Botrytis cinerea* isolates from a vineyard in Germany. J Plant Dis Protect 115:214–219

Kretschmer M, Leroch M, Mosbach A, Walker AS, Fillinger S, Mernke D, Schoonbeek HJ, Pradier JM, Leroux P, De Waard MA, Hahn M (2009) Fungicide-driven evolution and molecular basis of multidrug resistance in field populations of the grey mould fungus *Botrytis cinerea*. PLoS Pathog 5:e1000696

Kroken S, Glass NL, Taylor JW, Yoder OC, Turgeon BG (2003) Phylogenomic analysis of type I polyketide synthase genes in pathogenic and saprobic ascomycetes. Proc Natl Acad Sci USA 100:15670–15675

Kunz C, Vandelle E, Rolland S, Poinssot B, Bruel C, Cimerman A, Zotti C, Moreau E, Vedel R, Pugin A, Boccara M (2006) Characterization of a new, non-pathogenic mutant of *Botrytis cinerea* with impaired plant colonization capacity. New Phytol 170:537–550

Lanver D, Mendoza-Mendoza A, Brachmann A, Kahmann R (2010) Sho1 and Msb2-related proteins regulate appressorium development in the smut fungus *Ustilago maydis*. Plant Cell 22:2085–2101

Leroch M, Kleber A, Silva E, Coenen T, Koppenhöfer D, Shmaryahu A, Valenzuela PD, Hahn M (2013a) Transcriptome profiling of *Botrytis cinerea* conidial germination reveals upregulation of infection-related genes during the pre-penetration stage. Eukaryot Cell 12:614–626

Leroch M, Plesken C, Weber RW, Kauff F, Scalliet G, Hahn M (2013b) Gray mold populations in German strawberry fields are resistant to multiple fungicides and dominated by a novel clade closely related to *Botrytis cinerea*. Appl Environ Microbiol 79:159–167

Leroch M, Mernke D, Koppenhoefer D, Schneider P, Mosbach A, Doehlemann G, Hahn M (2011) Living colors in the gray mold pathogen *Botrytis cinerea*: codon-optimized genes encoding green fluorescent protein and mCherry, which exhibit bright fluorescence. Appl Environ Microbiol. 77:2887–2897

Li X, Kerrigan J, Chai W, Schnabel G (2012a) *Botrytis caroliniana*, a new species isolated from blackberry in South Carolina. Mycologia 104:650–658

Li B, Wang W, Zong Y, Qin G, Tian S (2012b) Exploring pathogenic mechanisms of *Botrytis cinerea* Secretome under different ambient pH based on comparative proteomic analysis. J Proteome Res 11:4249–4260

Liu W, Zhou X, Li G, Li L, Kong L, Wang C, Zhang H, Xu JR (2011a) Multiple plant surface signals are sensed by different mechanisms in the rice blast fungus for appressorium formation. PLoS Pathog 7:e1001261

Liu W, Leroux P, Fillinger S (2008) The HOG1-like MAP kinase Sak1 of *Botrytis cinerea* is negatively regulated by the upstream histidine kinase Bos1 and is not involved in dicarboximide- and phenylpyrrole-resistance. Fungal Genet Biol 45:1062–1074

Liu W, Soulié MC, Perrino C, Fillinger S (2011b) The osmosensing signal transduction pathway from *Botrytis cinerea* regulates cell wall integrity and MAP kinase pathways control melanin biosynthesis with influence of light. Fungal Genet Biol 48:377–387

Manteau S, Abouna S, Lambert B, Legendre L (2003) Differential regulation by ambient pH of putative virulence factor secretion by the phytopathogenic fungus *Botrytis cinerea*. FEMS Microbiol Ecol. 43:359–366

Martinez F, Blancard D, Lecomte P, Levis C, Dubos B, Fermaud M (2003) Phenotypic differences between vacuma and transposa subpopulations of *Botrytis cinerea*. Eur J Plant Pathol 109:479–488

Martinez F, Dubos B, Fermaud M (2005) The role of saprotrophy and virulence in the population dynamics of *Botrytis cinerea* in vineyards. Phytopathology 95:692–700

Massaroli M, Moraga J, Bastos Borges K, Ramirez-Fernández J, Viaud M, González Collado I, Durán-Patrón R, Hernández-Galán R (2013) A shared biosynthetic pathway for botcinins and botrylactones revealed through gene deletions. Chem Bio 14:132–136

Mengiste T (2012) Plant immunity to necrotrophs. Annu Rev Phytopathol. 50:267–294

Michielse CB, Becker M, Heller J, Moraga J, Collado IG Tudzynski P (2011) The *Botrytis cinerea* Reg1 protein, a putative transcriptional regulator, is required for pathogenicity, conidiogenesis, and the production of secondary metabolites. Mol Plant Microbe Interact 24:1074–1085

Morcx S, Kunz C, Choquer M, Assie S, Blondet E, Simond-Côte E, Gajek K, Chapeland-Leclerc F, Expert D, Soulie MC (2013) Disruption of Bcchs4, Bcchs6 or Bcchs7 chitin synthase genes in *Botrytis cinerea* and the essential role of class VI chitin synthase (Bcchs6). Fungal Genet Biol 52:1–8

Morschhäuser J (2010) Regulation of multidrug resistance in pathogenic fungi. Fungal Genet Biol 47:94–106

Mosbach A, Leroch M, Mendgen KW, Hahn M (2011) Lack of evidence for a role of hydrophobins in conferring surface hydrophobicity to conidia and hyphae of *Botrytis cinerea*. BMC Microbiol 11:10

Moyano C, Alfonso C, Gallegro J, Raposo R, Melgarejo P (2003) Comparison of RAPD and AFLP marker analysis as a means to study the genetic structure of *Botrytis cinerea* populations. Eur J Plant Pathol 109:515–522

Nielsen K, Yohalam DS (2001) Origin of a polyploidy Botrytis pathogen through interspecific hybridization between *Botrytis aclada* and *Botrytis byssoidea*. Mycologie 93:1064–1071

Nguyen QB, Itoh K, Van Vu B, Tosa Y, Nakayashiki H (2011) Simultaneous silencing of endo-$\beta$-1,4 xylanase genes reveals their roles in the virulence of *Magnaporthe oryzae*. Mol Microbiol 81:1008–1019

Noda J, Brito N, González C (2010) The *Botrytis cinerea* xylanase Xyn11A contributes to virulence with its necrotizing activity, not with its catalytic activity. BMC Plant Biol 10:38

Oliver RP, Solomon PS (2010) New developments in pathogenicity and virulence of necrotrophs. Curr Opin Plant Biol 13:415–419

Pihet M, Vandeputte P, Tronchin G, Renier G, Saulnier P, Georgeault S, Mallet R, Chabasse D, Symoens F, Bouchara JP (2009) Melanin is an essential component for the integrity of the cell wall of *Aspergillus fumigatus* conidia. BMC Microbiol 9:177

Pinedo C, Wang CM, Pradier JM, Dalmais B, Choquer M, Le Pêcheur P, Morgant G, Collado IG, Cane DE, Viaud M (2008) Sesquiterpene synthase from the botrydial biosynthetic gene cluster of the phytopathogen *Botrytis cinerea*. ACS Chem Biol 3:791–801

Prado-Cabrero A, Scherzinger D, Avalos J, Al-Babili S (2007) Retinal biosynthesis in fungi: characterization of the carotenoid oxygenase CarX from *Fusarium fujikuroi*. Eukaryot Cell 6:650–657

Quidde T, Osbourn AE, Tudzynski P (1998) Detoxification of $\alpha$-tomatine by *Botrytis cinerea*. Physiol Mol Plant Pathol 52:151–165

RajaguruShaw P, Shaw M (2010) Genetic differentiation between hosts and locations in populations of latent *Botrytis cinerea* in southern England. Plant Pathol 59:1081–1090

Rampitsch C, Tinker NA, Subramaniam R, Barkow-Oesterreicher S, Laczko E (2012) Phosphoproteome profile of *Fusarium graminearum* grown in vitro under nonlimiting conditions. Proteomics 12:1002–1005

Reis H, Pfiffi S, Hahn M (2005) Molecular and functional characterization of a secreted lipase from *Botrytis cinerea*. Mol Plant Pathol 6:257–267

Rolke Y, Liu S, Quidde T, Williamson B, Schouten A, Weltring KM, Siewers V, Tenberge KB, Tudzynski B, Tudzynski P (2004) Functional analysis of H(2)O(2)-generating systems in Botrytis cinerea: the major Cu-Zn-superoxide dismutase (BCSOD1) contributes to virulence on French bean, whereas a glucose oxidase (BCGOD1) is dispensable. Mol Plant Pathol 5:17–27

Rossall S, Mansfield JW, Hutson RA (1980) Death of *Botrytis cinerea* and *B. fabae* following exposure to wyerone derivatives *in vitro* and during infection development in broad bean leaves. Physiol Plant Pathol 16:135–146

Rowe HC, Kliebenstein DJ (2007) Elevated genetic variation within virulence-associated *Botrytis cinerea* polygalacturonase loci. Mol Plant-Microbe Interact 20:1126–1137

Rowe HC, Walley JW, Corwin J, Chan EK, Dehesh K, Kliebenstein DJ (2010) Deficiencies in jasmonate-mediated plant defense reveal quantitative variation in *Botrytis cinerea* pathogenesis. PLoS Pathog 6:e1000861

Rowe HC, Kliebenstein DJ (2010) All mold is not alike: the importance of intraspecific diversity in necrotrophic plant pathogens. PLOS Pathog 6:e1000759

Rui O, Hahn M (2007) The Slt2-type MAP kinase Bmp3 of *Botrytis cinerea* is required for normal saprotrophic growth, conidiation, plant surface sensing and host tissue colonization. Mol Plant Pathol 8:173–184

Saikia S, Nicholson MJ, Young C, Parker EJ, Scott B (2008) The genetic basis for indole-diterpene chemical diversity in filamentous fungi. Mycol Res 112:184–199

Schamber A, Leroch M, Diwo J, Mendgen K, Hahn M (2010) The role of mitogen-activated protein (MAP) kinase signalling components and the Ste12 transcription factor in germination and pathogenicity of *Botrytis cinerea*. Mol Plant Pathol 11:105–119

Schouten A, Wagemakers L, Stefanato FL, van der Kaaij RM, van Kan JA (2002) Resveratrol acts as a natural profungicide and induces self-intoxication by a specific laccase. Mol Microbiol 43:883–894

Schumacher J, Kokkelink L, Huesmann C, Jimenez-Teja D, Collado IG, Barakat R, Tudzynski P, Tudzynski B (2008a) The cAMP-dependent signaling pathway and its role in conidial germination, growth, and virulence of the gray mold *Botrytis cinerea*. Mol Plant-Microbe Interact 21:1443–1459

Schumacher J, Viaud M, Simon A, Tudzynski B (2008b) The G$\alpha$ subunit BCG1, the phospholipase C (BcPLC1), and the calcineurin phosphatase coordinately regulate gene expression in the grey mold fungus *Botrytis cinerea*. Mol Microbiol 67:1027–1050

Schumacher J, de Larrinoa IF, Tudzynski B (2008c) Calcineurin-responsive zinc finger transcription factor CRZ1 of *Botrytis cinerea* is required for growth, development, and full virulence on bean plants. Eukaryot Cell 7:584–601

Schumacher J, Pradier JM, Simon A, Traeger S, Moraga J, González Collado I, Viaud M, Tudzynski B (2012) Natural variation in the VELVET gene *bcvel1* affects virulence and light-dependent differentiation in *Botrytis cinerea*. PLoS One 7:e47840

Schumacher J (2012) Tools for *Botrytis cinerea*: New expression vectors make the gray mold fungus more accessible to cell biology approaches. Fungal Genet Biol 49:483–497

Schumacher J, Gautier A, Morgant G, Le Pêcheur P, Azeddine S, Fillinger S, Leroux P, Tudzynski B, Viaud M (2013) A functional bikaverin biosynthetic gene cluster in rare isolates of Botrytis cinerea is positively regulated by the Velvet complex PLoS One 8:e53729

Schulze Gronover C, Schorn C, Tudzynski B (2004) Identification of *Botrytis cinerea* genes up-regulated during infection and controlled by the Galpha subunit BCG1 using suppression subtractive hybridization (SSH). Mol Plant-Microbe Interact 17:537–546

Shah P, Gutierrez-Sanchez G, Orlando R, Bergmann C (2009) A proteomic study of pectin-degrading enzymes secreted by *Botrytis cinerea* grown in liquid culture. Proteomics 9:3126–3135

Shah P, Powell ALT, Orlando R, Bergmann C, Gutierrez-Sanchez G (2012) Proteomic analysis of ripening tomato fruit infected by *Botrytis cinerea*. J Proteome Res 11:2178–2192

Shipunov A, Newcombe G, Raghavendra AK, Anderson CL (2008) Hidden diversity of endophytic fungi in an invasive plant. Am J Bot 95:1096–1108

Siegmund U, Heller J, van Kan JA, Tudzynski P (2013) The NADPH oxidase complexes in *Botrytis cinerea*: Evidence for a close association with the ER and the tetraspanin Pls1. PLoS One 8:e55879

Shlezinger N, Minz A, Gur Y, Hatam I, Dagdas YF, Talbot NJ, Sharon A (2011a) Anti-apoptotic machinery protects the necrotrophic fungus *Botrytis cinerea* from host-induced apoptotic-like cell death during plant infection. PLoS Pathog 7:e1002185

Shlezinger N, Doron A, Sharon A (2011b) Apoptosis-like programmed cell death in the grey mould fungus *Botrytis cinerea*: genes and their role in pathogenicity. Biochem Soc Trans 39:1493–1498

Siewers V, Viaud M, Jimenez-Teja D, Collado IG, Gronover CS, Pradier JM, Tudzynski B, Tudzynski P (2005) Functional analysis of the cytochrome P450 monooxygenase gene *bcbot1* of *Botrytis cinerea* indicates that botrydial is a strain-specific virulence factor. Mol Plant-Microbe Interact 18:602–612

Siewers V, Kokkelink L, Smedsgaard J, Tudzynski P (2006) Identification of an abscisic acid gene cluster in the grey mold *Botrytis cinerea*. Appl Environ Microbiol 72:4619–4626

Simon A, Dalmais B, Morgant G, Viaud M (2013) Screening of a *Botrytis cinerea* one-hybrid library reveals a Cys2His2 transcription factor involved in the regulation of secondary metabolism gene clusters. Fung Genet Biol 52:9–19

Sowley ENK, Dewey FM, Shaw MW (2010) Persistent, symptomless, systemic, and seed-borne infection of lettuce by *Botrytis cinerea*. Eur J Plant Pathol 126:61–71

Spanu PD, Abbott JC, Amselem J, Burgis TA, Soanes DM, Stüber K, Loren Ver, van Themaat E, Brown JK, Butcher SA, Gurr SJ, Lebrun MH, Ridout CJ, Schulze-Lefert P, Talbot NJ, Ahmadinejad N, Ametz C, Barton GR, Benjdia M, Bidzinski P, Bindschedler LV, Both M, Brewer MT, Cadle-Davidson L, Cadle-Davidson MM, Collemare J, Cramer R, Frenkel O, Godfrey D, Harriman J, Hoede C, King BC, Klages S, Kleemann J, Knoll D, Koti PS, Kreplak J, López-Ruiz FJ, Lu X, Maekawa T, Mahanil S, Micali C, Milgroom MG, Montana G, Noir S, O'Connell RJ, Oberhaensli S, Parlange F, Pedersen C, Quesneville H, Reinhardt R, Rott M, Sacristán S, Schmidt SM, Schön M, Skamnioti P, Sommer H, Stephens A, Takahara H, Thordal-Christensen H, Vigouroux M, Wessling R, Wicker T, Panstruga R (2010) Genome expansion and gene loss in powdery mildew fungi reveal tradeoffs in extreme parasitism. Science 330:1543–1546

Staats M, van Baarlen P, van Kan JA (2005) Molecular phylogeny of the plant pathogenic genus *Botrytis* and the evolution of host specificity. Mol Biol Evol 22:333–346

Staats M, van Baarlen P, Schouten A, van Kan JA, Bakker FT (2007a) Positive selection in phytotoxic protein-encoding genes of *Botrytis* species. Fungal Genet Biol 44:52–63

Staats M, van Baarlen P, Schouten A, van Kan JA (2007b) Functional analysis of NLP genes from *Botrytis elliptica*. Mol Plant Pathol 8:209–214

Staats M, van Kan JA (2012) Genome update of *Botrytis cinerea* strains B05.10 and T4. Eukaryot Cell 11:1413–1414

Stefanato FL, Abou-Mansour E, Buchala A, Kretschmer M, Mosbach A, Hahn M, Bochet CG, Métraux JP, Schoonbeek HJ (2009) The ABC transporter BcatrB from *Botrytis cinerea* exports camalexin and is a virulence factor on *Arabidopsis thaliana*. Plant J 58:499–510

Stergiopoulos I, Collemare J, Mehrabi R, de Wit PJGM (2013) Phytotoxic secondary metabolites and peptides produced by plant pathogenic Dothideomycete fungi. FEMS Microbiol Rev 37:67–93

Sutherland A, Auclair K, Vederas JC (2001) Recent advances in the biosynthetic studies of lovastatin. Curr Opin Drug Discov Devel 4:229–236

Tan K-C, Oliver RP, Solomon PS, Moffat CS (2010) Proteinaceous necrotropic effectors in fungal virulence. Funct Plant Biol 37:907–912

Tan K-C, Ferguson-Hunt M, Rybak K, Waters ODC, Bond CS, Stukenbrock EH, Friesen TL, Faris JD, McDonald BA, Oliver RP (2012) Quantitative variation in effector activity of ToxA isoforms from *Stagonospora nodorum* and *Pyrenophora tritici-repentis*. Mol Plant-Microbe Interact 25:515–522

Tani H, Koshino H, Sakuno E, Cutler HG, Nakajima H (2006) Botcinins E and F and botcinolide from *Botrytis cinerea* and structural revision of botcinolides. J Nat Prod 69:722–725

Temme N, Tudzynski P (2009) Does *Botrytis cinerea* ignore H(2)O(2)-induced oxidative stress during infection? Characterization of botrytis activator protein 1. Mol Plant Microbe Interact. 22:987–998

Temme N, Oeser B, Massaroli M, Heller J, Simon A, González Collado I, Viaud M, Tudzynski P (2012) BcAtf1- A global regulator controls various differentiation processes and phytotoxin production in *Botrytis cinerea*. Mol Plant Pathol 3:704–718

Tenberge KB (2004) Morphology and cellular organization in *Botrytis* interaction with plants. In: Elad Y, Williamson B, Tudzynski P, Delen N (eds) Botrytis: biology, pathology and control. Kluwer Academic Publishers, Dordrecht, The Netherlands, pp 67–84

Valette-Collet O, Cimerman A, Reignault P, Levis C, Boccara M (2003) Disruption of *Botrytis cinerea* pectin methylesterase gene Bcpme1 reduces virulence on several host plants. Mol Plant-Microbe Interact 16:360–367

van Baarlen P, Staats M, van Kan JAL (2004) Induction of programmed cell death in lily by the fungal pathogen *Botrytis elliptica*. Mol Plant Pathol 5:559–574

van Baarlen P, Woltering EJ, Staats M, van Kan JAL (2007) Histochemical and genetic analysis of host and non-host interactions of Arabidopsis with three *Botrytis* species: an important role for cell death control. Mol Plant Pathol 8:41–54

van Kan JA (2006) Licensed to kill: the lifestyle of a necrotrophic plant pathogen. Trends Plant Sci 11:247–253

Van Vu B, Itoh K, Nguyen QB, Tosa Y, Nakayashiki H (2012) Cellulases belonging to glycoside hydrolase families 6 and 7 contribute to the virulence of *Magnaporthe oryzae*. Mol Plant-Microbe Interact 25:1135–1141

Viaud M, Brunet-Simon A, Brygoo Y, Pradier JM, Levis C (2003) Cyclophilin A and calcineurin functions investigated by gene inactivation, cyclosporin A inhibition and cDNA arrays approaches in the phytopathogenic fungus *Botrytis cinerea*. Mol Microbiol 50:1451–1465

Viaud M, Fillinger S, Liu W, Polepalli JS, Le Pêcheur P, Kunduru AR, Leroux P, Legendre L (2006) A class III histidine kinase acts as a novel virulence factor in *Botrytis cinerea*. Mol Plant-Microbe Interact 19:1042–1050

Villen J, Gygi SP (2008) The SCX/IMAC enrichment approach for global phosphorylation analysis by mass spectrometry. Nat Protoc 3:1630–1638

Vogt K, Bhabhra R, Rhodes JC, Askew DS (2005) Doxycycline-regulated gene expression in the opportunistic fungal pathogen *Aspergillus fumigatus*. BMC Microbiol 5:1

Walker AS, Gautier AL, Confais J, Martinho D, Viaud M, Le P, Cheur P, Dupont J, Fournier E (2011) *Botrytis pseudocinerea*, a new cryptic species causing gray mold in French vineyards in sympatry with *Botrytis cinerea*. Phytopathology 101:1433–1445

De Waard MA, Andrade AC, Hayashi K, Schoonbeek HJ, Stergiopoulos I, Zwiers LH (2006) Impact of fungal drug transporters on fungicide sensitivity, multidrug resistance and virulence. Pest Manag Sci 62:195–207

Weiberg A, Wang M, Lin FM, Zhao H, Zhang Z, Kaloshian I, Huang HD, Jin H (2013) Fungal small RNAs suppress plant immunity by hijacking host RNA interference pathways. Science 342:118–123

Williamson B, Tudzynski B, Tudzynski P, van Kan JAL (2007) Pathogen profile—*Botrytis cinerea*: the cause of grey mould disease. Mol Plant Pathol 8:561–580

Williams B, Kabbage M, Kim HJ, Britt R, Dickman MB (2011) Tipping the balance: Sclerotinia *sclerotiorum secreted* oxalic acid suppresses host defenses by manipulating the host redox environment. PLoS Pathog 7(6):e1002107

Windram O, Madhou P, McHattie S, Hill C, Hickman R, Cooke E, Jenkins DJ, Penfold CA, Baxter L, Breeze E, Kiddle SJ, Rhodes J, Atwell S, Kliebenstein DJ, Kim YS, Stegle O, Borgwardt K, Zhang C, Tabrett A, Legaie R, Moore J, Finkenstadt B, Wild DL, Mead A, Rand D, Beynon J, Ott S, Buchanan-Wollaston V, Denby KJ (2012) Arabidopsis defense against *Botrytis cinerea*: chronology and regulation deciphered by high-resolution temporal transcriptomic analysis. Plant Cell 24:3530–3537

Wong Sak Hoi J, Dumas B (2010) Ste12 and Ste12-like proteins, fungal transcription factors regulating development and pathogenicity. Eukaryot Cell 9:480–485

Yin W, Keller NP (2011) Transcriptional regulatory elements in fungal secondary metabolism. J Microbiol 49:329–339

Yang Q, Yan L, Gu Q, Ma Z (2012) The mitogen-activated protein kinase kinase kinase BcOs4 is required for vegetative differentiation and pathogenicity in *Botrytis cinerea*. Appl Microbiol Biotechnol 96:481–492

Yan L, Yang Q, Jiang J, Michailides TJ, Ma Z (2011) Involvement of a putative response regulator Brrg 1 in the regulation of sporulation, sensitivity to fungicides, and osmotic stress in *Botrytis cinerea*. Appl Microbiol Biotechnol 90:215–226

Yan L, Yang Q, Sundin GW, Li H, Ma Z (2010) The mitogen-activated protein kinase kinase BOS5 is involved in regulating vegetative differentiation and virulence in *Botrytis cinerea*. Fungal Genet Biol 47:753–760

Zhang J, Wu MD, Li GQ, Yang L, Yu L, Jiang DH, Huang HC, Zhuang WY (2010a) *Botrytis fabiopsis*, a new species causing chocolate spot of broad bean in central China. Mycologia 102:1114–1126

Zhang J, Zhang L, Li GQ, Yang L, Jiang DH, Zhuang WY, Huang HC (2010b) *Botrytis sinoallii*: a new species of the grey mould pathogen on Allium crops in China. Mycoscience 5:421–431

# 3 Alternaria Comparative Genomics: The Secret Life of Rots

Ha X. Dang and Christopher B. Lawrence

## 3.1 Introduction

*Alternaria* species are a major cause of necrotrophic diseases of plants. In addition, they are often clinically associated with allergic respiratory disorders in humans but are rarely found to cause invasive infections. They can be classified as belonging to kingdom Fungi, subkingdom Eumycotera, phylum *Fungi Imperfecti* (a non-phylogenetic or artificial phylum of fungi without known sexual stages whose members may be related; taxonomy does not reflect relationships), form class Hyphomycetes, form order Moniliales, form family Dematiaceae, genus *Alternaria*. Some species of *Alternaria* are the asexual anamorph of the ascomycete *Pleospora,* others are speculated to be anamorphs of *Leptosphaeria* (Rotem 1994; Morales et al. 1995; Dong et al. 1998). They are also broadly classified as members of the Dothideomycete group of fungi.

*Alternaria* species are some of the most common fungi encountered by humans. The number of species is estimated to range in the hundreds (Rotem 1994). Many species are common saprophytes and have been recovered from very diverse substrates: plant material, sewage, leather, wood pulp, paper, textiles, building supplies, stone monuments, optical instruments, cosmetics, computer disks, and even jet fuel (references too numerous to list here). This suggests that their genomes most likely contain many genes that encode proteins/enzymes that allow for the acquisition and utilization of quite diverse substrates as carbon and/or nitrogen sources for growth and reproduction.

*Alternaria* species are some of the most well known producers of diverse secondary metabolites especially toxins (Montemurro and Visconti 1992). Over 70 small molecule compounds have been reported from *Alternaria* (Montemurro and Visconti 1992). Some of these metabolites are potent mycotoxins (e.g., alternariol, alternariol methyl ether, tenuazonic acid, etc.) with mutagenic and teratogenic properties, and have been linked to certain forms of cancer, such as esophogeal cancer, because of their genotoxic properties (Liu et al. 1991). The occurrence of potentially harmful *Alternaria* metabolites in food and food products such as grains, peanuts, tomato products, fresh fruits, and vegetables is becoming an increasing environmental concern (Bottalico and Logrieco 1998). In addition, an emerging and critical area of research utilizes *A. alternata* f. sp. *lycopersici*, which produces the sphingolipid-like AAL toxin, as a model organism in the investigation of toxin-mediated programmed cell death (apoptosis) in animals and plants (Abbas et al. 1995; Wang et al. 1996).

H. X. Dang
Virginia Bioinformatics Institute, Virginia Tech, Blacksburg, VA 24061, US

C. B. Lawrence (✉)
Department of Biological Sciences, Virginia Bioinformatics Institute, Virginia Tech, Blacksburg, VA 24061, US
e-mail: lawrence@vbi.vt.edu

R. A. Dean et al. (eds.), *Genomics of Plant Associated Fungi and Oomycetes: Dicot Pathogens*,
DOI: 10.1007/978-3-662-44056-8_3, © Springer-Verlag Berlin Heidelberg 2014

Collectively these data strongly suggest that *Alternaria* is an ideal genus for studying the genome evolution and organization of genes involved in secondary metabolite biosynthesis and secretion.

## 3.2 Alternaria Plant Pathogens

The so-called "rots" are among the most destructive plant diseases caused by Necrotrophic fungi that includes *Alternaria* spp. They inflict substantial tissue damage on their hosts in advance of and in concert with hyphal colonization. Necrotrophs represent the largest class of plant pathogens, yet our understanding of host-parasite interactions involving this class of pathogens is still lacking to a great extent. These fungi are tremendously important economically, although they represent a small percentage of fungal diversity they cause up to 80 % of losses due to fungal diseases in some parts of the world (Rotem 1994). Besides being devastating foliar pathogens, as decomposers, necrotrophic *Alternaria* species are ubiquitous postharvest pathogens and contribute to the spoilage of a substantial portion of our agricultural output (Wilson and Wisniewski 1994). In contrast, several *Alternaria* species or their small molecules have shown promise as biocontrol agents of weeds and mycoparasites (Rotem 1994).

Although they are sometimes considered unsophisticated in comparison to the more elegant biotrophic pathogenic fungi, necrotrophic fungi are highly specialized in order to successfully avoid, or suppress, host resistance mechanisms. In general, necrotrophic fungi employ a variety of approaches to circumvent the host-defense response by interfering with the activation of the response or negating its effect(s). Some important necrotrophic fungi, such as *Cochliobolus* and *Alternaria*, accomplish this by the production of low molecular weight, host-specific or host-selective phytotoxins (HSTs). Other pathogens that have a major necrotrophic stage do not produce HSTs, although some produce non-HSTs. The molecular basis for pathogenicity in these organisms remains largely unknown. This group contains many important plant-pathogenic fungal genera, including *Botrytis*, *Sclerotinia*, *Mycosphaerella*, *Fusarium*, *Leptosphaeria*, and *Colletotrichum*. These "HST" or "non-HST" are diverse in chemical structure and include secondary metabolites, cyclic peptides, and even proteins. In some plant-pathogen systems these toxins have been shown to be a primary or major determinant of pathogenicity. In other cases, they only serve to increase virulence. In most scenarios, host plant resistance mechanisms to necrotrophic fungi are not well understood but appear to function by interfering with the ability of the pathogen to suppress defenses through toxins.

The majority of the research to date on *Alternaria* pathogenesis mechanism has been centered upon the role of phytotoxins in disease. More toxins have been characterized from this genus than from any other. All of the plant pathogenic *Alternaria* species to date have been reported to produce toxins with very diverse biochemical structures (Rotem 1994; Agrios 1997). For many of the pathogenic *Alternaria* species, toxin production has been shown to be essential in enabling disease development on particular hosts. *Alternaria alternata* is ubiquitous in nature and mainly leads a saprophytic lifestyle, however, the transition from a nonspecific and nonpathogenic saprophytic type to a host-specific pathogen requires the production of HSTs (Rotem 1994). There are at least seven known host-parasite interactions in which HSTs produced by *A. alternata* pathotypes are required for disease (Akamatsu et al. 1997; Tanaka et al. 1999; Johnson et al. 2000; Masunaka et al. 2000). Thus, *Alternaria*-plant interactions are ideal for studying the role of toxins in infection processes.

HSTs produced by *A. alternata* pathotypes are mainly low molecular weight secondary metabolites. Some of these toxins, such as the AAL toxin, produced by *A. alternata* f.sp. *lycopersici*, are sphingolipid-like molecules and are structurally similar to fumonisins (Wang et al. 1996; Gilchrist 1998). Other toxins of diverse structure include cyclic despipeptide- based molecules such as the AM toxin from *A. alternata* f.sp. *mali* (Johnson et al. 2000). *Alternaria solani* like many of the pathogenic *Alternaria* species has been reported to produce non-HSTs such as alternaric acid, alternariol, tenuazonic acid, and zinniol; however, there has also been a report of a HST being produced by this species (Rotem 1994; Maiero et al.). *Alternaria brassicicola* has been reported to produce several structurally diverse phytotoxic molecules. A proteinaceous host-specific toxin (AB toxin) and toxic metabolites including despipeptides and diterpenoid fucicoccin-like compounds (Cooke et al. 1997; Otani et al. 1998; MacKinnon et al. 1999; Otani et al. 2001; McKenzie et al. 1998).

One interesting metabolite produced by *A. brassicicola* is the histone deacetylase (HDAC) inhibitor, depudecin (Kwon et al. 1998). HDAC inhibitors are known to cause changes in gene expression, induce apoptosis, developmental arrest, and detransformation of oncogene-transformed cells (Baidyaroy et al. 2002). HDAC inhibitors are also being explored as potential anticancer agents (Jung 2001). It is noteworthy that another HDAC inhibitor, HC-toxin, is produced by the closely related phytopathogenic fungus *Cochliobolus carbonum* and recently another *Alternaria* spp. (Baidyaroy et al. 2002), (personal communication Jonathon Walton). This suggests that HDAC inhibitors may have a critical role in determining virulence and host specificity in pathosystems especially those with necrotrophs.

Besides toxic components in spore germination fluids from *A. brassicicola*, little was known about other genes employed during pathogenesis until fairly recently. Early work examined the role of cutinase and lipase in *A. brassicicola* pathogenesis (Yao and Köller 1994; Yao and Koller 1995). Biolistic transformation resulted in disruption of the CUTABI gene, and affected saprophytic growth (cutin was no longer able to be utilized as a sole carbon source). However, disruption of CUTABI had no significant affect on *A. brassicicola* pathogenicity (Yao and Koller 1995). An extracellular lipase was found to be produced by *A. brassicicola* in vitro (Berto et al. 1997). Anti-lipase antibodies were found to significantly decrease *A. brassicicola's* ability to cause disease on cauliflower leaves (Berto et al. 1999). The *A. brassicicola* genome project described later has recently enabled the completion of several larger scale functional genomics studies that have now resulted in the identification of many new genes that contribute to virulence especially secondary metabolite genes, mitogen-activated protein (MAP) kinases, and transcription factors (Cho et al. 2006, 2007, 2009, 2012; Craven et al. 2008; Wight et al. 2009; Kim et al. 2007, 2009; Srivastava et al. 2012, 2013).

## 3.3 Alternaria and Human Respiratory Health

Sensitivity to the fungus *A. alternata* is a common cause of allergic rhinitis and believed to be a common cause of atopic or allergic asthma. Epidemiological studies from a variety of locations worldwide indicate that *Alternaria* sensitivity is closely linked with the development of asthma and up to 70 % of mold-allergic patients have skin test reactivity to *Alternaria* (O'Hollaren et al. 1991; Gergen and Turkeltaub 1992; Halonen et al. 1997; Andersson et al. 2003; Salo et al. 2006). Additionally, *Alternaria* sensitization has been determined to be one of the most important factors in the onset of childhood asthma in the southwest deserts of the US and other arid regions (Peat et al. 1993; Halonen et al. 1997). *Alternaria* spores are routinely found in atmospheric surveys in the United States, and in other countries and are the most frequently encountered fungal spore type (Hoffman 1984). Fungal exposure differs from pollen exposure in quantity (airborne spore counts are often 1,000-fold greater than pollen counts) and duration (e.g., *Alternaria* exposure occurs for months, whereas ragweed exposure for example occurs for weeks). This prolonged intense exposure is similar to that of other asthma-associated allergens such as cat dander and dust mites. Indeed it has long been speculated that this type of exposure may be partially responsible for both the chronic nature and severity of asthma in *Alternaria* sensitive individuals (Van Leeuwen 1924).

Although some research has been performed on the physiological and molecular identification of *Alternaria* allergens, only three major and five minor allergenic proteins have been described to date (Sanchez and Bush 2001; Breitenbach and Simon–Nobbe 2002). The biological role of these allergens and other fungal products in the development of allergy and asthma is very poorly understood. Other than a few studies demonstrating binding of these allergens to IgE-specific antibodies in human sera from patients diagnosed as being *Alternaria* sensitive, virtually nothing is known about how these highly immunoreactive proteins interact with the host. It has been demonstrated in several recently published studies that protease activity in *A. alternata* culture filtrates has marked proinflammatory properties in treated lung cells (Kouzaki et al. 2009; Matsuwaki et al. 2009). Thus, there is clearly a need to identify and elucidate the role of *Alternaria* allergens, immunostimulatory proteins as well as perhaps small molecules (secondary metabolites) in the development of allergic airway diseases from both diagnostic and immunotherapeutic perspectives. Indeed in a recent survey by the National Institute of Environmental Health and Safety (NIEHS) of 831 homes containing 2,456 individuals it was found that the prevalence of current symptomatic asthma increased with increasing *Alternaria* concentrations. Higher levels of *Alternaria* antigens in the environment significantly increased odds of having had asthma symptoms during the preceding year, more so than other examined antigens (Salo et al. 2006).

In addition, *Alternaria* species are gaining attention as emerging human invasive pathogens, particularly in immuno-compromised patients (Anaissie et al. 1989; Rossmann et al. 1996). Several *Alternaria* species have been found associated with infections of the cornea, oral and sinus cavities, respiratory tract, skin, and nails (de Bievre 1991; de Hoog et al. 2000).

## 3.4 Current Status of Alternaria Genomics

At present multiple genomes of various *Alternaria* spp. have been sequenced and annotated. *A. brassicicola* (ATCC 96836) was the first genome to be sequenced at The Genome Institute at Washington University as part of a collaborative project with the Virginia Bioinformatics Institute (VBI) funded by the USDA Microbial Genome Sequencing Program in 2004. This genome assembly has been deposited into NCBI genbank and the annotation was performed at the VBI. The annotated genome can currently be accessed at the DOE-JGI genome portal (http://genome.jgi-psf.org/Altbr1). Following the completion of the *A. brassicicola* genome, several other *Alternaria* genomes have been sequenced at the VBI and at the University of Arizona Genome Center independently or as part of a project funded by the National Science Foundation Systematics Program. These include several isolates of saprophytic *Alternaria spp., A. alternata* isolates used for commercial allergen production, and various other plant -pathogenic species primarily found within the "alternata" and "porri" taxonomic clades of the genus.

We have comprehensively annotated and analyzed *A. brassicicola* and two *A. alternata* genomes. Other *Alternaria* genomes are currently being investigated in additional studies. In this chapter, we will discuss results related to the genome analyses of these species. For convenience, we abbreviate *A. brassicicola* as *Ab*, *A. alternata* ATCC 66891 as *Aa*1, and *A. alternata* ATCC 11680 as *Aa*2, and will interchangeably use these abbreviations and the full species names. It is important to also mention that although ATCC 11680 was deposited as *A. alternata* it exhibits several morphological features of the closely related species, *Alternaria tenuissima* also found within the same taxonomic clade (*alternata* clade) (Barry Pryor, University of Arizona, personal communication). Thus, until further refinement of species name can be made, we will refer to this isolate also as *A. alternata* or *Aa*2 in this chapter.

## 3.5 Alternaria Genome Annotation

To annotate *Alternaria* genomes, a custom pipeline for fungal genome annotation and comparative genomics was developed at VBI

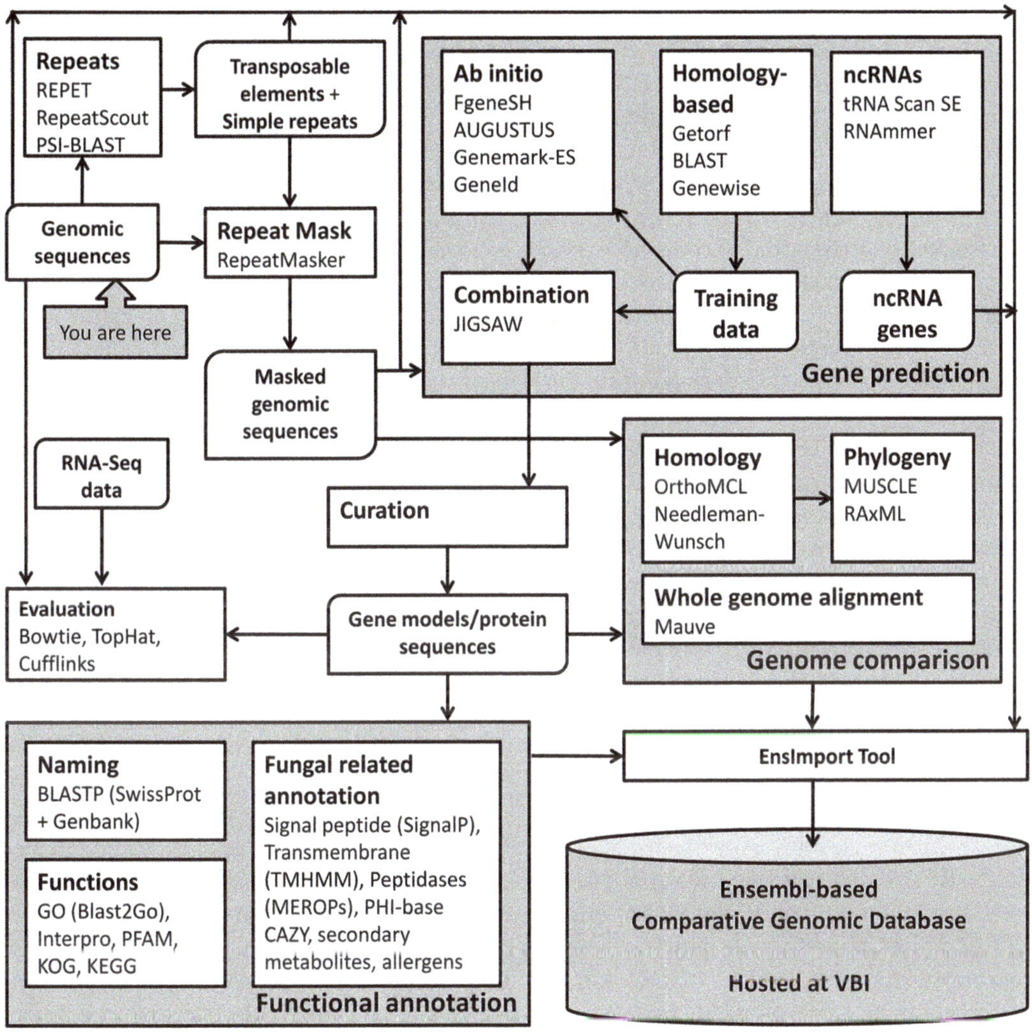


**Fig. 3.1** *Alternaria* genome annotation and comparison pipeline at the VBI

and is nearing completion (Fig. 3.1). The pipeline takes input from assembled genomes (i.e. genomic sequences, often in the form of contigs or supercontigs). These genomic sequences are first scanned for often in the form of contigs (both transposable elements and simple repeats). Repetitive sequences are then masked out from the original genomic sequences. Masked genomic sequences are the starting point for various subsequent analyses, including gene prediction, functional annotation, and genome comparison.

### 3.5.1 Repetitive Sequence Annotation

Transposable elements (and other types of repeats) in fungal genomes make up a portion from as small as ~1 % to as large as ~65 % (Galagan et al. 2005; Wicker et al. 2007; Spanu et al. 2010; Rouxel et al. 2011). Although most fungal genomes have relatively small amounts of repetitive DNA, it is thought to play important roles in fungal evolution (Daboussi and Capy 2003; Thon et al. 2006; Braumann et al. 2008).

For repetitive sequence discovery, an existing tool (REPET) is used as a part of the annotation pipeline at the VBI. REPET (Flutre et al. 2011) was developed specifically for fungi and uses BLAST (Altschul et al. 1997), RECON (Bao and Eddy 2002) and PILER (Edgar and Myers 2005) for de novo repetitive element prediction. It was first used for predicting repeats in the genome of another closely related Dothideomycete, *Leptosphaeria maculans* (Rouxel et al. 2011). In the annotation of *Alternaria* repetitive sequence, results from REPET are combined with repetitive sequences also discovered de novo by RepeatScout (Price et al. 2005). All consensus repetitive elements are then compared using BLAST and clustered to remove redundancy. In the *Alternaria* annotation pipeline, a consensus sequence is considered redundant and removed if $\geq$70 % of it is covered by another sequence in a pairwise BLAST alignment with an expected or e-value $\leq 10^{-10}$. After repeat families are identified, they are classified by both an automatic classification pipeline and human curation. REPET is used again to classify repeat elements in which their repeat structures are identified and their sequences are compared with known elements from Repbase (Jurka et al. 2005). After the initial annotation by REPET, another round is performed where repeat families are searched against Genbank using PSI-BLAST and BLASTX to identify homologous transposable elements (e-value $\leq 10^{-1}$). Repetitive sequences are then masked by RepeatMasker (http://repeatmasker.org) before further analyses are performed.

We found that repetitive sequences in *Alternaria* accounted for ~1–10 % of a genome. For example, the percentage of repetitive sequences in *A. brassicicola* and *A. alternata* genomes were ~9 and ~1 %, respectively. The larger repetitive content is a possible explanation for higher genome rearrangement rate in *A. brassicicola* (which will be discussed later).

### 3.5.2 Gene Prediction

The VBI fungal genome annotation pipeline uses multiple gene predictors including de novo and evidence-based gene prediction tools. To date, evidence-based gene predictors such as homology-based methods are among the best methods for eukaryotic gene structure annotation. Homology-based gene predictors, build gene models by aligning DNA sequences with known proteins, genes, or mRNAs. It is often a powerful approach to recover orthologs of known genes in genomes of closely related species that can also be used as training gene models for gene predictor software. In order to maximize the accuracy of predicted gene models, multiple gene predictors for eukaryotic and fungal genomes are used in the VBI *Alternaria* annotation pipeline, including homology-based tools like Genewise (Birney et al. 2004), as well as machine learning based tools such as FgeneSH, AUGUSTUS (Stanke and Waack 2003), Genemark-ES (Ter-Hovhannisyan et al. 2008), and GeneID (Blanco et al. 2007). When annotating *Alternaria* gene structure using the VBI pipeline, supervised gene predictors are re-trained (when needed) using known *Alternaria* gene models surveyed from Genbank and high scoring genes built from Genewise. To speed up predictions with Genewise, Emboss getorf software (Rice et al. 2000) is used to identify all possible open reading frames (ORFs) that are longer than 500 bp from genomic sequences and then BLASTed against Genbank to identify possible homologous proteins. ORFs that are found to match with known proteins (e.g., e-value $\leq 10^{-20}$) are paired with these proteins, and Genewise is then called to predict gene models for those ORFs from their paired proteins.

Gene models from individual predictors are then combined to produce the best models. The practice of combining results from different computational approaches is known as ENSEMBLE learning and has proven useful in many fields. Since eukaryotic gene prediction is still far from perfect, combination approaches could potentially produce more accurate gene models than individual methods (Bernal et al. 2012). There are many tools to combine multiple gene models into one model. The JIGSAW combiner (Allen and Salzberg 2005) is used for the VBI *Alternaria* annotation pipeline. JIGSAW uses decision trees to score different combinations of

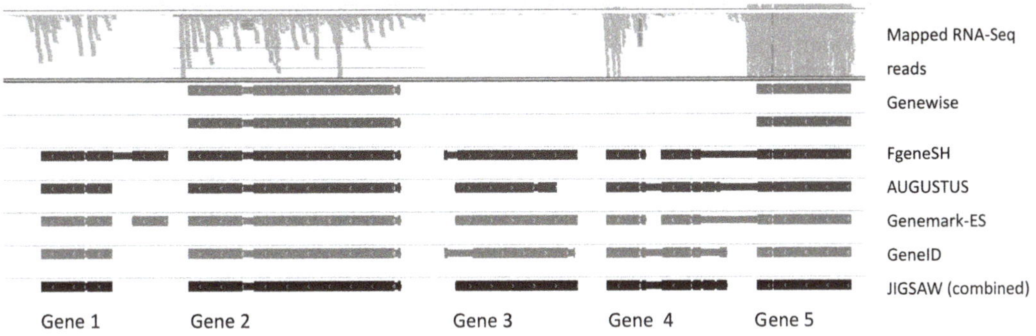


**Fig. 3.2** Examples of *Alternaria alternata* gene models. Gene models are predicted by individual gene predictors and combined by JIGSAW along with 10 million RNA-Seq reads mapped onto the genome. JIGSAW genes 1 and 5 are better supported by RNA-Seq while for gene 4, FgeneSH and AUGUSTUS models are better supported. Gene 2 is the same for all predictors and gene 3 is not expressed in this RNA-Seq sample

individual gene models derived from various predictors and selects the best scoring models. A subset of high quality gene models that is not used to train individual gene predictors is used to train JIGSAW for gene prediction. Beside protein coding genes, non-coding genes are predicted using RNAmmer (Lagesen et al. 2007) (rRNA genes) and tRNAScan-SE (Lowe and Eddy 1997) (tRNA genes). Lastly, RNA-sequencing reads are mapped to the genome (and quantified) using the Tuxedo tool set (Trapnell et al. 2012) as an additional annotation feature and aids in accurate gene prediction.

The total number of genes predicted using the combination approach for *A. brassicicola*, *A. alternata* ATCC 66891, and *A. alternata* ATCC 11680 are 10514, 11635, and 12323, respectively. These numbers are comparable with the number of predicted genes for other filamentous fungi of similar genome sizes (Galagan et al. 2003; Nierman et al. 2005; Hane et al. 2007). In a transcriptomic validation of the *Aa*1 predicted gene models produced by the pipeline, at least 2 % more RNA-Seq reads from a dataset of 10 million reads, were able to be mapped to a fewer number of the combined gene models compared to that mapped to each set of gene models produced by an individual method thus, suggesting some improvements using the JIG-SAW based approach. Several examples of gene models combined from multiple predictions are illustrated in Fig. 3.2.

### 3.5.3 Protein Annotation

Following gene prediction, gene models are transcribed and then translated using the protein module in the annotation pipeline. Various comparative computational tools are then used to infer the functions of the new proteins, based on their sequence under the assumption that proteins with similar sequences are possible orthologs that share the same functions. This comparative approach is widely used in annotation of protein function, and is relatively accurate for proteins that possess a high level of sequence similarity with known proteins or protein families.

In the VBI *Alternaria* annotation pipeline, new protein sequences are searched (using BLAST) against publicly available sequence databases such as the Swissprot (The UniProt Consortium 2008) and Genbank (Benson et al. 2011) to identify putative homologous proteins. Based upon results of these searches, new proteins are named according to the Broad Institute standard operating procedure and predicted functions, and other corresponding annotation are then assigned to the new proteins.

To gain more detailed information regarding putative functions of the proteins, domain annotation is performed in addition to whole sequence level BLAST-based approaches. In the VBI pipeline, InterproScan (Hunter et al. 2009) is used to query predicted proteins against protein signature databases, including PFAM, Superfamily, ProDom, Panther, TIGR, PIR, SMART, and PROSITE, and domain architectures of the predicted proteins are derived and then used to infer protein functions in addition to BLAST-based annotation. The predicted proteins are also assigned to the eukaryote orthologous groups (KOGs) by searching the KOG profiles from the NCBI conserved domain database (CDD) (Marchler-Bauer et al. 2011) using RPS-BLAST (Marchler-Bauer et al. 2002). Gene ontology (GO) terms are assigned with new proteins by a combination of results from Interpro and Blast2GO (Conesa et al. 2005).

### 3.5.4 Functional Annotation with Relevance to Saprophytic and Human/Plant Pathogenic Fungi

Specific annotation features of fungal proteins are important to predict especially those thought to play important roles in fungal pathogenicity and survival. These features may also include protein localization and secretion attributes, ability to acquire and digest different substrates, production of secondary metabolites (toxins, antibiotics, etc.), suppression of host defense, and features associated with the ability to detoxify antifungal chemicals and proteins from the host to name a few. For example, effector proteins are often small, cysteine-rich secreted virulence associated proteins (Stergiopoulos and de Wit 2009). PHI-base (http://www.phi-base.org) annotation also aids in identifying proteins that are similar to the known fungal plant pathogenicity-related proteins. BLAST searches against PHI-base database are used to link predicted proteins with experimentally verified pathogenicity, virulence and effector genes. Carbohydrate active enzymes (discussed in more detail later) are often secreted and important for the fungal saprophytic lifestyle but also may play a role in pathogenesis via plant cell wall breakdown.

In the VBI pipeline, secreted proteins are predicted by SignalP (Bendtsen et al. 2004) (using both HMM and neural network models), WoLF-psort (Horton et al. 2007), and Phobius (Käll et al. 2004). A protein is predicted to be secreted when a signal peptide is found by all three programs. This is quite strict and aims at improving confidence of predicted secretomes. Secreted proteins shorter than 300 amino acids, that have at least four cysteines in their sequences are classified as small cysteine-rich secreted proteins, and are candidates for further investigation as effectors. Transmembrane proteins are often involved in sensing the environment and subsequent signaling, and are predicted using TMHMM software (Krogh et al. 2001).

### 3.5.5 Carbohydrate Active Enzymes

Many fungi including *Alternaria spp.* are ubiquitous saprophytes that have the capability of acquiring nutrients from dead or decaying organic sources in the environment. Nutrients are often acquired via the action of carbohydrate active enzymes, proteinases, and lipases. In addition to BLAST against Genbankand/or Swissprot, carbohydrate -active enzymes are annotated using the CAZy annotation tool (Cantarel et al. 2009). Distribution of pectate lyases, carboxyl esterases, and glycoside hydrolases of 16 fungi including three *Alternaria* isolates according to their CAZy profile is depicted in Table 3.1. Interestingly, the *Alternaria* genomes investigated thus far harbor the largest number of carbohydrate- active enzymes suggesting one possible reason for the genus' success as a ubiquitous saprophyte, especially *A. alternata*. Pectate lyase type enzymes are also expanded in all *Alternaria* genomes compared to other fungi.

### 3.5.6 Allergen Homologs and Putative Proinflammatory Proteins Are Numerous in Alternaria

As described previously *A. alternata* is a clinically important allergenic fungus. Allergenic fungi often possess many proteins (allergens) that can trigger allergic hypersensitive type

**Table 3.1** Distribution of pectate lyases (PL), carboxyl esterases (CE), and glycoside hydrolases (GH) in 16 fungal genomes based on CAZy annotation

| Species | PLs | CEs | GHs | Total |
|---|---|---|---|---|
| *Ceriporiopsis subvermispora* | 6 | 15 | 175 | 196 |
| *Laccaria bicolor* | 7 | 20 | 181 | 208 |
| *Schizophyllum commune* | 16 | 38 | 245 | 299 |
| *Blumeria graminis f. sp. hordei* | 0 | 9 | 63 | 72 |
| *Trichoderma reesei* QM6a | 5 | 16 | 199 | 220 |
| *Myceliophthora thermophila* | 8 | 26 | 211 | 245 |
| *Thielavia terrestris* | 4 | 24 | 220 | 248 |
| *Neurospora crassa* | 4 | 21 | 179 | 204 |
| *Podospora anserina* | 7 | 39 | 232 | 278 |
| *Magnaporthe oryzae* | 5 | 52 | 276 | 333 |
| *Aspergillus fumigatus* Af293 | 14 | 31 | 272 | 317 |
| *Alternaria alternata* ATCC 11680 | 24 | 57 | 309 | 390 |
| *Alternaria alternata* ATCC 66981 | 24 | 57 | 305 | 386 |
| *Alternaria brassicicola* ATCC 96836 | 24 | 43 | 258 | 325 |
| *Phaeosphaeria nodorum* SN15 | 10 | 49 | 273 | 332 |
| *Leptosphaeria maculans* | 19 | 33 | 231 | 283 |

Note that *A. alternata* isolates possess the highest number of total enzymes compared to other fungi. Pectate lyase type enzymes are also expanded in all *Alternaria* genomes compared to other fungi. *GH* glycoside hydrolases, *GT* glycosyl transferases, *PL* polysaccharide lyases

I-IgE reactions in sensitized patients. Many allergens and allergen-like proteins also possess innate immunity stimulatory activity that may be important for the initial sensitization process. Previous genomic-scale studies have found that fungi such as the clinically relevant species *A. fumigatus* harbor many allergen homologs, including species- specific and cross-reactive isoforms (Bowyer et al. 2006; Bowyer and Denning 2007). At the VBI, a comparative genomics approach is used to annotate allergen homolog content of fungal genomes. The predicted proteins are compared with and classified according to Allergen Online database of known allergens from diverse sources (http://www.allergenonline.org) using BLAST-based approaches (e-value $\leq 10^{-10}$). Interestingly, the *A. alternata* genome contains the highest number of allergen-like proteins compared to any other fungal genome including *A. fumigatus* (Table 3.2). This may at least partially explain why *Alternaria* is one of the most clinically important allergenic fungi. Currently, machine learning based approaches (e.g., support vector machine) are being developed to more accurately predict potential allergens in fungal genomes at the VBI.

It has been shown that some allergens from fungi, and other organisms such as dust mites, as well as, other non-allergenic proteins that contribute to the overall inflammatory response in humans are potent proteases (Kouzaki et al. 2009; Matsuwaki et al. 2009). Thus, in addition to allergens, proteases in *Alternaria* genomes are identified and classified using an online batch BLAST search against the *MEROPS* database (Rawlings et al. 2012). The *MEROPS* database is a resource for information on peptidases (also termed proteases, proteinases, and proteolytic enzymes). *MEROPS* also includes the proteins that inhibit peptidase, and about 3,000 individual peptidases and inhibitors are included in the database. If a protein is found similar to a peptidase, it is assigned to the putative family and class of the matched peptidase. Using this approach it is clear that *A. alternata* has an expanded repertoire of unique putative protease genes, especially serine type proteases, compared to *A. brassicicola* that could contribute to its inflammatory potential (Fig. 3.3).

**Table 3.2** Distribution of allergen homologs in *Alternaria* and *Aspergillus*

| Allergen Type | *A. alternata* | *A. brassicicola* | *A. fumigatus* |
|---|---|---|---|
| Total | 292 | 219 | 233 |
| Aero fungi | 145 | 125 | 136 |
| Aero plant | 76 | 54 | 63 |
| Contact | 72 | 51 | 40 |
| Food plant | 23 | 16 | 27 |
| Aero mite | 16 | 9 | 10 |
| Bacteria airway | 11 | 5 | 3 |
| Venom or salivary | 12 | 12 | 10 |
| Aero animal | 8 | 6 | 7 |
| Aero insect | 4 | 3 | 2 |
| Unassigned | 1 | 1 | 1 |
| Worm (parasite) | 1 | 1 | 1 |
| Food animal | 0 | 0 | 1 |

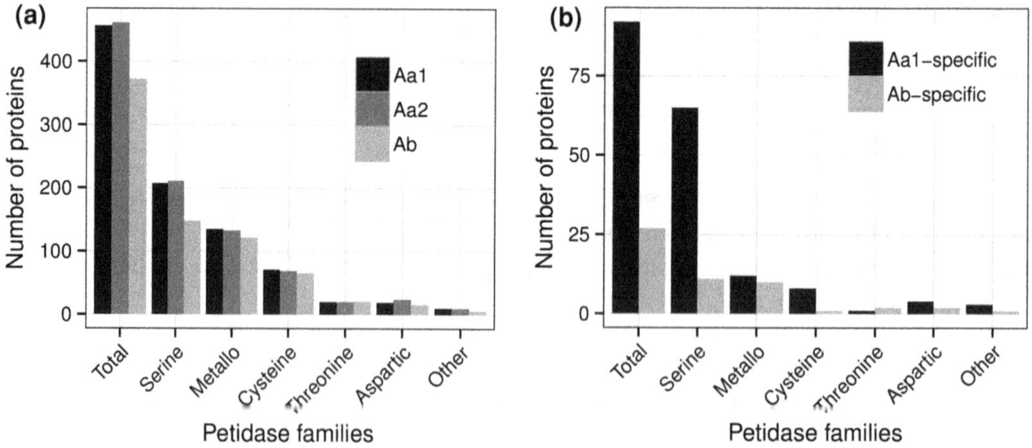


**Fig. 3.3** Peptidase distribution in *A. alternata ATCC 66891 (Aa*1*), A. alternata ATCC 11680 (Aa2), and A. brassicicola (Ab)* genomes (**a**), and in specific proteins from a pairwise comparison between *Aa*1 and *Ab*

### 3.5.7 Secondary Metabolites and Virulence Factors

Another class of genes that are important for fungal pathogenicity is secondary metabolite biosynthetic genes. In addition to BLAST, SMURF is used (Khaldi et al. 2010) to identify secondary metabolite gene clusters, in addition, to looking at domain architecture of the proteins to identify and characterize secondary metabolite production-related genes. Interestingly, *A. brassicicola* and *A. alternata* genomes share a core set of polyketide synthase (PKS) and non-ribosomal (NPS) genes, (6 PKS and 8 NPS) but each genome has their own unique sets of these types of genes/clusters suggesting different evolutionary pressures to acquire, and retain these types of interesting genes depending on lifestyle. *A. brassicicola* has seven unique PKS and NPS while Aa1 has 10 and Aa2 has 13. In addition to specific secondary metabolite biosynthetic genes, several other types of genes (transcription factors, map kinases, dehydrins, etc.) have been identified as virulence factors in *A. brassicicola*. Table 3.3 depicts the majority of virulence factors discovered to date in *A.*

**Table 3.3** *A. brassicicola* virulence factors

| Predicted Gene ID | Genbank Accession | Gene name | Description | Host | Pathogenicity of mutant | Reference |
|---|---|---|---|---|---|---|
| AB02513 | DY543081 | AbDhn1 | dehydrin-like protein | Cabbage | Reduced virulence | Pochon et al. (2013) |
| AB08993 | JX891381 | AbDhn2 | dehydrin-like protein | Cabbage | Reduced virulence | Pochon et al. Pochon et al. (2013) |
| AB05365 | JX891382 | AbDhn3 | dehydrin-like protein | Cabbage | Reduced virulence | (2013) |
| AB06533 | JQ899199 | AbPf2 | transcription factor | Cabbage Arabidopsis | Loss of pathogenicity | Cho et al. (2013) |
| AB08163 | JN835469 | AbVf19 | transcription factor | Cabbage | Reduced virulence | Srivastava et al. (2012) |
| AB02276 | JF487829 | Amr1 | melanin regulation | Cabbage | Increased virulence | Cho et al. (2012) |
| AB10176 | N/A | AbVf8 | transcription factor | Cabbage | Reduced virulence | Cho et al. (2012) |
| AB04090 | N/A | AbPacC | transcription factor | Cabbage | Loss of pathogenicity | Cho et al. (2012) |
| AB09132 | N/A | AbSte12 | transcription factor | Cabbage | Loss of pathogenicity | Cho et al. (2009) |
| AB05686 | AY987486 | AbHog1 | MAP kinase (HOG1) | Cabbage Arabidopsis | Reduced virulence | Joubert et al. (2011a) |
| AB07134 AB07135 | AY705975 | Amk2/ AbSlt2 | MAP kinase (SLT2) | Cabbage Arabidopsis | Reduced virulence | Joubert et al. (2011a) |
| | N/A | AbHacA | major UPR transcription regulator | Cabbage Arabidopsis | Loss of pathogenicity | Joubert et al. (2011b) |
| AB01916 | ACZ57548 | AbPKS9 | depudecin polyketide synthase | Cabbage | Reduced virulence | Wight et al. (2009) |
| AB04869 | EU223383 | TmpL | transmembrane protein 1 | Cabbage Arabidopsis | Reduced virulence | Kim et al. (2009) |
| AB07871 | N/A | AbPro1 | transcription factor | Cabbage | Reduced virulence | Cho et al. (2009) |
| AB03201 | AY700092 | AbNIK1 | two-component histidine kinase | Cabbage | Reduced virulence | Cho et al. (2009) |
| AB06909 | N/A | AbSNF1 | sucrose non-fermenting kinase | Cabbage | Reduced virulence | Cho et al. (2009) |
| AB10519 | EF989017 | AbAso1 | (anastomosis-1) | Cabbage | Loss of pathogenicity | Craven et al. (2008) |
| AB08356 AB08354 AB08355 | N/A | AbNPS2 | non-ribosomal peptide synthetase | Cabbage | Age-dependent virulence reduction | Kim et al. (2007) |
| AB09565 | AY515257 | Amk1 | MAP kinase (Fus3/ Kss) | Cabbage Oil seed rape Mustard | Loss of pathogenicity | Cho et al. (2007) |
| AB01758 | DQ860091 | AbNPS6 | nonribosomal peptide synthetase (NPS6, Siderophore) | Arabidopsis | Reduced virulence | Oide et al. (2006) |

*brassicicola* including several genes involved in secondary metabolite biosynthesis (PKS and NPS). In addition to those genes listed in Table 3.3, several additional PKS and NPS genes have now also been shown to be virulence factors in *A. brassicicola* (Lawrence, unpublished).

## 3.6 Comparative Genomics

Comparison of whole genomes helps in understanding evolution (e.g., genome rearrangement) as well as genetic similarities and differences between fungal species/isolates. Such comparative genomics have been used widely in many fungal genome projects to discover the evolutionary and pathological relationship between closely and/or distantly related species (Galagan et al. 2005; Ma et al. 2010). In the *Alternaria* genomes project, besides genome annotation, comparative genomics approaches are used in many cross-genome analyses such as studying genome rearrangement (whole genome alignment), comparing whole genome content, and analyzing homologous and species-specific genes.

### 3.6.1 Whole Genome Alignment and Genome Rearrangement

Whole genome alignment is an effective tool to study evolution events such as genome rearrangements. For example, in a pairwise comparison, the genomic sequences (contigs, supercontigs, etc.) of two fungal genomes of interest are aligned. Mauve progressive whole genome aligner (Darling et al. 2004) is used for alignment of *Alternaria* genomes. When aligning sequences of large contigs and supercontigs, local alignments are initiated and expanded that result in multiple aligned blocks. With respect to the direction (strand) of the aligned blocks, they are classified into two groups: non-inverted blocks (+/+ alignment, plus strand) and inverted blocks (+/− or −/+ alignment, reverse strand). This information is used to estimate genomic inversion events. Since the contigs/supercontigs has no strand information, in order to estimate genomic inversion ratio, it is assumed that when aligning two contigs/supercontigs, the total length of inverted aligned blocks is always smaller than the total length of the non-inverted aligned blocks. Therefore, the first aligned block is considered the plus strand, and strands assigned for other blocks depending on whether it is the same direction as the first block or not. The length of the same strand blocks are totaled, and a plus strand designation is reassigned to all the blocks that account for the larger total length, and the reverse strand assigned to the rest of the blocks. An example of whole genome alignment can be seen in Fig. 3.4. These analyses collectively suggest that genome rearrangements have occurred at a higher rate in *A. brassicicola* compared to *A. alternata*. For example, results of genome alignment and homology analyses between *A. brassicicola* and *A. alternata* show that 49 % of non-syntenic genes in *A. brassicicola* are homologous to *A. alternata* genes, while only 31 % of non-syntenic genes in *A. alternata* are homologous to *A. brassicicola* genes. Interestingly, 86 % of *A. brassicicola* specific genes are in syntenic regions (aligned regions between *A. alternata* and *A. brassicicola* genomes). Likewise, 70 % of *A. alternata* specific genes are in the syntenic regions. One might speculate that each species faced some similar but also unique evolutionary pressures prompting genome rearrangements. For example, co-evolutionary pressure between plant and pathogen in the case of *A. brassicicola* may have contributed to a higher overall level of genome rearrangement compared to *A. alternata* leading a primarily saprophytic lifestyle.

### 3.6.2 Genome General Function Comparison

General functional comparison provides an overview of the similarity and difference between genomes and is often done at high-level functional categories. The annotation pipeline assigned approximately 6,179 (53 %), 6,265 (51 %), and 5,406 (51 %) of *Aa1*, *Aa2*, and *Ab* predicted proteins to at least one KOG group. Compared to the *A. brassicicola* genome, high-level KOG classification for both *A. alternata* genomes showed very similar composition in most functional categories, except

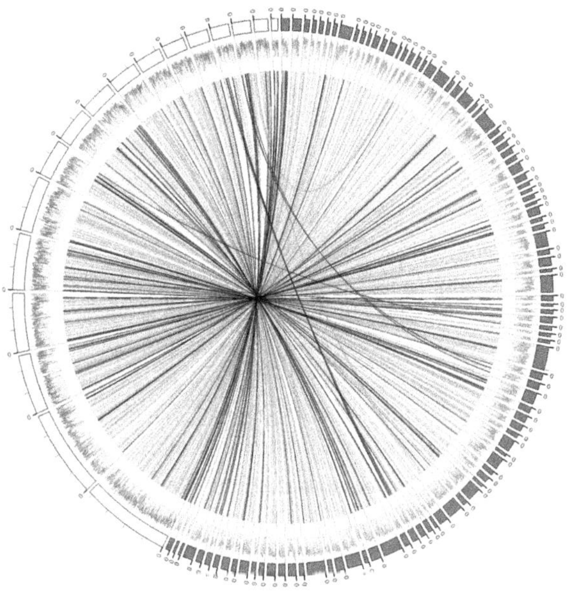

| | *A. alternata* | *A.brassicicola* |
|---|---|---|
| Genome Size | ~33.2Mb | ~32Mb |
| Aligned bases | 89% | 87% |
| Overall identity | 67% | 71% |
| Inverted blocks | 10% | 9% |
| Non-syntenic genes | 1,299 | 517 |
| Homologous genes | 8,632 | 8,612 |
| Specific genes | 3,003 | 1,902 |
| Non-syntenic homologous genes | 397 (31%) | 255 (49%) |
| Syntenic specific genes | 2,101 (70%) | 1,640 (86%) |

**Fig. 3.4** Whole genome pairwise alignment of *A. alternata* ATCC 66981 *(Aa1)* and *A. brassicicola*. In the left circos plot, the outmost track displays *Aa*1 contigs (*gray*) and *Ab* supercontigs (*white*). The ribbons link aligned blocks (*gray* indicates putative non-reversed strand alignment and black indicates putative reversed strand alignment). The histogram (second track from the outside) shows percent similarity between aligned blocks. Only contigs/supercontigs longer than 100 kb and alignment blocks longer than 10 kb are displayed. Alignment statistics is shown on the right

for significant overrepresentation of genes associated with secondary metabolites, and both lipid and carbohydrate transport, and metabolism. This result may partially explain why *A. alternata* is a more ubiquitous saprophyte compared to *A. brassicicola*.

In GO annotation, we were able to assign 2,103 GO terms to 6,447 genes (~61 % of genes) for *A. brassicicola,* making up 24,529 gene-GO term unique assignments. For *Aa*1, 29,680 gene-GO term pairs were assigned (2,170 GO terms and 7,093 genes). For *Aa*2, 30,523 gene-GO term pairs were assigned (2,232 GO terms and 7,319 genes). The ratios of total genes that were found to match with some GO terms (and also previously with KOG) are roughly equal among the three fungi (61 % for *Ab*, 61 % for *Aa*1, and 59 % for *Aa*2). This indicates the amounts of annotation information that we obtained for the three fungi are approximately equal, which made our comparative genomics approaches reliable despite the fact that different technologies were used to sequence the *A. alternata* (454-GS-Flx Titanium) and *A. brassicicola* (ABI-Sanger) genomes at different coverage levels (6× for *Ab*, ~20× for *Aa*1 and *Aa*2).

We then mapped those GO terms to higher level GO slim terms using the GO-perl package (http://search.cpan.org/~cmungall/go-perl/). We created an *Alternaria* GO slim subset using a slightly modifying subset developed previously for *Aspergillus* (Arnaud et al. 2010). We added 10 more high-level terms to the GO slim subset which are ancestors of most of the GO terms, that we were unable to map to the higher level terms other than the three top level terms (biological process, molecular function, and cellular component). The top GO categories (assigned to more than 15 % genes) for all species are metabolic process (GO:0008152), membrane (GO: 0016020), nucleotide binding (GO:0000166), ion binding (GO:0043167), oxidoreductase activity (GO:0016491), and hydrolase activity (GO: 0016787). The functional classifications of the two *A. alternata* and *A. brassicicola* genomes are very similar in each of the biological process, molecular function, and cellular component categories. The *A. brassicicola* genome had slightly more genes mapped to DNA, RNA, nucleotide, and nucleic acid binding categories, while *A. alternata* genomes had slightly higher numbers of genes mapped to GO terms related to metabolism, similar to results of KOG annotation.

Results from PFAM annotation showed that the two *A. alternata* strains have dramatically more heterokaryon incompatibility domain (HET, PF06985) containing proteins than *A. brassicicola*. Heterokaryon incompatibility genes are known to prevent hyphal fusion of genetically different filamentous fungi (Glass et al. 2000). Moreover, heterokaryon incompatibility genes have been proposed to represent a vegetative self/nonself recognition system preventing heterokaryon formation between unlike individuals to limit horizontal transfer of cytoplasmic infectious elements. Thus, it may be interesting in the future to examine the molecular evolution of the HET-like proteins in *A. alternata* to gain more insight into the possible reasons for this dramatic increase compared to *A. brassicicola*. Besides HET domain containing protein coding genes, we found that dipeptidyl peptidases (X-Pro dipeptidyl-peptidase) are missing in *A. brassicicola* while the two *A. alternata* genomes each have six copies. Interestingly, a dipeptidyl peptidase has been found to be a virulence factor during infection of immune-compromised mice by *Aspergillus* in addition to being an allergen (Beauvais et al. 1997). The over-abundance of X-Pro dipetidyl-petidases in *A. alternata,* as well as the increased number of total proteases and other allergen homologs described earlier when compared with *A. brassicicola* suggests that *A. alternata* possesses important genes that may contribute to inflammation in humans.

### 3.6.3 Homology and Specific Gene Analysis

Homology analysis (ortholog and paralog identification) for *Alternaria* genomes was performed using a bi-directional best BLAST hit and Markov clustering method via OrthoMCL software (Li et al. 2003). Four Dothideomycete fungi chosen for the analysis were: *A. alternata ATCC 66981 (Aa1),* *A. brassicicola, Leptosphaeria maculans* (closely related brassica pathogen), and *Stagonospora nodorum* (closely related wheat pathogen).

Among four fungi, 6,402 orthologous groups were identified. Interestingly, *Aa*1 had 1,690 specific protein families and *Ab* had 1,589 specific protein families. We found 73 gene/protein families that were specific to the two brassica pathogens *A. brassicicola* and *L. maculans* (*Ab/Lm* specific proteins). Most of these protein families are of unknown function. However, three CAZY-type enzymes were among the *Ab/Lm* specific proteins including a predicted secreted alpha-L-fucosidase (glycoside hydolase family 29), a glycosyl transferase family 90, and a predicted sercreted glycoside hydrolase family 105. It is possible these enzymes are important for degrading cell wall components that are more specific or abundant in plants found in the Brassicaceae family. Most of the genes that encode those protein families, were found to be highly expressed during infection of cabbage by *Ab* (top 10 % expressed genes, FPKM > 170), according to RNA-seq data from Srivastava et al. (2012). GH105 is among the top 100 highly expressed genes in *Ab* during infection with FPKM > 1,000 for both wild type and mutant of a virulence factor *Abvf*19.

Results of four-species ortholog analysis were also used to infer ortholog groups for *Aa*1 and *Ab*. The two species share 8,255 ortholog groups, *Aa*1 has 2,872 specific gene/protein families (3,003 genes) and *Ab* has 1,875 specific gene/protein families (1,902 genes). Gene ontology analysis between the unique genes of the two species shows that *Ab* possesses many more specific genes related to nucleotide binding while *Aa*1 has more specific genes that facilitate metabolism. For example, *Ab* is enriched in GO:0003676 (nucleic acid binding), GO:0003723 (RNA binding) while *Aa*1 is enriched in GO:0008152 (metabolic process), GO:0016491 (oxidoreductase activity), and GO:0016787 (hydrolase activity).

KOG comparison between *Aa*1 specific genes, *Ab* specific genes, and *Aa*1/*Ab* orthologous genes revealed that *Aa*1 has greater gene expansion in many categories, including secondary metabolites, carbohydrate, and lipid metabolism that may again explain the reason why *A. alternata* is so ubiquitous due to an increased capacity to utilize diverse carbon sources in the environment. For example, 11 % (1,119 genes) of the *Aa*1/*Ab* homologous genes that were able to be characterized by KOG were annotated with carbohydrate transport and metabolism, while the ratios for *Aa*1 and *Ab* are 16 % (173 genes, increased) and 9 % (42 genes, reduced). For the lipid transport and metabolism, the proportion of genes for the homologous gene set and *Aa*1, *Ab* specific gene sets are 14 % (1,361 genes), 22 % (231 genes), and 16 % (76 genes), respectively. This is again evidenced that *Aa*1 has more genes related to metabolism agreeing with GO classification, PFAM, and CAZY analysis. Interestingly, in a few KOG categories, *Aa*1 showed gene expansion while in *Ab*, gene reduction or no change was observed, including carbohydrate transport and metabolism, defense mechanism, cell wall biogenesis—inorganic ion transport and metabolism, and energy production and conversion.

## 3.7 Housing, Visualization, and Distribution of Genomic Data

It is very important to make results from genome annotation, genome comparison, and various other computational analyses easily accessible to scientists. Web-based genome browsers are the most commonly used tool to present, visualize, and distribute such data. A number of web-based genome browsers have been developed and widely adopted, among which the most popular and feature rich platforms are UCSC (Karolchik et al. 2003), EBI Ensembl (Flicek et al. 2011), NCBI Map Viewer (Wolfsberg 2010), JGI VISTA (Dubchak 2008), and Gbrowse (Donlin 2002). For housing and displaying *Alternaria* genome sequences and annotation features, the Ensembl platform was ultimately chosen because of its strength in visualizing comparative genomics data.

An Ensembl-independent tool (named EnsImport) was developed at the VBI that allows one to easily import various types of genome annotation and comparison data to an Ensembl-based genomic core database and comparative database. These data include genome assemblies, gene models/transcripts/proteins, multiple genome functional annotations, gene/protein/whole genome alignments, and homology data. This tool has been used, along with genome annotation and comparison pipelines, to integrate *Alternaria* fungal genome annotations and comparative genomics data into an installation of Ensemble (version 68.0) at the VBI. The EnsImport tool is written in Perl/BioPerl and it supports standard formats such as FASTA, AGP, GFF3 and XMFA, and outputs from widely used tools such as BLAST, Interpro, RepeatMasker, OrthoMCL, and Blast2GO. EnsImport provides a "point and-click" solution for ones who prefer to use only the browser features of Ensembl to present genomes and comparative data analyzed outside of Ensembl. For those who mainly rely on Ensembl pipelines, this tool can assist with integrating additional analyses not available in Ensembl into Ensembl databases. EnsImport has been extensively used to port *Alternaria* genome annotation and comparison to the customized installation of Ensembl at the VBI (*Alternaria* Genomes Database) that will be publicly available in the near future.

## References


Abbas HK, Tanaka T, Shier WT (1995) Biological activities of synthetic analogues of Alternaria alternata toxin (AAL-toxin) and fumonisin in plant and mammalian cell cultures. Phytochemistry 40:1681–1689

Agrios GN (1997) Plant pathology, 4th edn. Academic Press, San Diego

Akamatsu H, Itoh Y, Kodama M et al (1997) AAL-toxin-deficient mutants of Alternaria alternata tomato pathotype by restriction enzyme-mediated integration. Phytopathology 87:967–972. doi:10.1094/PHYTO.1997.87.9.967

Allen JE, Salzberg SL (2005) JIGSAW: integration of multiple sources of evidence for gene prediction. Bioinformatics 21:3596–3603. doi:10.1093/bioinformatics/bti609

Altschul SF, Madden TL, Schäffer AA et al (1997) Gapped BLAST and PSI-BLAST: a new generation of protein database search programs. Nucleic Acids Res 25:3389–3402. doi:10.1093/nar/25.17.3389

Anaissie EJ, Bodey GP, Rinaldi MG (1989) Emerging fungal pathogens. Eur J Clin Microbiol Infect Dis 8:323–330. doi:10.1007/BF01963467

Andersson M, Downs S, Mitakakis T et al (2003) Natural exposure to Alternaria spores induces allergic rhinitis symptoms in sensitized children. Pediatr Allergy Immunol 14:100–105

Arnaud MB, Chibucos MC, Costanzo MC et al (2010) The aspergillus genome database, a curated comparative genomics resource for gene, protein and sequence information for the aspergillus research community. Nucleic Acids Res 38:D420–D427. doi:10.1093/nar/gkp751

Baidyaroy D, Brosch G, Graessle S et al (2002) Characterization of inhibitor-resistant histone deacetylase activity in plant-pathogenic fungi. Eukaryot Cell 1:538–547

Bao Z, Eddy SR (2002) Automated de novo identification of repeat sequence families in sequenced genomes. Genome Res 12:1269–1276. doi:10.1101/gr.88502

Beauvais A, Monod M, Wyniger J et al (1997) Dipeptidyl-peptidase IV secreted by Aspergillus fumigatus, a fungus pathogenic to humans. Infect Immun 65:3042–3047

Bendtsen JD, Nielsen H, von Heijne G, Brunak S (2004) Improved prediction of signal peptides: SignalP 3.0. J Mol Biol 340:783–795. doi:10.1016/j.jmb.2004.05.028

Benson DA, Karsch-Mizrachi I, Clark K et al (2011) GenBank. Nucleic Acids Res 40:D48–D53. doi:10.1093/nar/gkr1202

Bernal A, Crammer K, Pereira F (2012) Automated gene-model curation using global discriminative learning. Bioinformatics 28:1571–1578. doi:10.1093/bioinformatics/bts176

Berto P, Belingheri L, Dehorter B (1997) Production and purification of a novel extracellular lipase from Alternaria brassicicola. Biotechnol Lett 19:533–536. doi:10.1023/A:1018333219304

Berto P, Comménil P, Belingheri L, Dehorter B (1999) Occurrence of a lipase in spores of Alternaria brassicicola with a crucial role in the infection of cauliflower leaves. FEMS Microbiol Lett 180:183–189

De Bievre C de (1991) Alternaria spp. pathogenic to man: epidemiology. J Mycol Med 118:50–58

Birney E, Clamp M, Durbin R (2004) Genewise and genomewise. Genome Res 14:988–995. doi:10.1101/gr.1865504

Blanco E, Parra G, Guigó R (2007) Using geneid to identify genes. Curr Protoc Bioinforma Ed Board Andreas Baxevanis Al Chapter 4:Unit 4.3. doi:10.1002/0471250953.bi0403s18

Bottalico A, Logrieco A (1998) Toxigenic Alternaria species of economic importance. Mycotoxins Agric Food Saf 65:108

Bowyer P, Denning DW (2007) Genomic analysis of allergen genes in Aspergillus spp.: the relevance of genomics to everyday research. Med Mycol 45:17–26. doi:10.1080/13693780600972907

Bowyer P, Fraczek M, Denning DW (2006) Comparative genomics of fungal allergens and epitopes shows widespread distribution of closely related allergen and epitope orthologues. BMC Genom 7:251. doi:10.1186/1471-2164-7-251

Braumann I, van den Berg M, Kempken F (2008) Strain-specific retrotransposon-mediated recombination in commercially used *Aspergillus niger* strain. Mol Genet Genomics 280:319–325. doi:10.1007/s00438-008-0367-9

Breitenbach M, Simon-Nobbe B (2002) The allergens of Cladosporium herbarum and Alternaria alternata. Chem Immunol 81:48–72

Cantarel BL, Coutinho PM, Rancurel C et al (2009) The carbohydrate-active enzymes database (CAZy): an expert resource for glycogenomics. Nucleic Acids Res 37:D233–D238. doi:10.1093/nar/gkn663

Cho Y, Cramer RA Jr, Kim K-H et al (2007) The Fus3/Kss1 MAP kinase homolog Amk1 regulates the expression of genes encoding hydrolytic enzymes in Alternaria brassicicola. Fungal Genet Biol 44:543–553. doi:10.1016/j.fgb.2006.11.015

Cho Y, Davis JW, Kim K-H et al (2006) A high throughput targeted gene disruption method for Alternaria brassicicola functional genomics using linear minimal element (LME) constructs. Mol Plant-Microbe Interact (MPMI) 19:7–15. doi:10.1094/MPMI-19-0007

Cho Y, Kim K-H, La Rota M et al (2009) Identification of novel virulence factors associated with signal transduction pathways in Alternaria brassicicola. Mol Microbiol 72:1316–1333

Cho Y, Ohm RA, Grigoriev IV, Srivastava A (2013) Fungal-specific transcription factor AbPf2 activates pathogenicity in Alternaria brassicicola. Plant J 75:498–514. doi:10.1111/tpj.12217

Cho Y, Srivastava A, Ohm RA et al (2012) Transcription factor Amr1 induces melanin biosynthesis and suppresses virulence in Alternaria brassicicola. PLoS Pathog 8:e1002974. doi:10.1371/journal.ppat.1002974

Conesa A, Gotz S, Garcia-Gomez JM et al (2005) Blast2GO: a universal tool for annotation, visualization and analysis in functional genomics research. Bioinformatics 21:3674–3676. doi:10.1093/bioinformatics/bti610

Cooke DEL, Jenkins PD, Lewis DM (1997) Production of phytotoxic spore germination liquids by Alternaria brassicae and A. brassicicola and their effect on species of the family brassicaceae. Ann Appl Biol 131:413–426. doi:10.1111/j.1744-7348.1997.tb05169.x

Craven KD, Vélëz H, Cho Y et al (2008) Anastomosis is required for virulence of the fungal necrotroph Alternaria brassicicola. Eukaryot Cell 7:675–683

Daboussi M-J, Capy P (2003) Transposable elements in filamentous fungi. Annu Rev Microbiol 57:275–299. doi:10.1146/annurev.micro.57.030502.091029

Darling ACE, Mau B, Blattner FR, Perna NT (2004) Mauve: multiple alignment of conserved genomic sequence with rearrangements. Genome Res 14:1394–1403. doi:10.1101/gr.2289704

Dong J, Chen W, Crane JL (1998) Phylogenetic studies of the Leptosphaeriaceae, pleosporaceae and some other loculoascomycetes based on nuclear ribosomal DNA sequences. Mycol Res 102:151–156. doi:10.1017/S0953756297004826

Donlin MJ (2009) Using the Generic Genome Browser (GBrowse). Curr Protoc Bioinforma. Chapter 9: Unit 9.9. doi:10.1002/0471250953.bi0909s28

Dubchak I (2008) Comparative analysis and visualization of genomic sequences using VISTA browser and associated computational tools. In: Bergman NH (ed) Computational and Genomics. Humana Press, pp 3–16

Edgar RC, Myers EW (2005) PILER: identification and classification of genomic repeats. Bioinformatics 21:i152–i158. doi:10.1093/bioinformatics/bti1003

Flicek P, Amode MR, Barrell D et al (2011) Ensembl 2012. Nucleic Acids Res 40:D84–D90. doi:10.1093/nar/gkr991

Flutre T, Duprat E, Feuillet C, Quesneville H (2011) Considering transposable element diversification in de novo annotation approaches. PLoS ONE 6:e16526. doi:10.1371/journal.pone.0016526

Galagan JE, Calvo SE, Borkovich KA et al (2003) The genome sequence of the filamentous fungus neurospora crassa. Nature 422:859–868. doi:10.1038/nature01554

Galagan JE, Calvo SE, Cuomo C et al (2005) Sequencing of Aspergillus nidulans and comparative analysis with A. fumigatus and A. oryzae. Nature 438:1105–1115. doi:10.1038/nature04341

Gergen PJ, Turkeltaub PC (1992) The association of individual allergen reactivity with respiratory disease in a national sample: data from the second national health and nutrition examination survey, 1976–1980 (NHANES II). J Allergy Clin Immunol 90:579–588

Gilchrist DG (1998) Programmed cell death in plant disease: the purpose and promise of cellular suicide. Annu Rev Phytopathol 36:393–414. doi:10.1146/annurev.phyto.36.1.393

Glass NL, Jacobson DJ, Shiu PK (2000) The genetics of hyphal fusion and vegetative incompatibility in filamentous ascomycete fungi. Annu Rev Genet 34:165–186. doi:10.1146/annurev.genet.34.1.165

Halonen M, Stern DA, Wright AL et al (1997) Alternaria as a major allergen for asthma in children raised in a desert environment. Am J Respir Crit Care Med 155:1356–1361

Hane JK, Lowe RGT, Solomon PS et al (2007) Dothideomycete plant interactions illuminated by genome sequencing and EST analysis of the wheat pathogen stagonospora nodorum. Plant Cell 19:3347–3368. doi:10.1105/tpc.107.052829

Hoffman DR (1984) Mould allergens. In: Yousef A-D and Joanne FD (eds) Mould Allergy, Philadelphia, Lea and Febiger, pp 104–116

De Hoog GS, Guarro J, Figueras M, Gené J (2000) Atlas of clinical fungi. In: De Hoog GS, Guarro J, Figueras M, Gené J (eds) Centraalbureau voor schimmelcultures utrecht, the Netherlands

Horton P, Park K-J, Obayashi T et al (2007) WoLF PSORT: protein localization predictor. Nucleic Acids Res 35:W585–W587. doi:10.1093/nar/gkm259

Hunter S, Apweiler R, Attwood TK et al (2009) InterPro: the integrative protein signature database. Nucleic Acids Res 37:D211–D215. doi:10.1093/nar/gkn785

Johnson RD, Johnson L, Itoh Y et al (2000) Cloning and characterization of a cyclic peptide synthetase gene from Alternaria alternata apple pathotype whose product is involved in AM-toxin synthesis and pathogenicity. Mol Plant-Microbe Interact (MPMI) 13:742–753. doi:10.1094/MPMI.2000.13.7.742

Joubert A, Bataille-Simoneau N, Campion C et al (2011a) Cell wall integrity and high osmolarity glycerol pathways are required for adaptation of Alternaria brassicicola to cell wall stress caused by brassicaceous indolic phytoalexins. Cell Microbiol 13:62–80. doi:10.1111/j.1462-5822.2010.01520.x

Joubert A, Simoneau P, Campion C et al (2011b) Impact of the unfolded protein response on the pathogenicity of the necrotrophic fungus Alternaria brassicicola. Mol Microbiol 79:1305–1324. doi:10.1111/j.1365-2958.2010.07522.x

Jung M (2001) Inhibitors of histone deacetylase as new anticancer agents. Curr Med Chem 8:1505–1511. doi:10.2174/0929867013372058

Jurka J, Kapitonov VV, Pavlicek A et al (2005) Repbase update, a database of eukaryotic repetitive elements. Cytogenet Genome Res 110:462–467. doi:10.1159/000084979

Käll L, Krogh A, Sonnhammer EL (2004) A combined transmembrane topology and signal peptide prediction method. J Mol Biol 338:1027–1036. doi:10.1016/j.jmb.2004.03.016

Karolchik D, Baertsch R, Diekhans M et al (2003) The UCSC genome browser database. Nucleic Acids Res 31:51–54. doi:10.1093/nar/gkg129

Khaldi N, Seifuddin FT, Turner G et al (2010) SMURF: genomic mapping of fungal secondary metabolite clusters. Fungal Genet Biol FG B 47:736–741. doi:10.1016/j.fgb.2010.06.003

Kim K-H, Cho Y, LA Rota M et al (2007) Functional analysis of the Alternaria brassicicola non-ribosomal peptide synthetase gene AbNPS2 reveals a role in conidial cell wall construction. Mol Plant Pathol 8:23–39. doi:10.1111/j.1364-3703.2006.00366.x

Kim K-H, Willger SD, Park S-W et al (2009) TmpL, a transmembrane protein required for intracellular redox homeostasis and virulence in a plant and an animal fungal pathogen. PLoS Pathog 5:e1000653. doi:10.1371/journal.ppat.1000653

Kouzaki H, O'Grady SM, Lawrence CB, Kita H (2009) Proteases induce production of thymic stromal

lymphopoietin by airway epithelial cells through protease-activated receptor-2. J Immunol 183:1427–1434
Krogh A, Larsson B, von Heijne G, Sonnhammer EL (2001) Predicting transmembrane protein topology with a hidden markov model: application to complete genomes. J Mol Biol 305:567–580. doi:10.1006/jmbi.2000.4315
Kwon HJ, Owa T, Hassig CA et al (1998) Depudecin induces morphological reversion of transformed fibroblasts via the inhibition of histone deacetylase. Proc Natl Acad Sci USA 95:3356–3361
Lagesen K, Hallin P, Andreas Rodland E et al (2007) RNAmmer: consistent and rapid annotation of ribosomal RNA genes. Nucl Acids Res 160 gkm. doi:10.1093/nar/gkm160
Van Leeuwen WS (1924) Bronchial asthma in relation to climate. Proc R Soc Med 17:19
Li L, Stoeckert CJ, Roos DS (2003) OrthoMCL: Identification of ortholog groups for eukaryotic genomes. Genome Res 13:2178–2189. doi:10.1101/gr.1224503
Liu GT, Qian YZ, Zhang P et al (1991) Relationships between Alternaria alternata and oesophageal cancer. IARC Sci Publ 105:258–262
Lowe T, Eddy S (1997) tRNAscan-SE: a program for improved detection of transfer RNA genes in genomic sequence. Nucl Acids Res 25:955–964. doi:10.1093/nar/25.5.955
Ma L-J, van der Does HC, Borkovich KA et al (2010) Comparative genomics reveals mobile pathogenicity chromosomes in Fusarium. Nature 464:367–373. doi:10.1038/nature08850
MacKinnon SL, Keifer P, Ayer WA (1999) Components from the phytotoxic extract of Alternaria brassicicola, a black spot pathogen of canola. Phytochemistry 51:215–221. doi:10.1016/S0031-9422(98)00732-8
Maiero M, Bean GA, Ng TJ (1991) Toxin production by Alternaria solani and its related phytotoxicity to tomato breeding lines. Phytopathology 81:1030–1033
Marchler-Bauer A, Lu S, Anderson JB et al (2011) CDD: a conserved domain database for the functional annotation of proteins. Nucleic Acids Res 39:D225–D229. doi:10.1093/nar/gkq1189
Marchler-Bauer A, Panchenko AR, Shoemaker BA et al (2002) CDD: a database of conserved domain alignments with links to domain three-dimensional structure. Nucleic Acids Res 30:281–283. doi:10.1093/nar/30.1.281
Masunaka A, Tanaka A, Tsuge T et al (2000) Distribution and characterization of AKT homologs in the tangerine pathotype of alternaria alternata. Phytopathology 90:762–768. doi:10.1094/PHYTO.2000.90.7.762
Matsuwaki Y, Wada K, White TA et al (2009) Recognition of fungal protease activities induces cellular activation and eosinophil-derived neurotoxin release in human eosinophils. J Immunol 183:6708–6716
McKenzie KJ, Robb J, Lennard JH (1998) Toxin production by Alternaria pathogens of oilseed rape (Brassica napus). Crop research 28:67−81
Montemurro N, Visconti A (1992) Alternaria metabolites-chemical and biological data. Alternaria: Biol Plant Dis Metab, In: Topics in secondary metabolism, J. Chelkowski and A. Visconti (eds) Amsterdam, Elsevier, 449–557
Morales VM, Jasalavich CA, Pelcher LE et al (1995) Phylogenetic relationship among several Leptosphaeria species based on their ribosomal DNA sequences. Mycol Res 99:593–603. doi:10.1016/S0953-7562(09)80719-3
Nierman WC, Pain A, Anderson MJ et al (2005) Genomic sequence of the pathogenic and allergenic filamentous fungus Aspergillus fumigatus. Nature 438:1151–1156. doi:10.1038/nature04332
O'Hollaren MT, Yunginger JW, Offord KP et al (1991) Exposure to an aeroallergen as a possible precipitating factor in respiratory arrest in young patients with asthma. N Engl J Med 324:359–363
Oide S, Moeder W, Krasnoff S et al (2006) NPS6, Encoding a Nonribosomal peptide synthetase involved in siderophore-mediated iron metabolism, is a conserved virulence determinant of plant pathogenic ascomycetes. Plant Cell Online 18:2836–2853. doi:10.1105/tpc.106.045633
Otani H, Kohnobe A, Kodama M, Kohmoto K (1998) Production of a host-specific toxin by germinating spores of Alternaria brassicicola. Physiol Mol Plant Pathol 52:285–295. doi:10.1006/pmpp.1998.0147
Otani H, Kohnobe A, Narita M, et al. (2001) A new type of host-selective toxin, a protein from Alternaria brassicicola. Deliv Percept Pathog Signals Plants APS Press St Paul Minn 68–76
Peat J, Tovey E, Mellis C et al (1993) Importance of house dust mite and Alternaria allergens in childhood asthma: an epidemiological study in two climatic regions of Australia. Clin Exp Allergy 23:812–820
Pochon S, Simoneau P, Pigné S et al (2013) Dehydrin-like Proteins in the Necrotrophic fungus alternaria brassicicola have a role in plant pathogenesis and stress response. PLoS ONE 8:e75143. doi:10.1371/journal.pone.0075143
Price AL, Jones NC, Pevzner PA (2005) De novo identification of repeat families in large genomes. Bioinforma Oxf Engl 21(Suppl 1):i351–i358. doi:10.1093/bioinformatics/bti1018
Rawlings ND, Barrett AJ, Bateman A (2012) MEROPS: the database of proteolytic enzymes, their substrates and inhibitors. Nucleic Acids Res 40:D343–D350. doi:10.1093/nar/gkr987
Rice P, Longden I, Bleasby A (2000) EMBOSS: the European molecular biology open software suite. Trends Genet TIG 16:276–277
Rossmann SN, Cernoch PL, Davis JR (1996) Dematiaceous fungi are an increasing cause of human disease. Clin Infect Dis 22:73–80
Rotem J (1994) The genus Alternaria: biology, epidemiology, and pathogenicity. APS Press, St. Paul, Minn
Rouxel T, Grandaubert J, Hane JK et al (2011) Effector diversification within compartments of the Leptosphaeria maculans genome affected by repeat-induced point mutations. Nat Commun 2:202. doi:10.1038/ncomms1189

Salo PM, Arbes SJ Jr, Sever M et al (2006) Exposure to Alternaria alternata in US homes is associated with asthma symptoms. J Allergy Clin Immunol 118:892–898
Sanchez H, Bush RK (2001) A review of Alternaria alternata sensitivity. Rev Iberoam Micol 18:56–59
Spanu PD, Abbott JC, Amselem J et al (2010) Genome expansion and gene loss in powdery mildew fungi reveal tradeoffs in extreme parasitism. Science 330:1543–1546. doi:10.1126/science.1194573
Srivastava A, Cho IK, Cho Y (2013) The Bdtf1 gene in Alternaria brassicicola is important in detoxifying brassinin and maintaining virulence on Brassica species. Mol Plant-Microbe Interact (MPMI). doi:10.1094/MPMI-07-13-0186-R
Srivastava A, Ohm RA, Oxiles L et al (2012) A zinc-finger-family transcription factor, AbVf19, is required for the induction of a gene subset important for virulence in Alternaria brassicicola. Mol Plant-Microbe Interact (MPMI) 25:443–452. doi:10.1094/MPMI-10-11-0275
Stanke M, Waack S (2003) Gene prediction with a hidden Markov model and a new intron submodel. Bioinformatics 19:ii215–ii225. doi:10.1093/bioinformatics/btg1080
Stergiopoulos I, de Wit PJGM (2009) Fungal effector proteins. Annu Rev Phytopathol 47:233–263. doi:10.1146/annurev.phyto.112408.132637
Tanaka A, Shiotani H, Yamamoto M, Tsuge T (1999) Insertional mutagenesis and cloning of the genes required for biosynthesis of the host-specific AK-toxin in the Japanese pear pathotype of Alternaria alternata. Mol Plant Microbe Interact 12:691–702
Ter-Hovhannisyan V, Lomsadze A, Chernoff YO, Borodovsky M (2008) Gene prediction in novel fungal genomes using an ab initio algorithm with unsupervised training. Genome Res 18:1979–1990. doi:10.1101/gr.081612.108
The UniProt Consortium (2008) The universal protein resource (UniProt). Nucl Acids Res 36:D190–D195. doi:10.1093/nar/gkm895
Thon MR, Pan H, Diener S et al (2006) The role of transposable element clusters in genome evolution and loss of synteny in the rice blast fungus magnaporthe oryzae. Genome Biol 7:R16. doi:10.1186/gb-2006-7-2-r16
Trapnell C, Roberts A, Goff L et al (2012) Differential gene and transcript expression analysis of RNA-seq experiments with top hat and cufflinks. Nat Protoc 7:562–578. doi:10.1038/nprot.2012.016
Wang H, Li J, Bostock RM, Gilchrist DG (1996) Apoptosis: a functional paradigm for programmed plant cell death induced by a host-selective phytotoxin and invoked during development. Plant Cell Online 8:375–391
Wicker T, Sabot F, Hua-Van A et al (2007) A unified classification system for eukaryotic transposable elements. Nat Rev Genet 8:973–982. doi:10.1038/nrg2165
Wight WD, Kim K-H, Lawrence CB, Walton JD (2009) Biosynthesis and role in virulence of the histone deacetylase inhibitor depudecin from Alternaria brassicicola. Mol Plant Microbe Interact 22:1258–1267
Wilson CL, Wisniewski ME (1994) Biological control of postharvest diseases: theory and practice. CRC press
Wolfsberg TG (2010) Using the NCBI map viewer to browse genomic sequence data. Curr Protoc Bioinforma. Chapter 1:Unit 1.5.1–25. doi:10.1002/0471250953.bi0105s29
Yao C, Koller W (1995) Diversity of cutinases from plant pathogenic fungi: different cutinases are expressed during saprophytic and pathogenic stages of Alternaria brassicicola. MPMI-Mol Plant Microbe Interact 8:122–130
Yao C, Köller W (1994) Diversity of cutinases from plant pathogenic fungi: cloning and characterization of a cutinase gene from Alternaria brassicicola. Physiol Mol Plant Pathol 44:81–92

# 4 Verticillium alfalfae and V. dahliae, Agents of Verticillium Wilt Diseases

Patrik Inderbitzin, Bart P. H. J. Thomma, Steve J. Klosterman, and Krishna V. Subbarao

## 4.1 Introduction

### 4.1.1 Agricultural Relevance of *Verticillium*

Verticillium wilts are vascular wilt diseases caused by species of *Verticillium,* and are among the most devastating fungal diseases worldwide. Over 400 different plant hosts, including major agricultural crops and ornamentals (Pegg and Brady 2002), are susceptible to Verticillium wilt mainly in temperate, less frequently in subtropical, and rarely in tropical climates of the world (Hawksworth and Talboys 1970a, b; Hawksworth 1970b, c; Pegg and Brady 2002; Inderbitzin et al. 2011a). Economic losses in excess of 50 % have been reported on many high value crops, including cotton (Friebertshauser and DeVay 1982), lettuce (Subbarao et al. 1997; Atallah et al. 2011a) and potato (Rowe and Powelson 2002). In strawberry, losses of up to 75 % have been recorded in absence of soil fumigation (Wilhelm and Paulus 1980). Verticillium wilt is a significant problem on many more important crops and ornamentals (Pegg and Brady 2002), but precise loss estimates are generally unavailable.

Management of Verticillium wilt diseases is challenging. *Verticillium* species are soilborne and are able to survive in the soil in absence of a host for at least 14 years (Wilhelm 1955). Host infection is initiated through the roots, and slowly progresses through the vasculature into the shoots (Fradin and Thomma 2006). Distinct external symptoms, including wilting and tissue necrosis, typically only appear near the end of the disease cycle, often coinciding with host maturity (Isaac and Harrison 1968). *Verticillium* species are best controlled by soil fumigation, but the most effective and most widely used soil fumigant, methyl bromide, has been phased out due to environmental concerns (Enebak 2012), compounding the problem caused by *Verticillium* in agriculture.

There are currently ten *Verticillium* species (Inderbitzin et al. 2011a), of which *V. dahliae* is the most economically important infecting over 200 host species worldwide (Pegg and Brady 2002). The other species have more restricted global distributions and narrower host ranges (Inderbitzin et al. 2011a; Hawksworth and Talboys 1970a), but *V. albo-atrum* (Platt et al. 2000), *V. alfalfae* (Heale and Isaac 1963), *V. longisporum* (Dunker et al. 2008; Babadoost et al. 2004), *V. nonalfalfae* (Down et al. 2007; Radišek et al. 2006), *V. tricorpus* (Platt et al. 2000) and *V. zaregamsianum* (Usami et al. 2011) may also produce significant losses. *Verticillium nubilum* has historically been considered a

P. Inderbitzin · K. V. Subbarao (✉)
Department of Plant Pathology, University of California, Davis, CA, USA
e-mail: kvsubbarao@ucdavis.edu

B. P. H. J. Thomma
Laboratory of Phytopathology, Wageningen University, Wageningen, The Netherlands

S. J. Klosterman
USDA-ARS, Salinas, CA, USA

R. A. Dean et al. (eds.), *Genomics of Plant-Associated Fungi and Oomycetes: Dicot Pathogens*,
DOI: 10.1007/978-3-662-44056-8_4, 

minor pathogen, and causes disease on tomato and potato in pathogenicity tests (Isaac 1953). The agricultural importance of *V. isaacii* and *V. klebahnii* (Inderbitzin et al. 2011a) is unknown, but they were isolated from lettuce and artichoke, respectively (Qin et al. 2008; Inderbitzin et al. 2011a).

This chapter focuses mainly on the two agriculturally important species, *V. alfalfae* and *V. dahliae*, for which annotated genome sequences are currently available, with a tangential focus on *V. longisporum* for which a partial genome sequence has been generated.

### 4.1.2 Taxonomy and Relationship to Other Plant Pathogens

*Verticillium* is a small genus in the family *Plectosphaerellaceae* (Zare et al. 2007) that belongs to the subclass *Hypocreomycetidae,* which is part of the class *Sordariomycetes* in the phylum *Ascomycota* (Zhang et al. 2006). The *Plectosphaerellaceae* harbor several additional, generally minor plant pathogens, including *Lectera colletotrichoides* (Cannon et al. 2012), *Musicillium theobromae* (Hawksworth and Holliday 1970), and *Plectosphaerella cucumerina* (Ramos et al. 2013). Also closely related to *Verticillium* is the family *Glomerellaceae* (Zhang et al. 2006), which contains *Colletotrichum*, the causal agents of the important anthracnose diseases (Latunde-Dada 2001). Among the more distant relatives are several agents of vascular wilt, including *Ceratocystis fagacearum* (Juzwik et al. 2011) and *Fusarium oxysporum* (Michielse and Rep 2009), both of the *Hypocreomycetidae,* and *Ophiostoma novo-ulmi* (Brasier 1991) of the *Sordariomycetidae*. Since *Ceratocystis, Fusarium* and *Ophiostoma* are not closely related to each other or to *Verticillium* (Zhang et al. 2006), the ability to colonize the xylem and trigger symptoms of vascular wilt may have evolved multiple times in the *Sordariomycetes.*

*Verticillium* species are characterized morphologically by the elongate conidiogenous cells that are arranged in whorls along a vertical axis, and by the presence of resting structures that consist of thick walled, pigmented cells (Fig. 4.1). However, unrelated fungi that used to be placed in *Verticillium* share these morphological features, including *Gibellulopsis nigrescens,* the former *V. nigrescens,* generally considered a saprotroph (Hawksworth 1970a), and *M. theobromae,* the former *V. theobromae,* a nonvascular wilt pathogen of banana (Hawksworth and Holliday 1970; Zare et al. 2007; Pegg and Brady 2002).

Species boundaries and phylogenetic relationships in *Verticillium* are well resolved, and were determined by multilocus phylogenetic analyses and comparisons to herbarium type material (Inderbitzin et al. 2011a). These studies resolved longstanding controversies in *Verticillium* taxonomy (Barbara and Clewes 2003). *Verticillium dahliae* was confirmed to be a species distinct from the diploid hybrid *V. longisporum,* which originated at least three different times by separate hybridization events that involved four different parental lineages, including *V. dahliae* (Inderbitzin et al. 2011b). The fact that "*Verticillium albo-atrum*" consisted of three different lineages led to the description of two new species, *V. alfalfae* and *V. nonalfalfae* that are sister species and close relatives of *V. dahliae* and *V. longisporum.* The name *V. albo-atrum* was maintained for the lineage that is more closely related to *V. tricorpus* than to *V. alfalfae* and *V. nonalfalfae* (Inderbitzin et al. 2011a). The correct species name for the "*V. albo-atrum*" strain whose genome is described in this chapter (Klosterman et al. 2011), is *V. alfalfae.*

*Verticillium* species are challenging to identify and differentiate based on morphology. The species treated in this chapter differ by their resting structures and sizes of conidia. The resting structures of *V. dahliae* are rounded to elongate and are named microsclerotia, and the resting structures of *V. longisporum* are similar to the ones in *V. dahliae,* whereas *V. alfalfae* has resting mycelium (Inderbitzin et al. 2011a). A morphological key for species identification is available (Inderbitzin et al. 2011a), but results should be confirmed using species specific PCR assays (Inderbitzin et al. 2013), or DNA sequencing and phylogenetic analyses.

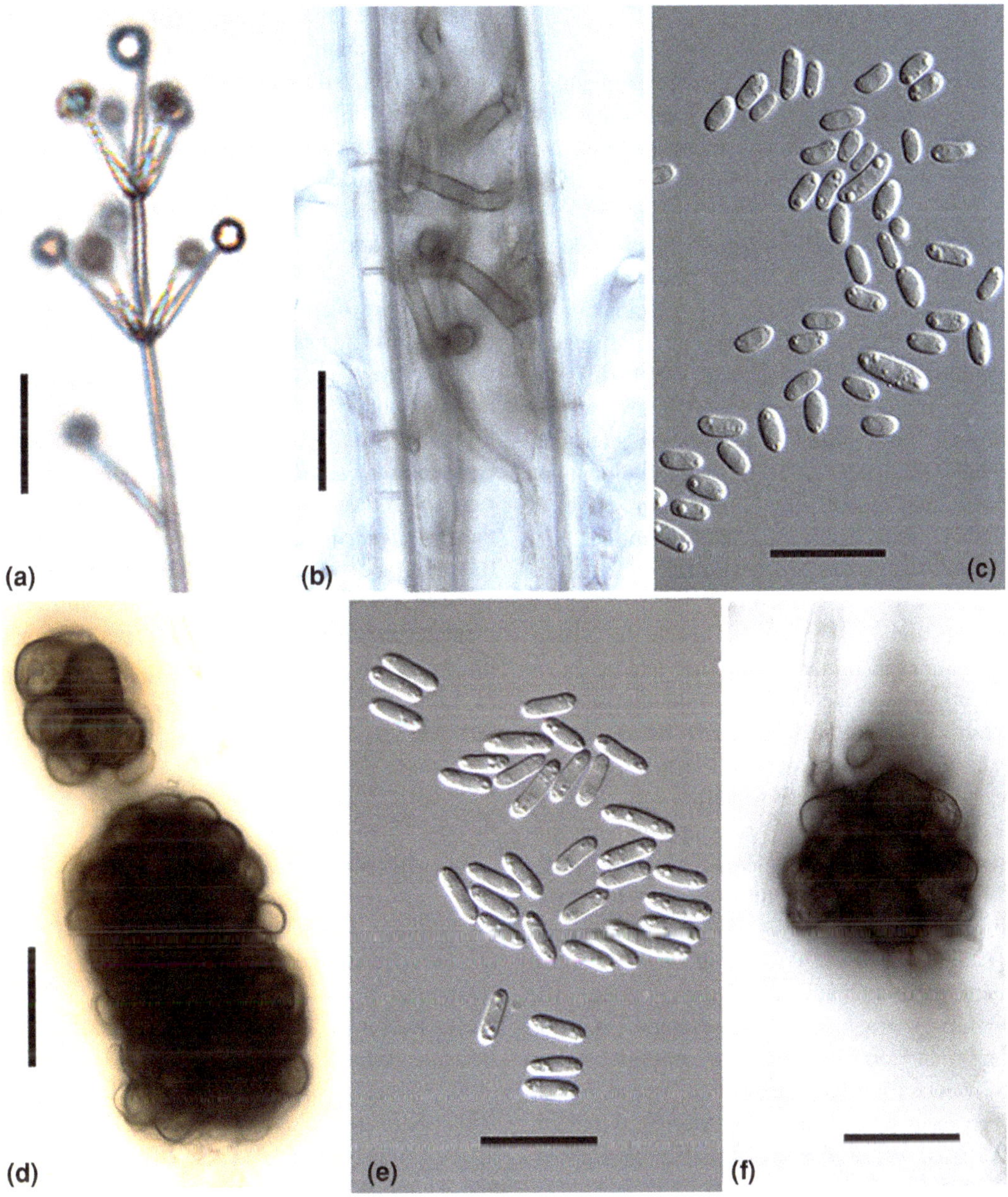


**Fig. 4.1** Select morphological features of *Verticillium alfalfae*, *V. dahliae* and *V. longisporum*. **a** Typical *Verticillium* conidiophore, from *V. alfalfae* strain PD682 after 31 days on WA-p. **b** Resting mycelium of *V. alfalfae* strain PD683 in lumen of a host cell after 32 days on WA-p. **c** Typical *Verticillium* conidia, from *V. dahliae* strain PD322 after 9 days on PDA. **d** Microsclerotia of *V. dahliae* strain PD322 after 12 days on WA-p. **e** Conidia of *V. longisporum* strain PD348 after 35 days on PDA, *V. longisporum* conidia tend to be longer than the ones of other *Verticillium* species. **f** Rounded microsclerotium of *V. longisporum* strain PD356 after 35 days on PDA, microsclerotia of *V. longisporum* are similar to the ones in *V. dahliae*. *Scale bar* **a** = 30 μm; **b–f** = 20 μm. Imaging method: **a** = DS; **b**, **d**, **f** = BF; **c**, **e** = DIC. Modified from Inderbitzin et al. (2011a), see this publication for strain information, growth media and imaging methods

### 4.1.3 Life Cycle

The life cycle of *Verticillium* species begins in the soil with dormant resting structures that in *V. dahliae* are referred to as microsclerotia. Microsclerotia are clusters of specialized, thick-walled and highly melanized fungal cells that survive in the soil or associated with plant debris (Harrison and Isaac 1969), and plant infection is initiated when microsclerotia germinate and produce one to several hyphae that extend toward host roots (Fitzell et al. 1980). How and where root infections are initiated differs between hosts, although penetration sites may include root tips as well as root elongation zones, and appressoria have been observed (Vallad and Subbarao 2008). Intracellular growth through the cortex and across the endodermis is followed by entry into the xylem (Bishop and Cooper 1983). Inside the xylem, conidiation is initiated and the conidia move upward through the xylem with the transpiration stream. Once trapped in pits or by vessel end walls (Garber and Houston 1966), the conidia germinate, and hyphae penetrate into adjacent vessels. The first wilting symptoms are displayed as the fungus moves into the shoots. Coinciding with the maturity of the susceptible host (Isaac and Harrison 1968), *V. dahliae* invades the parenchyma, conidiates, and forms microsclerotia (Fradin and Thomma 2006). The duration over which wilting symptoms develop varies. In tomato, extensive wilting appears 6–8 weeks after inoculation (Chen et al. 2004), whereas in lettuce, the first external symptoms take 8–10 weeks to appear (Vallad and Subbarao 2008).

### 4.1.4 Host Range

Verticillium wilts overwhelmingly affect dicotyledonous plants in temperate regions, more than 400 different hosts are known (Pegg and Brady 2002). It is not possible to conclusively compile host ranges for individual *Verticillium* species from literature surveys (Walker 1990; Inderbitzin and Subbarao 2014), because of inconsistent application of species names due to past taxonomic disagreements (Hawksworth and Talboys 1970b), and the inherent unreliability of species identification based on morphological characters (Inderbitzin and Subbarao 2014). But it is clear that *V. dahliae* has by far the widest host range among *Verticillium* species, affecting plants in at least fourteen different families, and maybe more than 200 hosts (Pegg and Brady 2002). Hosts include the economically important and widely planted cotton, lettuce, olives, pistachios, potato, strawberry, tomato, and sunflower. *Verticillium alfalfae* has so far only been isolated from *Medicago sativa* (alfalfa), and *V. longisporum* is generally restricted to hosts in the *Brassicaceae,* and is a major pest of canola (oilseed rape) in Europe (Heale and Karapapa 1999). Expansion of host range has been documented, one example is lettuce, where susceptibility to *V. dahliae* first developed in California in 1995 (Subbarao et al. 1997; Atallah et al. 2011a).

### 4.1.5 Genetic Transformation of *Verticillium*

*Verticillium* is amenable to *Agrobacterium tumefaciens*-mediated transformation (ATMT) (Dobinson et al. 2004), which has successfully been used for functional characterization of *Verticillium* pathogenicity related genes (Dobinson et al. 2004; Rauyaree et al. 2005; Klimes et al. 2006; Klimes and Dobinson 2006; Gao et al. 2010; Tzima et al. 2010, 2011, 2012; Santhanam and Thomma 2013; Santhanam et al. 2013; de Jonge et al. 2012), and the generation of random mutants by non target transfer-DNA (T-DNA) insertion (Knight et al. 2009; Maruthachalam et al. 2011; Santhanam 2012). These studies demonstrated that the frequency of homologous recombination in *Verticillium* is relatively high as compared to other filamentous fungi, possibly because *Verticillium* genomes contain relatively low amounts of repetitive DNA, such as transposable elements (Klosterman et al. 2011;

Amyotte et al. 2012; de Jonge et al. 2013). The more traditional protoplast-mediated transformation method (Olmedo-Monfil et al. 2004) was also performed in *Verticillium* (Dobinson 1995), but ATMT often leads to larger numbers of stable transformants and single copy T-DNA insertions than conventional transformation methods (Weld et al. 2006). Protocols for the construction of transformation vectors and ATMT transformation for *V. dahliae* were described in detail by Garcia-Pedrajas et al. (2013), Paz et al. (2011) and Santhanam (2012).

## 4.2 Verticillium Genomes and Bioinformatics Resources at the Broad Institute

The genome sequences of *V. alfalfae* strain Ms.102 (=Va.Ms.102), *V. dahliae* strain Ls.17 (=VdLs.17) and *V. longisporum* strain Bob.70 were generated by the Broad Institute, and are part of the *Verticillium* comparative genomics database where they can be accessed and are available for download (http://www.broadinstitute.org/annotation/genome/verticillium_dahliae/MultiHome.html). The *V. alfalfae* Ms.102 and *V. dahliae* Ls.17 genome sequences were analyzed by Klosterman et al. (2011).

The Broad Institute *Verticillium* site houses complete genome draft sequence, annotation, and map data for *V. dahliae* Ls.17 and *V. alfalfae* Ms.102, which on the site used to be featured under its previous name, *V. albo-atrum* Ms.102. The sequences differ in coverage and thus quality. The *V. dahliae* Ls.17 genome sequence was of the highest quality among the three genomes at 7.48-fold coverage, the *V. alfalfae* Ms.102 genome was covered 4.08-fold (Table 4.1), and the *V. longisporum* Bob.70 genome sequence was a partial draft at 0.5-fold coverage. All three genomes were produced via a shotgun strategy and Sanger sequencing methodology at the Broad Institute (Klosterman et al. 2011). Background information on the sequenced strains is given at the Broad Institute website, except for *V. longisporum* Bob.70 which originated from diseased cauliflower in California (Qin et al. 2006). Contig base quality scores and other characteristics for *V. alfalfae* Ms.102 and *V. dahliae* Ls.17 are shown in Table 4.1. All data, including contigs, assemblies and gene and protein predictions, are available to the research community under the download tab at the Broad Institute's *Verticillium* Group website.

The infrastructure at the Broad Institute website for genome analyses includes search tools that enable identification of gene homology using the local database and NCBI-linked BLAST, and offers various options to perform gene and feature searches. Genes or features can be localized on the genome synteny map, which is viewable with a maximum window size of 100 kb. The Broad Institute's ARGO genome browser (Altschul et al. 1997) that is also provided as an open source software download, allows scanning of larger windows. The feature maps allow analysis of gene organization at sites of interest across the three species, and display the Fgenesh (Salamov and Solovyev 2000), GeneID (Parra et al. 2000), and GeneMark-ES (Ter-Hovhannisyan et al. 2008) gene prediction models, expressed sequence tags (ESTs), BLASTX hits (Altschul et al. 1997), and *V. dahliae* Ls.17 and *V. longisporum* Bob.70 fosmids.

## 4.3 Genome Structure

### 4.3.1 Genome Assembly

The *V. dahliae* Ls.17 DNA sequence read data was assembled into scaffolds, which were positioned onto chromosomes by comparison to an optical map (Schwartz et al. 1993) prepared at ~300X physical coverage with restriction enzyme *Afl*II. The map showed that *V. dahliae* Ls.17 contained eight chromosomes (linkage groups) (Table 4.1) (Klosterman et al. 2011), which was similar to the seven chromosomes that were detected in a different strain of *V. dahliae* (Pantou and Typas 2005). Of the 52 *V.*

**Table 4.1** Comparative genome statistics of *V. dahliae* and *V. alfalfae*

| Assembly statistics | *V. dahliae*—Ls.17 | *V. alfalfae*—Ms.102 |
|---|---|---|
| Total scaffold length (Mb) | 33.8 | 32.8 |
| Average base coverage (Fold) | 7.48 | 4.08 |
| Quality score (% Q40) | 95.21 | 84.58 |
| $N_{50}$ contig (kb) | 43.31 | 14.31 |
| $N_{50}$ scaffold (Mb) | 1.27 | 2.31 |
| Linkage groups | 8 | Nd[a] |
| GC-content (%) | 55.85 | 56.06 |
| Protein-coding genes | 10,535 | 10,221 |
| tRNA genes | 230 | 223 |

[a] *Nd* No data, table adapted from Klosterman et al. (2011)

*dahliae* Ls.17 scaffolds, 30 were positioned on the chromosomes, while 22 scaffolds corresponding to approximately 3 Mb could not be positioned (Fig. 4.2). No optical map was available for *V. alfalfae* Ms.102, but the high sequence identity of 92 % to *V. dahliae* Ls.17 enabled assembly of the *V. alfalfae* Ms.102 genome into 26 scaffolds using *V. dahliae* Ls.17 as a reference. The *V. alfalfae* Ms.102 scaffold N50 length, a measure for the quality of the genome assembly, was 2.31 Mb, indicating that half of all bases were contained in scaffolds of at least 2.31 Mb in length, and the *V. dahliae* Ls.17 scaffold N50 length was 1.27 Mb. More than 95 % of the *V. dahliae* Ls.17 and 85 % of the *V. alfalfae* Ms.102 genome sequence had a quality score of at least 40, suggesting one error per $10^4$ bp in those regions (Table 4.1) (Klosterman et al. 2011). PCR amplification and sequencing of six protein coding loci of *V. dahliae* Ls.17 (Inderbitzin et al. 2011b) implied an error rate across these loci of 0.6 %.

The genome assemblies of *V. alfalfae* Ms.102 and *V. dahliae* Ls.17 were similar in length, with 32.8 Mb for *V. alfalfae* Ms.102, and 33.8 Mb for *V. dahliae* Ls.17, which was close to the expected 35 Mb determined by optical mapping for *V. dahliae* Ls.17 (Table 4.1) (Klosterman et al. 2011). The *Verticillium* genomes are among the smallest in the *Hypocreomycetidae*, resembling in size the genomes of the saprotroph *Trichoderma reesei* with 34 Mb (Martinez et al. 2008), and the plant pathogen *F. graminearum* with 36.1 Mb (Cuomo et al. 2007). The closest relatives of *Verticillium* that have been sequenced, *Colletotrichum graminicola* and *Co. higginsianum,* have larger genomes, with 57.4 and 53.4 Mb, respectively (O'Connell et al. 2012). The genome size of the vascular wilt pathogen *O. ulmi,* in the *Sordariomycetidae,* is smaller with 31.5 Mb (Khoshraftar et al. 2013).

A global genome alignment between *V. alfalfae* Ms.102 and *V. dahliae* Ls.17 revealed the presence of four regions of approximately 300–350 kb each in *V. dahliae* Ls.17 that were absent in *V. alfalfae* Ms.102 (Fig. 4.2) (Klosterman et al. 2011). The four regions were named lineage-specific (LS) regions 1 to 4, and are discussed in more detail in Sect. 4.4.1.

## 4.3.2 Protein Coding Genes

### 4.3.2.1 Numbers of Genes and Functional Categories

Gene candidates in the *V. alfalfae* Ms.102 and *V. dahliae* Ls.17 genomes were predicted by the gene calling pipeline of the Broad Institute (Haas et al. 2011). The approach took into consideration previously characterized genes, and three different EST libraries of *V. dahliae* Ls.17 cultured in complete medium (~10,000 ESTs), lettuce root extract medium (~5,000), or low nitrogen medium (~5,000), that were used in computational and manual approaches to

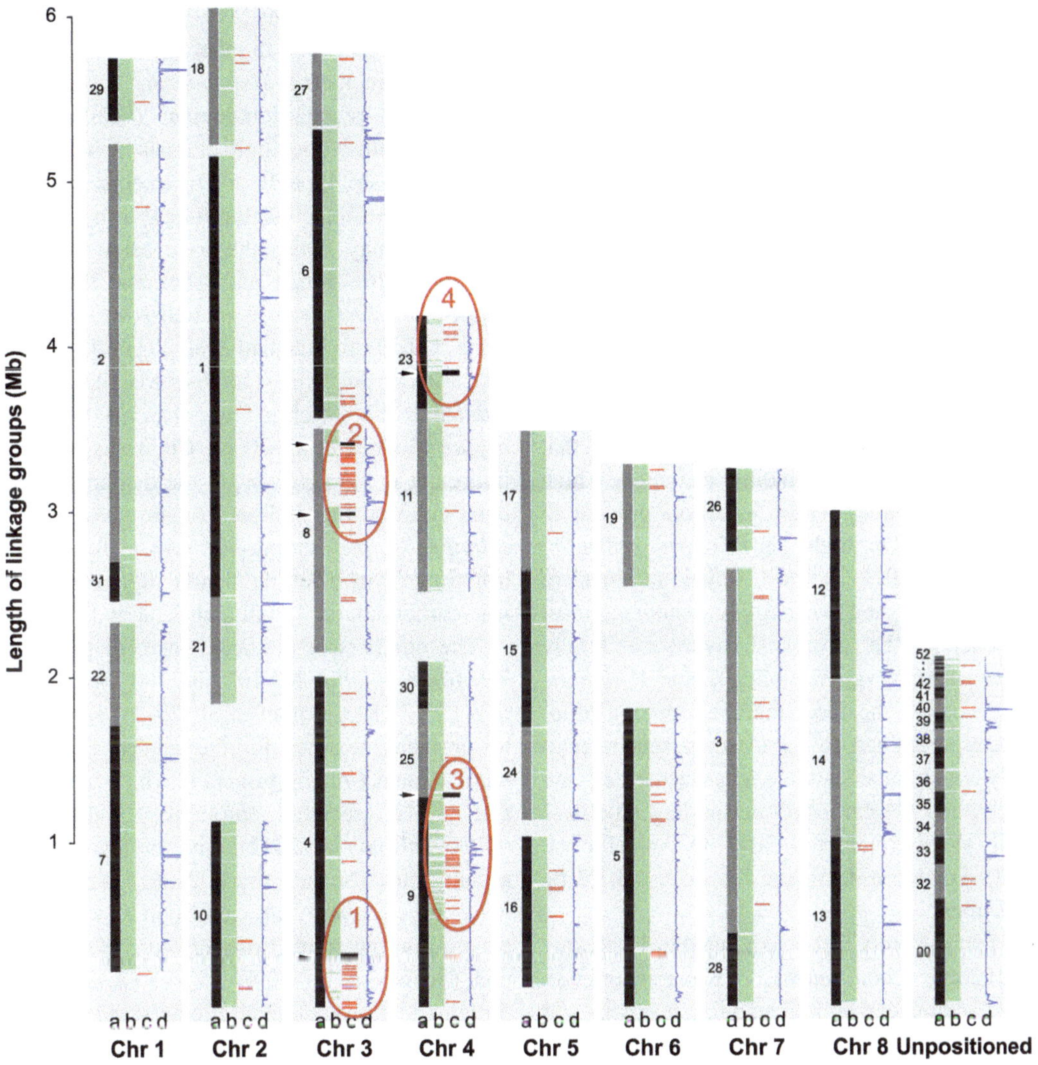


**Fig. 4.2** Syntenic alignments between *V. alfalfae* Ms.102 and *V. dahliae* Ls.17 genome sequences, and the distribution of transposable elements and EST alignments. *Verticillium dahliae* linkage groups (*black bars*) are shown as the reference, and the length of the *light grey* background to the *left* of each linkage group (in the scale of Mb) is defined by the *V. dahliae* optical map. For each chromosome, column **a** represents the *V. dahliae* genomic scaffolds positioned on the optical linkage groups separated by scaffold breaks. Scaffold numbers are adjacent to the blocks; column **b** displays the syntenic mapping of *V. alfalfae* scaffolds; column **c**, color *red* shows the density of transposable elements calculated with a 10 kb window, and color *black* represents the AT-rich regions; column **d** represents the density of ESTs calculated with a 10 kb window. Four LS regions that lack similarity to the genome of *V. alfalfae*, but are enriched for TEs are highlighted in red ovals and numbered as LS1, 2, 3, and 4. From Klosterman et al. (2011)

develop gene models for *V. alfalfae* Ms.102 and *V. dahliae* Ls.17 (Klosterman et al. 2011). There were 10,221 predicted protein-encoding genes for *V. alfalfae* Ms.102, and 10,535 for *V. dahliae* Ls.17 (Table 4.1). This is similar to other plant pathogens in the *Sordariomycetes*, including *O. ulmi* with 8639 genes (Khoshraftar et al. 2013), *Magnaporthe oryzae* with 11,109 genes (Dean et al. 2005), or the saprotroph *Neurospora crassa* with 10,082 genes (Galagan et al. 2003). More genes were projected for the *Verticillium* relatives *Co. graminicola* and *Co. higginsianum*, with 12,006 and 16,172, respectively (O'Connell et al. 2012). Functional categories have not been assigned for the majority of *V. alfalfae* Ms.102 and *V. dahliae* Ls.17 genes (Klosterman et al. 2011).

#### 4.3.2.2 Gene Family Expansions and Losses

Gene families with potential roles in pathogenicity and virulence were examined in detail in *V. alfalfae* Ms.102 and *V. dahliae* Ls.17 (Klosterman et al. 2011). These included the carbohydrate-active enzymes (CAZymes) that comprise cell wall-degrading enzymes, which are involved in the host infection process in *Verticillium* (Durrands and Cooper 1988a, b; Tzima et al. 2011). Some CAZyme families are expanded in the *Verticillium* genomes, most prominently the glycoside hydrolase family GH88 where *V. alfalfae* Ms.102 and *V. dahliae* Ls.17 contain the most members among ascomycetes examined so far. Also present in relatively high numbers are the polysaccharide-lyase family, and the carbohydrate-binding module-1 family (Klosterman et al. 2011; O'Connell et al. 2012). For more detail see the Sect. 4.4.1.2 on CAZymes.

Another family that is expanded in *V. alfalfae* Ms.102 and *V. dahliae* Ls.17 (Klosterman et al. 2011), is the gene family encoding necrosis- and ethylene-inducing-like proteins (NLPs) (Pemberton and Salmond 2004). In *V. dahliae* Ls.17, the number of NLPs is highly conserved between different isolates, but only two of the seven members in *V. dahliae* Ls.17 had direct roles in causing disease (Santhanam et al. 2013). For more, see Sect. 4.4.1.1 on effectors below.

Similar to other ascomycete plant pathogens (Soanes et al. 2008), the genomes of *V. dahliae* Ls.17 and *V. alfalfae* Ms.102 are enriched in proteins containing chitin-recognition motifs (Pfam ID: PF00187) and cutinases (Pfam ID: PF01083). There are 20 chitin-recognition domains in *V. alfalfae* Ms.102 and 28 in *V. dahliae* Ls.17, which are part of nine different genes in *V. alfalfae* Ms.102 and ten in *V. dahliae* Ls.17. The numbers of chitin-recognition domains in *V. alfalfae* Ms.102 and *V. dahliae* Ls.17 are higher than in other ascomycete plant pathogens, including *Magnaporthe oryzae* that has 18 chitin-recognition domains (Soanes et al. 2008). Chitin-recognition domains, including those of LysM type B or D proteins, may be beneficial in protecting the fungus from chitinases, and may contribute to evasion of host perception (de Jonge and Thomma 2009; de Jonge et al. 2010; van Esse et al. 2007; van den Burg et al. 2006; Marshall et al. 2011). There are 14 cutinases in both *V. alfalfae* Ms.102 and *V. dahliae* Ls.17, which is fewer than the 17 of *M. oryzae* (Soanes et al. 2008). Cutinases may be important in *Verticillium* for the degradation of plant cutin during the saprotrophic phase. In *F. solani* f. sp. *pisi*, cutinase is critical for saprotrophic growth when cutin was supplied as the sole carbon source (Stahl and Schäfer 1992).

The numbers of predicted secreted proteins, 759 for *V. alfalfae* Ms.102 and 780 for *V. dahliae* Ls.17 (Klosterman et al. 2011), are much lower than the ones for the related *Co. graminicola* and *Co. higginsianum,* which had 1650 and 2142 secreted proteins, respectively (O'Connell et al. 2012), and is low among ascomycetes (Soanes et al. 2008). Within the *Hypocreomycetidae,* the saprotroph *T. reesei* had a similarly low number of secreted proteins with 663 (Soanes et al. 2008). Small cysteine-rich proteins that are potential effectors and are part of the secretome are discussed in Sect. 4.4.1.1.

Comparative analyses of 47 transcription factor families between *V. alfalfae* Ms.102 and *V. dahliae* Ls.17 and six related pathogenic and nonpathogenic ascomycetes showed that in general, the transcription factor families were of similar sizes across all the species (Klosterman et al. 2011). But for several families, there were fewer members in *V. alfalfae* Ms.102 and *V. dahliae* Ls.17 than in related plant pathogens. For instance, the bHLH transcription factors had four members in *V. alfalfae* Ms.102, ten in *V. dahliae* Ls.17, and 46 in *F. oxysporum,* whereas *M. oryzae* was similar to *Verticillium* with nine members. The numbers of bZIP and $C_2H_2$ transcription factors followed a similar pattern in these species, but for the homeobox and MADS-box transcription factors, *V. dahliae* Ls.17 had the most members among the fungi examined with 15 and 7, respectively, as compared to 14 and 4, respectively, for *F. oxysporum* (Klosterman et al. 2011).

There were some major differences in gene content between the two *Verticillium* species, as 1102 genes were unique to *V. alfalfae* Ms.102, and 1357 were unique to *V. dahliae* Ls.17 (Klosterman et al. 2011). The full extent of the differences in gene content has not been explored, and may be partly due to the lower sequencing coverage in *V. alfalfae* Ms.102. However, some of the additional genes present in *V. dahliae* Ls.17 but absent in *V. alfalfae* Ms.102, including bZIP transcription factors, phospholipases, and ferric reductases, were significantly enriched within the Lineage-Specific (LS regions) (Fig. 4.2) (Klosterman et al. 2011), see Sect. 4.4.2 for more detail.

### 4.3.3 Repetitive DNA

#### 4.3.3.1 Overview

*Verticillium alfalfae* Ms.102 and *V. dahliae* Ls.17 differed in repetitive DNA content. Whereas less than 200 kb or 0.6 % of the *V. alfalfae* Ms.102 genome was repetitive, 1.68 Mb or 5 % of *V. dahliae* Ls.17 was repetitive. This difference was partly due to transposable elements that were more abundant in *V. dahliae* Ls.17 than in *V. alfalfae* Ms.102 (Klosterman et al. 2011), but may also partly reflect the lower sequencing coverage of the *V. alfalfae* Ms.102 genome with respect to *V. dahliae* Ls.17. Among the *Sordariomycetes*, *F. graminearum* and *M. oryzae* had similar levels of repetitive DNA, with 0.7 and 9.7 %, respectively (Cuomo et al. 2007; Altschul et al. 1997).

#### 4.3.3.2 Transposable Elements

Transposable elements or transposons are divided into two major classes, Class I and Class II (Kapitonov and Jurka 2008; Wicker et al. 2007). The *V. dahliae* Ls.17 genome encodes 56 complete Class I retrotransposons of the Gypsy, Copia, and LINE-like types, and 34 full-length Class II transposons of the "cut-and-paste" superfamilies Tc1/mariner, Activator, and Mutator (Amyotte et al. 2012). The *V. alfalfae* Ms.102 genome contains only one full-length Class I Copia-like retrotransposon, as well as one degenerate and one partial Copia-like element, and only short, degenerate sequences of the Gypsy-like type. Neither complete nor partial Class II transposons were found in *V. alfalfae* Ms.102 (Amyotte et al. 2012). All transposons were expressed in at least one of three growth conditions that were used for EST generation (Amyotte et al. 2012; Klosterman et al. 2011).

The transposons in *V. dahliae* Ls.17 were unevenly distributed, as more than half of all transposons were clustered within the four LS regions (Fig. 4.2), which may be related to the gene family expansions in those regions (Sect. 4.4.2) (Klosterman et al. 2011).

#### 4.3.3.3 Evidence of Repeat-Induced Point Mutations

Fungi have evolved Repeat-Induced Point Mutations (RIP) as a defense mechanism to guard against genome infiltration of potentially deleterious repetitive DNA, including transposons. RIP is linked to meiosis, and introduces C:G to T:A transitions into duplicate sequences (Cambareri et al. 1989). There is evidence for the presence of RIPed sequences in *V. dahliae* Ls.17. Long tracts of AT-rich DNA (1–5 kb) flank the LS regions in *V. dahliae* Ls.17, suggesting that a similar defense mechanism is operational in the genome of *V. dahliae* Ls.17. Class I transposons of the Gypsy-type showed RIP-like mutations (Klosterman et al. 2011). However, there were no signs of RIP in Copia or LINE-like retrotransposons, or any of the Class II transposons of *V. dahliae* Ls.17 (Amyotte et al. 2012). This may be due to several reasons, including increased susceptibility of one class of transposons to RIP, and introgression, possibly via horizontal transfer, of previously RIPed sequences or of non-RIPed transposons after RIP ceased to function or *V. dahliae* ceased to have sex. The *V. alfalfae* Ms.102 and *V. dahliae* Ls.17 genomes encode orthologs of RID (Klosterman et al. 2011), which functions in the RIP machinery of *N. crassa* (Freitag et al. 2002). However, sexual reproduction has not been observed in *V. alfalfae* Ms.102 and *V. dahliae* Ls.17, and it is unclear whether RIP currently occurs in these fungi.

## 4.4 Comparative Genomics

### 4.4.1 Similarities to Other Plant Pathogenic Fungi

#### 4.4.1.1 Effectors

Effectors are molecules secreted by pathogens during colonization of their hosts that function to modulate host physiology, often through suppression of host defenses, or to protect the pathogen from host defense responses employed to halt pathogen growth (de Jonge et al. 2011). Intense research efforts have focused on effectors over recent years (Thomma et al. 2011; de Jonge et al. 2011). In *Verticillium*, where few bona fide pathogenicity and virulence genes had been known (Fradin and Thomma 2006; Klosterman et al. 2009), genomics has significantly increased our knowledge of effector biology. Major advances include the characterization of Ave1, which is the determinant of race 1 in *V. dahliae* (de Jonge et al. 2012), as well as the discovery of the expansion in *Verticillium* of the LysM and the necrosis- and ethylene-inducing-like protein (NLP) families (Klosterman et al. 2011; Santhanam et al. 2013), known to be involved in virulence in fungi (de Jonge and Thomma 2009; Pemberton and Salmond 2004).

Many fungal effectors belong to a group of small cysteine-rich proteins that often lack recognizable protein domains, and thus have unknown functions. Bioinformatics analyses of the *V. alfalfae* Ms.102 and *V. dahliae* Ls.17 genomes identified 119 and 127 effector candidates, respectively (Klosterman et al. 2011), that were shorter than 400 amino acids, and also cysteine-rich, because they contained more than four cysteine residues. Of these, only two were orthologous to previously characterized effectors, the *Cladosporium fulvum* LysM effectors Ecp2 and Ecp6 (Klosterman et al. 2011), and none had been functionally characterized. Follow-up research using a population genomics approach that involved ten additional *V. dahliae* genomes, led to the discovery of Ave1, the first true small cysteine-rich effector protein in *Verticillium* (de Jonge et al. 2012).

Not all isolates of *V. dahliae* contain *Ave1*. There are two different races of *V. dahliae,* race 1 and race 2, and only *V. dahliae* race 1 isolates harbor *Ave1*. The gene encoding the corresponding cell surface receptor of tomato, *Ve1*, has been cloned (Kawchuk et al. 2001; Fradin et al. 2009, 2011). Ve1 recognizes Ave1, which triggers a defence response resulting in resistance toward *V. dahliae* race 1 (de Jonge et al. 2012). Since *V. dahliae* race 2 lacks Ave1, its interaction with *Ve1* tomato lines results in disease. *Ave1* encodes a secreted protein of 134 amino acids that is absent in *V. dahliae* Ls.17, which is race 2. An *Ave1* ortholog was detected in *V. alfalfae* Ms.102, where it was initially overlooked, likely due to low sequencing coverage (de Jonge et al. 2012).

Ave1 not only mediates avirulence on tomato plants that encode Ve1, but it also contributes to virulence in susceptible tomato. When tomato plants that lacked *Ve1* were inoculated with *V. dahliae* race 1 and race 2 isolates, the *V. dahliae* race 1 isolates carrying Ave1 were more virulent than the *V. dahliae* race 2 isolates (de Jonge et al. 2012), but details are not known.

Another noteworthy aspect of Ave1 is that it appears to be acquired in *V. dahliae* via horizontal transfer. Ave1 homologs are ubiquitous in plants, where they are generally annotated as plant natriuretic peptides (PNPs) (Gehring and Irving 2003), and are known from bacteria, where an Ave1 homolog was identified in the citrus canker pathogen *Xanthomonas axonopodis* pv. *citri,* as a bacterial virulence factor that mimics endogenous PNPs of the host (Gottig et al. 2008). Further orthologs were found in plant pathogenic fungi, namely *Co. higginsianum*, *Cercospora beticola*, and *F. oxysporum* f. sp. *lycopersici* (de Jonge et al. 2012), and more recently in *Co. orbiculare* (Gan et al. 2013) and in *V. nubilum* (Bart Thomma and Krishna Subbarao, unpublished data). Remarkably, several of the Ave1 orthologs can activate Ve1-mediated resistance (de Jonge et al. 2012).

Among the small cysteine-rich proteins found in *V. alfalfae* Ms.102 and *V. dahliae* Ls.17, there are also LysM effectors (Klosterman et al. 2011) that are widespread in fungi (de Jonge and Thomma 2009). *Verticillium alfalfae* Ms.102 and *V. dahliae* Ls.17 contain six and seven LysM effectors, respectively, which is more than the one to three LysM effectors detected in other fungi (Klosterman et al. 2011). Although LysM effectors have been characterized as virulence factors that act in the suppression of chitin-triggered immunity in various fungal plant pathogens, including the tomato leaf mold fungus *C. fulvum* (de Jonge et al. 2010), the *Septoria tritici* blotch fungus *Zymoseptoria tritici* (synonym of *Mycosphaerella graminicola*) (Marshall et al. 2011), and the rice blast fungus *M. oryzae* (Mentlak et al. 2012), their role in *Verticillium* remains enigmatic, as is the reason why this gene family is expanded in *Verticillium*.

Another group of small cysteine-rich proteins that is expanded in *Verticillium* is the necrosis- and ethylene-inducing-like protein (NLP) family that contain seven members, whereas most filamentous fungi carry only up to three of such genes (Klosterman et al. 2011; Santhanam et al. 2013). It has previously been shown that NLPs display cytotoxic activity toward dicotyledonous, but not monocotyledonous plant cells through plasma membrane permeabilization (Qutob et al. 2006; Gijzen and Nürnberger 2006; Ottmann et al. 2009). Only two of the seven *Verticillium* NLP family members displayed cytotoxic activity and were required for pathogenicity on tomato, Arabidopsis and *Nicotiana benthamiana*, but not on cotton (Wang et al. 2004; Zhou et al. 2012; Santhanam et al. 2013), the other NLP genes were noncytolytic. Until the discovery of noncytolytic NLPs in *Verticillium*, noncytolytic NLPs were only known from oomycete species, which generally contain drastically expanded NLP families with up to 70 members (Dong et al. 2012).

Little is known on how effector gene expression is regulated in planta (de Jonge et al. 2011), but Sge1 and its orthologs play a central role in regulation of effector gene expression in various fungi including *V. dahliae* Ls.17 (Santhanam and Thomma 2013). In the root-invading tomato wilt fungus *F. oxysporum* f. sp. *lycopersici*, Sge1 is a transcriptional regulator that is important for early infection, is localized in the nucleus, and is required for pathogenicity and in planta expression of various effector genes (Michielse et al. 2009). The *Botrytis cinerea* Sge1 homolog Reg1 (BcReg1) was required for pathogenicity on bean leaves, as Sge1 knockout mutants were unable to produce sesquiterpene and polyketide toxins (Michielse et al. 2011). Similarly, the ortholog Fgp1 of the wheat and barley pathogen *F. graminearum* was required for pathogenicity, and knockout mutants were deficient in trichothecene mycotoxin production (Jonkers et al. 2012). In *V. dahliae*, VdSge1 was required for pathogenicity on tomato (Santhanam and Thomma 2013). However, in contrast to its *F. oxysporum* ortholog, VdSge1 does not regulate expression of all *V. dahliae* effector genes, as VdSge1 is not required for induction of *Ave1*, nor for induction of the genes encoding cytotoxic NLPs (Santhanam et al. 2013). Thus, VdSge1 is not a central regulator of *Verticillium* effector gene expression in planta.

#### 4.4.1.2 Carbohydrate-Active Enzymes

Fungal cell walls consist of 80–90 % carbohydrates (Bartnicki-Garcia 1968), and carbohydrate-active enzymes (CAZymes) are important virulence factors for plant pathogenic fungi (Pietro et al. 2009), including *Verticillium* (Durrands and Cooper 1988a, b; Tzima et al. 2011). There are approximately 300 CAZyme families (Cantarel et al. 2009), some of which have more members in *V. alfalfae* Ms.102 and *V. dahliae* Ls.17 with respect to other fungi. Whereas *Co. higginsianum* overall has the most CAZymes of any ascomycete studied so far (O'Connell et al. 2012), *V. alfalfae* Ms.102 and *V. dahliae* Ls.17

have the most representatives of the glycoside hydrolase (GH) family GH88 with four genes, whereas *F. oxysporum* had three (Klosterman et al. 2011). Also more abundant in *Verticillium* were the polysaccharide lyases (PLs) that include pectin-degrading enzymes (Garron and Cygler 2010). The *Verticillium* genomes contain 34 and 35 PL genes for *V. alfalfae* Ms.102 and *V. dahliae* Ls.17, respectively, including pectate lyases in the PL1, PL3, PL9 families, and rhamnogalacturonan lyases in the PL4 and PL11 families (Klosterman et al. 2011). Only *Co. higginsianum* had more PL genes with 39 (O'Connell et al. 2012). Contributing to the pectinolytic machinery in *Verticillium* are the GHs mentioned above, including the GH88 and GH105 families, which act on the reaction products of PL family members PL4 and PL11 (van den Brink and de Vries 2011). *Colletotrichum* and *Verticillium* are part of the same lineage in the *Sordariomycetes,* but whether an increase in the number of PL genes in *Verticillium* and *Co. higginsianum* is due to common ancestry is unclear, since *Co. graminicola* only has 15 PL genes (O'Connell et al. 2012). Pectinolytic enzymes are virulence factors in *Verticillium* (Durrands and Cooper 1988a, b) where they may serve the degradation of the pectin-containing pit membranes. The gels that result from degradation of the pit membranes can obstruct xylem vessels, which correlates with appearance of wilting symptoms (Cooper and Wood 1980). Similar processes were described in the bacterial plant pathogen and causal agent of Pierce's disease, *Xylella fastidiosa* (Ellis et al. 2010; Roper et al. 2007). A regulator of pectin lyases and other cell wall-degrading enzymes in *V. dahliae,* the sucrose nonfermenting 1 protein (SNF1), is required for effective colonization of xylem vessels in host plants (Tzima et al. 2011).

The *Verticillium* genomes also contain a high number of CAZymes with a carbohydrate-binding module 1 (CBM1), 26 for *V. alfalfae* Ms.102 and 30 for *V. dahliae* Ls.17 (Klosterman et al. 2011). CBMs recognize and bind polysaccharides (Guillén et al. 2010), which may promote host colonization and survival during the saprotrophic phase. High numbers of CBM1s were also found in the white rot basidiomycete, *Phanerochaete chrysosporium* (Klosterman et al. 2011). *Phanerochaete chrysosporium* (Martinez et al. 2004) and *Verticillium* are distant relatives and differ in life style, but share the ability to grow as saprotrophs where the high numbers of CBM1s may profit the two fungi.

It has been suggested that a reduction in CAZymes in plant pathogenic fungi including *Ustilago maydis* and *Z. tritici,* may be part of a stealth strategy to avoid detection during biotrophic growth (Kämper et al. 2006; Goodwin et al. 2011). The fact that *V. alfalfae* Ms.102 and *V. dahliae* Ls.17 have an extensive biotrophic phase, and are enriched in CAZymes, suggests that having few CAZymes may not represent a stealth strategy, or that *V. alfalfae* Ms.102 and *V. dahliae* Ls.17 compensate by other means to avoid detection.

#### 4.4.1.3 Signaling Pathways

Signal transduction pathways transmit stimuli from outside the cell to transcription factors inside the nucleus, and allow a pathogen to initiate and adapt its infection strategy depending on the environment and host response. Among the best-studied signaling pathways are the cyclic AMP (cAMP) and the mitogen-activated protein (MAP) kinase pathways that are required for pathogenicity in both ascomycetes and basidiomycetes (Mehrabi et al. 2009), and control many of the same processes (Kronstad et al. 1998).

Comparative analyses of signal transduction pathways in fungi indicate conservation in the core networks of signaling proteins (Dean et al. 2005; Klosterman et al. 2007; Hamel et al. 2012). In *Saccharomyces cerevisiae,* the cAMP signaling pathway is involved in filamentous growth, glycogen accumulation and response to stress (Kronstad et al. 1998), and in *M. oryzae,* it is required for appressorium formation and invasive growth inside the host (Xu and Hamer 1996). *V. dahliae* Ls.17 contains all the components of the cAMP pathway of *S. cerevisiae,* except the ortholog of the target gene *Ime1*, and the *Ste18* and *Bcy1* orthologs, but the latter two were present in *V. alfalfae* Ms.102 (Fig. 4.3). VdPKAC1, the *S.*

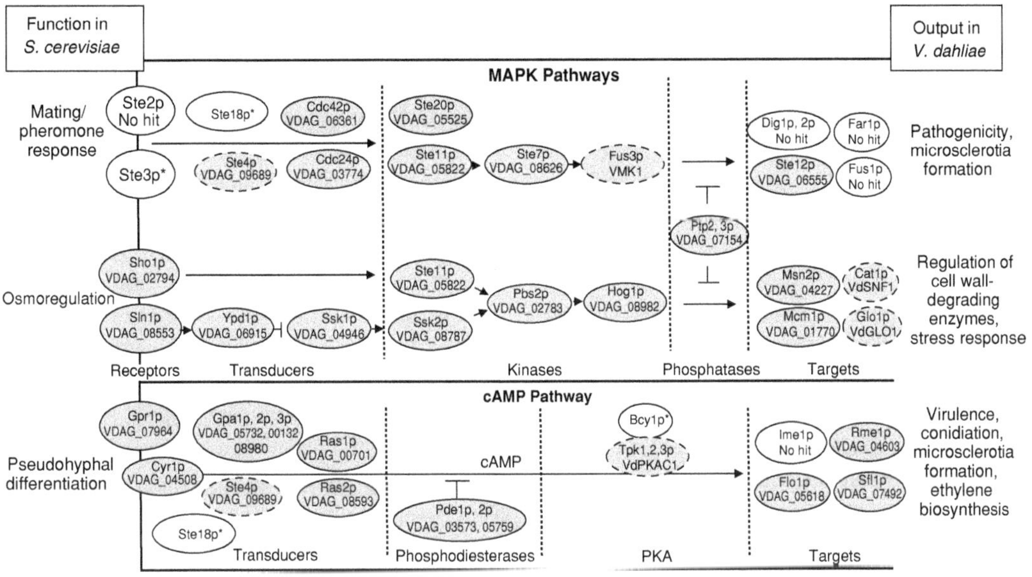


**Fig. 4.3** Conservation in three signal transduction pathways in *Saccharomyces cerevisiae* and *Verticillium dahliae*. Shaded ovals with solid outlines represent proteins that share homology between *S. cerevisiae* and *V. dahliae* based on BLASTp cut off e-value $<1 \times 10^{-8}$. *Shaded ovals* with *dashed* outlines represent genes functionally characterized in *V. dahliae*, see text for details and references, PKA is cAMP-dependent protein kinase A. No hit indicates not detectable in BLASTp searches. *BLASTp hits for Ste3p, Ste18p and Bcy1p were not identified among the *V. dahliae* proteins, but orthologs of each were identified encoded in *V. alfalfae* as VDBG_08677, VDBG_08761 and VDBG_00367, respectively. Orthologs of the proteins used in the searches were from *S. cerevisiae*, strain S288c present in NCBI. The overall relationship of the *S. cerevisiae* proteins in these signaling pathways was from Dean et al. (2005) and Xu (2000)

*cerevisiae* TPK1p ortholog was functionally characterized in *V. dahliae* Ls.17; it has a role in virulence, conidiation, microsclerotia formation, and ethylene biosynthesis (Tzima et al. 2010).

There are five MAP kinase pathways in *S. cerevisiae* (Chen and Thorner 2007), three of which are present in *M. oryzae* (Dean et al. 2005). The two MAP kinase pathways that have been partly characterized in *V. dahliae*, are illustrated in Fig. 4.3. In *S. cerevisiae*, they are involved in mating and osmoregulation, respectively (Dean et al. 2005), whereas in *V. dahliae*, the two pathways function in pathogenicity and microsclerotia formation, and regulation of cell wall-degrading enzymes and response to stress, respectively (Fig. 4.3). Almost all components of the two pathways in *S. cerevisiae* (Dean et al. 2005) were found in *V. dahliae* Ls.17 or *V. alfalfae* Ms.102, only four lacked orthologs in any of the two *Verticillium* genomes (Fig. 4.3).

The following *S. cerevisiae* MAP kinase-pathway or pathway-target orthologs were previously characterized in *V. dahliae*. The Fus3p ortholog VMK1 in *V. dahliae* Ls.17 is required for microsclerotia formation and virulence (Rauyaree et al. 2005). Likewise, the Cat1p ortholog VdSNF1 was identified as an important virulence factor and regulator of the expression of multiple cell wall-degrading enzymes in *V. dahliae* Ls.17 (Tzima et al. 2011), including pectate lyases mentioned in Sect. 4.4.1.2. There were three homologs of the *S. cerevisiae Glo1* in *V. alfalfae* Ms.102, and one in *V. dahliae* Ls.17. Disruption of the single *VdGLO1* gene in *V. dahliae* did not affect pathogenicity (Klimes et al. 2006), but a *Glo1* ortholog was required for maintenance of filamentous growth and pathogenic development of *U. maydis* (Leuthner et al. 2005). Glo1p oxidizes methylglyoxal (Inoue and Kimura 1996) and other small aldehydes and produces $H_2O_2$

(Leuthner et al. 2005). The expansion of the *Glo1* family in *V. alfalfae* Ms.102 may signify an adaptation to detoxify methylglyoxal, which is widely produced during normal metabolism in plants and other organisms.

An additional signaling pathway, the calcium signaling pathway (Nguyen et al. 2008) has not been investigated in detail, but BLAST analyses revealed four phospholipase C genes and a single calmodulin gene in *V. dahliae* Ls.17 (not shown), similar to the numbers reported for other filamentous ascomycetes (Amselem et al. 2011).

While the cAMP and MAP kinase signaling pathways are conserved in *V. dahliae* Ls.17, there is one major difference to other fungi. An additional homolog of the high osmolarity glycerol response MAP kinase Hog1p is encoded in LS region 1 of *V. dahliae* Ls.17 (Fig. 4.2). Hog1p and its orthologs in fungi regulate responses to osmostress (Xu 2000), and are typically single copy genes. The Hog1p encoding-gene in the *V. dahliae* Ls.17 LS region (VDAG_02354), and its paralog (VDAG_08982) are both 361 aa in length and share 87 % amino acid identity, but phylogenetic analyses revealed that the LS copy has diverged from related kinases (Klosterman et al. 2011). EST alignments indicate that the LS *Hog1* is expressed, but its function remains to be determined.

#### 4.4.1.4 Signatures of Horizontal Gene Transfer

Horizontal gene transfer plays an important role in the evolution of plant pathogenic fungi (Mehrabi et al. 2011; Rosewich and Kistler 2000; Oliver and Solomon 2008). In *V. alfalfae* Ms.102 and *V. dahliae* Ls.17, there is evidence for horizontal gene transfer from other fungal species, bacteria, and plants. Examples include a glucan glucosyltransferase (GT) of *V. alfalfae* Ms.102 and *V. dahliae* Ls.17, which is a virulence factor of *V. dahliae* on *Nicotiana benthamiana* and appears to have a bacterial origin (Klosterman et al. 2011). The GTs of *Verticillium* showed a high amino acid similarity of 38 % to GTs in the proteobacteria, which function in the generation of osmoregulated periplasmic glucans (Bohin 2000). GT orthologs were also found in *F. oxysporum* and the insect pathogen *Metarhizium anisopliae*. In phylogenetic analyses, the fungal GTs formed a single group that was most closely related to bacterial order *Rhizobiales* (Klosterman et al. 2011). This suggests that the GT orthologs of filamentous ascomycetes were integrated into an ancestor of *Fusarium, Metarhizium* and *Verticillium,* via horizontal transfer from bacteria in the *Rhizobiales.*

Another *V. dahliae* Ls.17 virulence factor, the effector Ave1 (Sect. 4.4.1.1), has a complicated evolutionary history (de Jonge et al. 2012). A phylogenetic tree of *Ave1* orthologs obtained from the fungal plant pathogens *Ce. beticola, Co. higginsianum, F. oxysporum,* and *V. dahliae,* various plant species, and the bacterium *Xanthomonas axonopodis*, showed that the *Xanthomonas*, fungal, and plant *Ave1* orthologs were intermixed. The *Ave1* orthologs differed by numerous substitutions among one another (de Jonge et al. 2012), suggesting that potentially multiple transfers of *Ave1* involving bacteria, filamentous fungi, and plants may have occurred a sufficiently long time ago for the substitutions to accumulate.

Retrotransposons may also have been acquired via horizontal transfer. This is because RIP was found to affect members of the Gypsy, but not the Copia retrotransposon superfamilies in *V. dahliae* Ls.17, indicating that either the Copia retrotransposons were resistant to RIP, or that already RIPed Gypsy retrotransposons were introduced into *V. dahliae* Ls.17 by horizontal transfer (Klosterman et al. 2011; Amyotte et al. 2012).

### 4.4.2 Lineage-Specific Regions

Alignment of the *V. alfalfae* Ms.102 and *V. dahliae* Ls.17 genome sequences revealed the presence of four regions of approximately 300–350 kb each in *V. dahliae* Ls.17 that were absent in *V. alfalfae* Ms.102 (Klosterman et al. 2011). DNA hybridization experiments with probes derived from each of the four regions and DNAs from 13 different strains of *V. alfalfae, V. dahliae,* and *V. longisporum,* showed that approximately half of the strains did not hybridize with any of the probes, and that hybridization patterns differed among many of the remaining strains (Klosterman et al. 2011). This suggested high variability between species and strains at the four regions, which were named lineage-specific (LS) regions 1 to 4 (Klosterman et al. 2011; Ma et al. 2010).

The size of LS regions is highly variable between different strains of *V. dahliae*. Whereas Klosterman et al. (2011) delineated LS regions based on the four major gaps in synteny between the chromosomes of *V. dahliae* Ls.17 and *V. alfalfa* Ms.102, which resulted in an LS region size estimate of 1.2–1.4 Mb, comparative genome analyses of 11 strains of *V. dahliae* suggested LS region sizes of 2.8–4.5 Mb depending on the isolate (de Jonge et al. 2013).

The LS regions contained genes with putative or known functions in pathogenicity and virulence of plant pathogenic fungi. Most prominently, Ave1, the effector and race determinant of *V. dahliae* Ls.17, was encoded in the LS3 region (de Jonge et al. 2013) (see Sect. 4.4.1.1). There were also two copies of the *vdt1* gene that was associated with host range specificity in *V. dahliae* isolates from tomato (Usami et al. 2001), one LysM type protein-encoding gene, and at least one member of the NLP family (Sect. 4.4.1.1) (Klosterman et al. 2011).

Several gene families were expanded in the LS regions, these encoded six bZIP transcription factors (PF00170), four ferric reductase-like proteins (PF01794), and three patatin-like phospholipases (PF01734) (Klosterman et al. 2011). The bZip transcription factor genes of the LS regions were under purifying or positive selection indicating maintenance of original and acquisition of new functions, respectively (Klosterman et al. 2011), and may play critical roles in pathogenesis including mediation of the oxidative stress response (Guo et al. 2011) and iron homeostasis (López-Berges et al. 2012). Ferric reductase-like proteins are also involved in iron metabolism, by facilitating iron uptake (Roman et al. 1993). The role of patatin-like phospholipases in fungal pathogenesis is unknown (Köhler et al. 2006), but orthologs in potato tubers contain general esterase activity and function in response to pathogen attack and other environmental stresses (Köhler et al. 2006; Hirschberg et al. 2001).

Phylogenetic analyses of the expanded LS region gene families and relatives showed that the families of bZIP transcription factors, ferric reductase-like encoding genes, and patatin-like phospholipases each contained a clade consisting exclusively of genes from LS regions (Klosterman et al. 2011), which indicated that gene duplication (Zhang 2003) occurred within the LS regions. The gene families were not confined to a single LS region, but were distributed among two or three of the LS regions depending on the gene family, and had thus moved between LS regions (Klosterman et al. 2011). Alternatively, the LS regions may represent a small chromosome that was broken up and integrated into larger chromosomes as suggested for the *Tox* loci of *Cochliobolus heterostrophus* (Turgeon and Lu 2000). The gene family phylogenies (Klosterman et al. 2011) do not resolve whether the LS region genes were acquired by *V. dahliae* Ls.17 via horizontal transfer, as proposed for LS region genes in *F. oxysporum* f. sp. *lycopersici* (Ma et al. 2010).

The LS regions of *F. oxysporum* comprised 40 % of the genome assembly including four entire chromosomes (Ma et al. 2010), much more than the less than 10 % in *V. dahliae* Ls.17, but in many other aspects the LS regions of the two fungi were similar. Many LS region genes seemed to be absent in other strains of the same species, and in *F. oxysporum*, only 20 % of the 1285 predicted LS region genes comprised a functional domain, as compared to 39 % of the 354 LS region genes in *V. dahliae* Ls.17, and in both fungi housekeeping genes were lacking in the LS regions (Ma et al. 2010; Klosterman et al. 2011). Gene families that were enriched in both fungi included phospholipases and transcription factors, such as those containing the bZIP domain (Ma et al. 2010; Klosterman et al. 2011). The *F. oxysporum* LS regions were also enriched in genes involved in fungal cell wall degradation and in signal transduction (Ma et al. 2010), but there were also representatives of chitinases and kinases among the 354 *V. dahliae* Ls.17 predicted LS proteins (Klosterman et al. 2011). The *F. oxysporum* secreted proteins were enriched on LS chromosome 14 with 9.5 % (Ma et al. 2010), which was not the case for any of the *V. dahliae* Ls.17 LS regions where 7 % of all encoded proteins were projected to be secreted (Klosterman et al. 2011). Genes of the *V. dahliae* Ls.17 LS regions were disproportionally represented in the transcript pools, as 23 % of the LS genes could be matched with transcripts, as opposed to 12 % of *V. dahliae* Ls.17 genes overall, which perhaps reflected the functional importance of the LS regions (Klosterman et al. 2011).

The LS regions were rich in transposable elements that play important roles in gene evolution. More than half of all transposable elements detected in *V. dahliae* Ls.17 were clustered within the LS regions (Fig. 4.2) (Klosterman et al. 2011; Amyotte et al. 2012) relative to 74 % in *F. oxysporum* (Ma et al. 2010). Clustering of transpo sable elements was also observed in other plant pathogenic fungi, including *M. oryzae* (Dean et al. 2005) and *Leptosphaeria maculans* (Rouxel et al. 2011). In these two fungi, and in *F. oxysporum* and *V. dahliae* Ls.17, the regions that were rich in transposable elements, also encoded effectors (Dean et al. 2005; Rouxel et al. 2011; Ma et al. 2010). The proximity of transposable elements and effectors increases the evolutionary potential for host range expansion by transposon-mediated inactivation of effector genes (Dean et al. 2005), or by changes to effector gene expression through chromosome rearrangement and alteration of chromatin structure (Rep and Kistler 2010). Diversification of effector and other genes may also be driven by transposable elements (Rouxel et al. 2011), as illustrated by a gene fusion identified in the *V. dahliae* Ls.17 LS2 region, which encodes both nitrilase and transposase domains (Amyotte et al. 2012). The only bona fide effector gene of *V. dahliae* Ls.17, *Ave1* is located in the transposon-rich LS3 region, immediately adjacent to a Ty1-copia retrotransposon (de Jonge et al. 2013).

## 4.5 Impact

### 4.5.1 Basic Research

#### 4.5.1.1 Population Genomics

**Markers for Assessing Genetic Variation**

Knowledge of the genetic diversity of *Verticillium* has long been recognized as a critical factor for effective disease management (Puhalla and Hummel 1983; Corsini et al. 1985; Rowe 1995). The variation within *Verticillium* species and populations has been evaluated by assessment of vegetative compatibility groups (VCGs) (Puhalla 1979; Joaquim and Rowe 1990), and by genetic profiling methods including restriction fragment length polymorphisms (RFLPs) (Carder and Barbara 1991), DNA sequencing (Nazar et al. 1991), random amplification of polymorphic DNA (RAPD) (Koike et al. 1996; Zeise and von Tiedemann 2002), and amplified fragment length polymorphisms (AFLPs) (Collins et al. 2003; Fahleson et al. 2003; Barasubiye et al. 1995).

However, much of the history and the dynamics of *Verticillium* populations remained unknown. Microsatellite markers, also referred to as simple sequence repeats (SSRs), offer exceptional resolution and are ideally suited for analysis of fine-scale population processes (Selkoe and Toonen 2006). Microsatellite markers have to be developed anew for each species or groups of species (Sunnucks 2000), and only became available for *Verticillium* when Atallah and Subbarao (2012) took advantage of the *V. dahliae* Ls.17 genome sequence (Klosterman et al. 2011) and developed 22 microsatellite loci (Almany et al. 2009). The microsatellite loci were deployed successfully, and revealed the population structure and gene flow pattern of *V. dahliae* in parts of Europe, North America and Chile, provided evidence of recombination in *V. dahliae* (Atallah et al. 2010; Atallah et al. 2012), and helped explore the linkage between *V. dahliae* race structure and genotype (Maruthachalam et al. 2010). However, many more

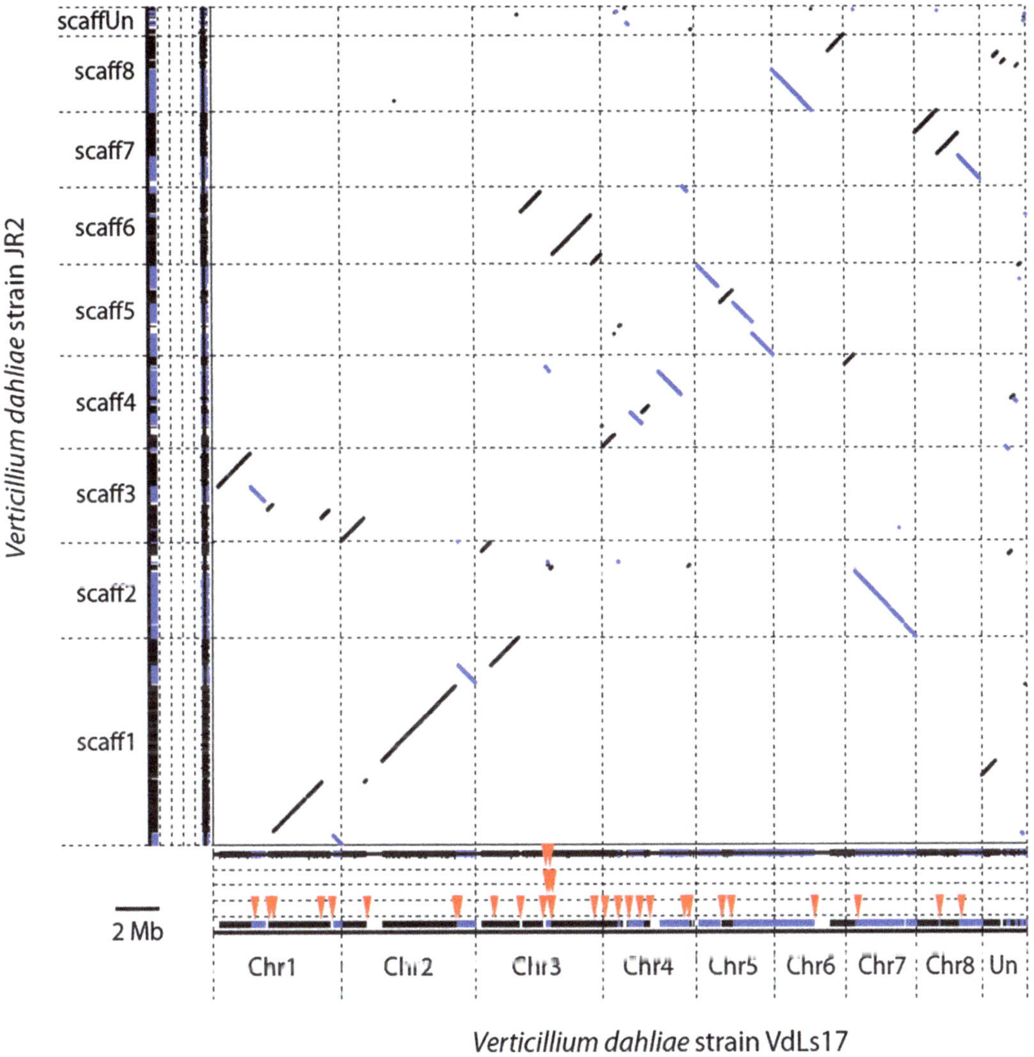


**Fig. 4.4** Whole-genome alignment of *Verticillium dahliae* strains Ls.17 (=VdLs17) and JR2 reveals extensive chromosomal rearrangements. Whole-genome dot-plot comparison with forward-forward alignments (*black*) and inversions (*blue*). *Red triangles* mark syntenic breakpoints. *Un* unplaced contigs. Adapted from de Jonge et al. (2013)

populations of *V. dahliae* and other *Verticillium* species remain unexplored. In addition to the microsatellite loci based on the *V. dahliae* Ls.17 genome, Berbegal et al. (2011) independently designed five additional microsatellite loci, thereby increasing the total of available microsatellite loci for the study of *V. dahliae* to 27. The loci are probably transferrable to *V. alfalfae* (Almany et al. 2009), and possibly also to the other close relatives, *V. longisporum* and *V. nonalfalfae* (Inderbitzin et al. 2011a).

Microsatellites may soon be replaced by the latest genomic marker, single nucleotide polymorphisms (SNPs), which can now be developed with relative ease (Elshire et al. 2011; Baird et al. 2008). However, SNPs may have their own drawbacks as summarized by Atallah and Subbarao (2012). Individual SNPs may be less informative than microsatellites for population analyses, are restricted to only four possible character states (A, C, T, or G), and may also be open for "ascertainment bias" in the selection of

SNP loci, especially when transferring SNPs between species, or using SNPs identified in one organism on another species (Atallah and Subbarao, 2012). Additionally, the 0.6 % error rate of the *V. dahliae* Ls.17 genome at six loci that were re-sequenced by Inderbitzin, Davis et al. (2011b), may result in an overestimation of the number of SNPs when the *V. dahliae* Ls.17 genome sequence is used as a reference in SNP discovery.

### Comparative Population Genomics

Comparative analyses of large numbers of genome sequences will soon be routine. In *Verticillium,* comparisons between ten *V. dahliae* genome sequences and the *V. dahliae* Ls.17 reference strain have already generated novel insight into the biology of *Verticillium* (de Jonge et al. 2012; de Jonge et al. 2013). The ten sequenced *V. dahliae* isolates originated from four different hosts and from soil, and represented both races (de Jonge et al. 2012). This allowed for genomic comparisons between multiple *V. dahliae* race 1 and race 2 genomes, which resulted in the identification of Ave1, the first fully characterized effector and race determinant of *V. dahliae* Ls.17 (Sect. 4.4.1.1) (de Jonge et al. 2012).

The second study involving the same eleven *V. dahliae* genomes focused on the lineage-specific regions (LS regions), which are regions with elevated proportions of isolate-specific DNA, as discussed in detail in Sect. 4.4.2. LS regions contain genes with established or putative roles in pathogenesis, and are thought to provide genetic flexibility to adapt to new environments (Klosterman et al. 2011). De Jonge et al. (2013) found that the eleven *V. dahliae* strains shared a core set of 9471 genes on approximately 30 Mb, and that the LS regions of each strain were up to 4 Mb in length, and contained up to 1000 genes that were strain-specific, or were shared by only a subset of strains. Analyses of gene expression during plant infection showed that upregulated *V. dahliae* genes were disproportionally located in the LS regions (de Jonge et al. 2013). The upregulated genes included four previously unknown putative effectors, two of which were knocked out and shown to be required for virulence (de Jonge et al. 2013). Taken together, and in addition to the already characterized LS region effector Ave1 (de Jonge et al. 2012), these data solidify the importance of the LS regions to pathogenesis in *V. dahliae*.

The LS regions may be an important driver of genetic diversity in *Verticillium.* No sexual state has ever been reported for *V. dahliae,* and the genetic structure of *V. dahliae* populations is mainly clonal as described in Sect. 4.5.1.2. Clonal plant pathogens have a low evolutionary potential (McDonald and Linde 2002), and are disadvantaged in the arms race with the host (Anderson et al. 2010). But *V. dahliae* is an extraordinarily successful plant pathogen with a very wide host range (Pegg and Brady 2002), and is capable of generating new races and expanding its host range (Atallah et al. 2011a; Pegg and Brady 2002). This suggests that the low level of sexual reproduction generates sufficient genetic diversity, that sex is more common than currently thought, or that there is another mechanism that generates genetic diversity, potentially involving the LS regions. This last possibility is supported by the disproportionate number of pathogenesis-related genes in the LS regions that are also enriched with transposable elements. More than half of all transposable elements in *V. dahliae* Ls.17 are associated with the LS regions (Klosterman et al. 2011; Amyotte et al. 2012). Transposable elements can promote diversity in various ways as described in Sect. 4.4.2, independent of sexual reproduction, and are an attractive explanation to account for the elevated genetic diversity at the LS regions of *V. dahliae*.

Transposable elements may also promote chromosomal rearrangements (Gray 2000; Mieczkowski et al. 2006), which were observed between most of the eleven *V. dahliae* strains, and were examined in detail in *V. dahliae* Ls.17 and *V. dahliae* JR2 (de Jonge et al. 2013). The two strains were 99.98 % similar in their core genomes, but their karyotypes differed by eleven intra- and seventeen interchromosomal rearrangements (de Jonge et al. 2013) (Fig. 4.4). As expected given the clustering of transposons in LS regions, the breakpoints of the chromosomal rearrangements correlated with LS regions (de Jonge et al. 2013). The extent of chromosomal polymorphism is negatively correlated to the frequency of meiosis (Kistler and Miao 1992), and therefore it is hard to imagine that karyotypes that differ by as many chromosomal rearrangements as *V. dahliae* Ls.17 and JR2, are easily sexually compatible (de Jonge et al. 2013). Thus, LS regions appear to be a double-edged sword for *V. dahliae* diversity. On the one hand there is evidence that LS regions generate genetic diversity without the need of sex and are potentially beneficial to *V. dahliae*, but on the other hand, LS regions are the sites of chromosomal rearrangements that interfere with meiosis, and reduce the likelihood for sexual recombination to generate novel genotypes. Further analysis of the two potentially contradictory roles of LS regions is needed to estimate their relative importance as drivers of genetic diversity in *V. dahliae.*

### 4.5.1.2 Evolution and Taxonomy

#### Contributions to Verticillium Phylogenetics and Taxonomy

Recent phylogenetic and taxonomic studies resolved the contemporary controversies about species boundaries and evolutionary histories and resulted in a stable taxonomic framework for *Verticillium* (Inderbitzin et al. 2011a, b). The availability of the *Verticillium* genome sequences (Klosterman et al. 2011) enabled PCR amplification and DNA sequencing of nuclear loci across a diverse collection of *Verticillium* isolates for use in phylogenetic analyses. The analyses showed that there are currently ten different species in *Verticillium*, and revealed the relationships among them. Each species was defined as a single group in the phylogenetic tree, and each group contained only one representative of a fungal type specimen from a herbarium, except for *V. longisporum*, a diploid hybrid. Using this approach, species-level phylogenetic groups were linked directly to type specimens that define species morphologically, and to which species names are permanently tied. The result was a taxonomic system where the species are delimited and named based on clearly defined criteria, thereby enabling consistent, expedient, and accurate species identification, which increases our knowledge of the biology of individual *Verticillium* species. An example of the practical implications of past taxonomic confusion in *Verticillium* is provided by Walker (1990).

The major previously unresolved issues in *Verticillium* systematics and evolution concerned *V. longisporum* and *V. albo-atrum* (Klosterman et al. 2009; Barbara and Clewes 2003). It is now settled that *Verticillium longisporum* is likely diploid, and a hybrid that evolved at least three different times by hybridization involving four different parental lineages representing three different species, including *V. dahliae*, and two other species that remain unknown, and also that *V. longisporum* is indeed a separate species from *V. dahliae* (Inderbitzin et al. 2011a, b). Regarding the name "*V. albo-atrum,*" which corresponds to at least two unrelated lineages (Barbara and Clewes 2003), it has been resolved that only the lineage related to *V. tricorpus* is *V. albo-atrum,* and the two groups that are related to *V. dahliae,* correspond to the new species *V. alfalfae* and *V. nonalfalfae* (Inderbitzin et al. 2011a). *Verticillium tricorpus* was split into three species including the two new species *V. isaacii* and *V. klebahnii* and each species corresponds to a major phylogenetic lineage (Inderbitzin et al. 2011a). A morphological key was provided and allows approximate identification of species (Inderbitzin et al. 2011a).

The major issue that hampers our understanding of *Verticillium* diversity is that all of the currently known species were described and collected mainly from agricultural crops, and we do not know much about the diversity of *Verticillium* on non-agricultural hosts, where many new species may await discovery.

#### Clonal Versus Outcrossing Population Structure

Sexual states are unknown in *Verticillium* and sexual recombination has never been observed, but there is evidence that at least *V. dahliae* may be able to mate and sexually recombine.

Sexual processes in ascomycetes are governed by the mating type locus *MAT* that encodes transcription factors regulating the expression of mating-related genes (Herskowitz 1989; Turgeon et al. 1993). In *V. dahliae*, each isolate either has a *MAT1-1* or a *MAT1-2* idiomorph at *MAT* (Usami et al. 2008, 2009), and in general in ascomycetes, only isolates that differ at *MAT* are sexually compatible (Debuchy and Turgeon 2006). Analyses of the *V. alfalfae* Ms.102 and *V. dahliae* Ls.17 genome sequences showed that *Verticillium MAT* is flanked by a DNA lyase gene (APN1) and a cytoskeleton assembly control gene (SLA2) (Klosterman et al. 2011), similarly to other fungi (Rydholm et al. 2007). In regularly recombining populations, *MAT* idiomorphs are expected to be present in approximately equal proportions (Milgroom 1996). However, in *V. dahliae*, *MAT1-2* isolates outnumber *MAT1-1* isolates by more than 10:1 (Usami et al. 2009; Inderbitzin et al. 2011b), suggesting that regular outcrossing does not occur in *V. dahliae*. Additional evidence for a history of sexual recombination is the occurrence of Repeat-Induced Point (RIP)-like mutation patterns in the *Verticillium* genomes that may be associated with meiosis (Klosterman et al. 2011; Cambareri et al. 1991). Alternatively or concurrently to sexual recombination, *Verticillium* species may be undergoing parasexual recombination. Parasexual recombination involves the fusion of somatic hyphae (Caten and Jinks 1966) and has been demonstrated in the laboratory in several species of *Verticillium* (Hastie 1962; Typas and Heale 1976). Evidence for recombination in *V. dahliae*, sexual or parasexual, includes microsatellite data that showed a relatively high genotypic diversity, linkage equilibrium in one population, and a low proportion of compatible loci in another population (Atallah et al. 2010). Neighbor-net network topology analyses (Atallah et al. 2011b), discordance between phylogenetic trees from different gene sequences (Inderbitzin et al. 2011b) and the lack of correlation between race structure and genotype (Maruthachalam et al. 2010), and VCGs and pathogenicity (Rowe 1995) also suggest that *V. dahliae* has at least a recent history of recombination.

The relative contributions of sexual and parasexual processes to recombination in *V. dahliae* and *Verticillium* in general are unknown, but parasexual recombination is not thought to be common in nature (Debets 1998). Nevertheless, it cannot be dismissed in *Verticillium* as long as sexual recombination has not been demonstrated.

## 4.5.2 Applied Research

### 4.5.2.1 Diagnostics

#### Microsatellite and Gene Flow Analyses

*V. dahliae* is a widespread agricultural pathogen. Surveys of the population structure of *V. dahliae* using microsatellite markers that were developed based on the *V. dahliae* Ls.17 genome sequence, have revealed surprising facts about gene flow and host range expansion in this species.

*V. dahliae* is the major cause of Verticillium wilt that affects lettuce in California (Subbarao et al. 1995). Verticillium wilt of lettuce is a relatively new disease first discovered in 1995 in coastal California near the Salinas Valley (Subbarao et al. 1997), the major lettuce growing area in the United States (Atallah et al. 2011a). In the 15 years since its discovery, Verticillium wilt of lettuce had spread throughout the entire Salinas Valley, resulting in significant losses (Atallah et al. 2011a). Verticillium wilt is also present in other areas of California, including in the neighboring Santa Clara Valley where lettuce seed is produced (Atallah et al. 2012), or the more distant Central Valley where Verticillium wilt impacts tomato production (Atallah et al. 2010). Surveys using microsatellite loci revealed that the *V. dahliae* population from the Salinas Valley was significantly differentiated from the lettuce seed population from the adjacent Santa Clara Valley, or the tomato population from the Central Valley with no significant gene flow occurring between the valleys (Atallah et al. 2010, 2012). Instead, the Salinas Valley population did not differ significantly from the populations recovered from infested spinach seed from Denmark and Washington State. Spinach is an important crop in the Salinas Valley and is often planted in rotation with lettuce. An increase in spinach production there coincided with the first detection of Verticillium wilt of lettuce near Salinas in 1995. This suggests that Verticillium wilt of lettuce was triggered by an increased level of *V. dahliae* inoculum in the soil derived from infested spinach seed. Lettuce has a relatively high tolerance to *V. dahliae* inoculum levels, and the threshold to disease may have been reached in 1995 (Atallah et al. 2011a). Spinach is not affected by Verticillium wilt, because it is harvested before symptoms develop. There is an ongoing significant gene flow via spinach seed into the Salinas Valley (Atallah et al. 2012), and the spinach strains have also been found on other crops including diseased peppers and strawberry (Atallah et al. 2010). These data suggest that the spinach seed infestation should be decreased in order to reduce the risk of importing novel and potentially even more devastating *V. dahliae* strains that could further harm the production of lettuce and other crops in the Salinas Valley.

#### Race-Specific Markers

*Verticillium* species do not show the distinct subspecific groupings according to host preference as do formae speciales in *Fusarium*. But preference toward a single host may exist, and a race structure has been reported in *V. dahliae* for tomato (Alexander 1962; Schaible et al. 1951) and lettuce (Vallad et al. 2006). The two races can be differentiated by a PCR assay that is based on a DNA region that is unique to race 1 (Usami et al. 2007; Maruthachalam et al. 2010). The *V. dahliae* Ave1 effector is only present in race 1 isolates of *V. dahliae* (de Jonge et al. 2012) (Sect. 4.4.1.1), which offers the possibility to design *Ave1*-specific primers that can be used to screen for the presence of the *Ave1* gene in *V. dahliae* and other *Verticillium* species. A PCR assay directly tied to the race determinant is useful, as the overall genetic diversity of *V. dahliae* does not correlate with race structure in *V. dahliae* (Maruthachalam et al. 2010; Atallah et al. 2011b).

### 4.5.2.2 *Verticillium* Species Specific Markers

*Verticillium* has been intensely studied for more than 100 years and the accumulated knowledge about its biology is considerable (Pegg and Brady 2002). But the characteristics of individual species and the differences between species is less well understood, as species boundaries have been controversial (Isaac 1967; Klosterman et al. 2009; Walker 1990), and consistent species identification has been complicated by the morphological plasticity of *Verticillium* strains in culture (Heale 2000). Linking biological information from the literature to current concepts of *Verticillium* species is often not possible (Inderbitzin et al. 2011a), because the information needed to confirm species identification, including morphological and DNA data, or the strains themselves, is unavailable.

Taking advantage of the *Verticillium* genomes (Klosterman et al. 2011), significant progress has been made in the study of *Verticillium* diversity (Inderbitzin et al. 2011a, b), and accurate and consistent means of species identifications are now available. This is because *Verticillium* species have been well defined, their phylogenetic relationships are known, and intraspecific variation has been assessed. This allows for species identification by means of DNA sequencing and phylogenetic analyses using reference DNA datasets, or more widely accessible, by means of species-specific PCR assays (Inderbitzin et al. 2013). Over time, consistent and accurate identification will increase our knowledge of the biological characteristics of *Verticillium* species and allow for a more effective disease management.

## 4.6 Future Perspectives

### 4.6.1 Basic Research

The genus *Verticillium* is an ideal subject for the study of fundamental aspects of fungal and general biology in the post genomics era. *Verticillium* is economically important and widely distributed, there are differences in host range and nutritional modes, the phylogenetic relationships between the species are known, the taxonomy is up to date, isolation, culturing, and transformation follow standard protocols, and a high quality, annotated genome sequence of *V. dahliae* Ls.17 is available as a reference.

One area of current major interest is the identification of effector genes using population genomics and RNA-seq, and the determination of the mechanistic bases of host-pathogen interactions (de Jonge et al. 2012, 2013; Faino et al. 2012). These studies have so far focused on a small number of *V. dahliae* genome sequences, but could be significantly advanced by examination of additional *V. dahliae* isolates, as well as isolates from other species that differ in host range and nutritional mode. These include *V. alfalfae* whose host range is restricted to alfalfa, its sister species *V. nonalfalfae* that has a wider host range that excludes alfalfa, *V. longisporum,* the diploid hybrid, whose host range is limited to *Brassicaceae,* possibly as a consequence of genetic or epigenetic genomics that are a consequence of hybridization, and *V. nubilum,* probably a saprotroph and a weak pathogen at best (Inderbitzin et al. 2011a, b). The biological roles of the expanded gene families in the LS regions, CAZymes, and other relevant genes including the glucosyl transferase gene acquired through horizontal gene transfer, should also be examined in greater detail. A related issue is whether there are any *Verticillium* toxins that are deployed during host infection (Pegg and Brady 2002), and gene clusters including the ones encoding polyketide synthases, could be a focus of these investigations.

Another area of current major interest is the question how the genetic diversity in *Verticillium* is being generated. Various mechanisms and genomic sites have been described in fungal plant pathogens that facilitate rapid adaptation in the arms race with plant hosts, including genomic locations enriched for transposons (Laurie et al. 2012; Raffaele et al. 2010; Klosterman et al. 2011; Van de Wouw et al. 2010; Rouxel et al. 2011; de Jonge et al. 2012), mutation and recombination in subtelomeric regions (Chuma et al. 2011), coregulated gene clusters (Palmer and Keller 2010), small dispensable chromosomes (Ma et al. 2010; Coleman et al. 2009; Stukenbrock et al. 2010; Goodwin et al. 2011), gene sparse regions (Raffaele et al. 2010), AT-rich isochore-like regions (Rouxel et al. 2011; Inderbitzin et al. 2010; O'Connell et al. 2012), genome hybridization and horizontal gene transfer (Ma et al. 2010; Goodwin et al. 2011; Coleman et al. 2009; Stukenbrock et al. 2012; Friesen et al. 2006; Gardiner et al. 2012). Many of these mechanisms operate in genomes that were shaped by repeat-driven expansion in sexually reproducing species. *V. dahliae* genomes only contain 4 % of repetitive DNA (Klosterman et al. 2011; Amyotte et al. 2012; de Jonge et al. 2013), which are primarily clustered in the LS regions. Since *V. dahliae* populations do not regularly recombine (Usami et al. 2009; Inderbitzin et al. 2011b), possibly due to chromosomal rearrangements driven by the clusters of transposable elements, the diversity generated in the LS regions may be crucial for adaptation to changing environments in *V. dahliae* (de Jonge et al. 2013). Future research will focus on the relative contributions to *Verticillium* genetic diversity of the LS regions, as well as sexual and parasexual recombination. These studies should encompass the induction of the sexual states in culture, and genomic analysis for meiosis and other sex-related genes. Also, we would like to point out that our knowledge of *Verticillium* depends almost entirely on isolates collected from agricultural areas. A major effort to find and study *Verticillium* isolates from natural habitats and possibly new species should be undertaken, and may be facilitated by an environmental metagenomics approach that could pinpoint novel substrates.

### 4.6.2 Applied Research

The impact of the human genome project on healthcare has been gradual and is taking many years to be fully realized (Green and Guyer 2011; Anonymous 2011). Similarly, the contributions of *Verticillium* genomics to management of Verticillium wilt diseases are likely to take some time to be fully translated into significant improvements in disease management. Contributions have so far included the design of PCR primers to differentiate *Verticillium* species and lineages (Tran et al. 2013; Inderbitzin et al. 2013), and the elucidation of global migration patterns of *V. dahliae* (Atallah et al. 2012), which may result in new quarantine regulations. Maybe most significantly, genomics-driven advances in our understanding of how *Verticillium* species interact with their hosts, will open the door for the design of novel approaches to prevent and reduce the impact of Verticillium wilt, both by classical breeding, genetic engineering and the design of novel fungicides.

## References


Alexander LJ (1962) Susceptibility of certain *Verticillium*-resistant tomato varieties to an Ohio isolate of the pathogen. Phytopathology 52(10):998–1000

Almany GR, De Arruda MP, Arthofer W, Atallah ZK, Beissinger SR, Berumen ML, Bogdanowicz SM, Brown SD, Bruford MW, Burdine C, Busch JW, Campbell NR, Carey D, Carstens BC, Chu KH, Cubeta MA, Cuda JP, Cui Z, Datnoff LE, Dávila JA, Davis ES, Davis RM, Diekmann OE, Eizirik E, Fargallo JA, Fernandes F, Fukuda H, Gale LR, Gallagher E, Gao Y, Girard P, Godhe A, Gonçalves EC, Gouveia L, Grajczyk AM, Grose MJ, Gu Z, Halldén C, Härnström K, Hemmingsen AH, Holmes G, Huang CH, Huang C-C, Hudman SP, Jones GP, Kanetis L, Karunasagar I, Karunasagar I, Keyghobadi N, Klosterman SJ, Klug PE, Koch J, Koopman MM, Köppler K, Koshimizu E, Krumböck S, Kubisiak T, Landis JB, Lasta ML, Lee C-Y, Li Q, Li S-H, Lin R-C, Liu M, Liu NA, Liu WC, Liu Y, Loiseau A, Luan W, Maruthachalam KK, McCormick HM, Mellick R, Monnahan PJ, Morielle-Versute E, Murray TE, Narum SR, Neufeld K, De Nova PJG, Ojiambo PS, Okamoto N, Othman AS, Overholt WA, Pardini R, Paterson IG, Patty OA, Paxton RJ, Planes S, Porter C, Pratchett MS, Püttker T, Rasic G, Rasool B, Rey O, Riegler M, Riehl C, Roberts JMK, Roberts PD, Rochel E, Roe KJ, Rossetto M, Ruzzante DE, Sakamoto T, Saravanan V, Sarturi CR, Schmidt A, Schneider MPC, Schuler H, Serb JM, Serrão ETA, Shi Y, Silva A, Sin YW, Sommer S, Stauffer C, Strüssmann CA, Subbarao KV, Syms C, Tan F, Tejedor ED, Thorrold SR, Trigiano RN, Trucco MI, Tsuchiya-Jerep MTN, Vergara P, Van De Vliet MS, Wadl PA, Wang A, Wang H, Wang RX, Wang X, Wang YAN, Weeks AR, Wei F, Werner WJ, Wiley EO, Williams DA, Wilkins RJ, Wisely SM, With KA, Wu D, Yao C-T, Yau C, Yeap B-K, Zhai B-P, Zhan X, Zhang G Y, Zhang SY, Zhao RU, Zhu L (2009) Permanent genetic resources added to Molecular Ecology Resources Database 1 May 2009–31 July 2009. Mol Ecol Resour 9(6):1460–1466. doi:10.1111/j.1755-0998.2009.02759.x

Altschul SF, Madden TL, Schäffer AA, Zhang J, Zhang Z, Miller W, Lipman DJ (1997) Gapped BLAST and PSI-BLAST: a new generation of protein database search programs. Nucleic Acids Res 25:3389–3402. doi:10.1093/nar/25.17.3389

Amselem J, Cuomo CA, van Kan JAL, Viaud M, Benito EP, Couloux A, Coutinho PM, de Vries RP, Dyer PS, Fillinger S, Fournier E, Gout L, Hahn M, Kohn L, Lapalu N, Plummer KM, Pradier J-M, Quévillon E, Sharon A, Simon A, ten Have A, Tudzynski B, Tudzynski P, Wincker P, Andrew M, Anthouard V, Beever RE, Beffa R, Benoit I, Bouzid O, Brault B, Chen Z, Choquer M, Collémare J, Cotton P, Danchin EG, Da Silva C, Gautier A, Giraud C, Giraud T, Gonzalez C, Grossetete S, Güldener U, Henrissat B, Howlett BJ, Kodira C, Kretschmer M, Lappartient A, Leroch M, Levis C, Mauceli E, Neuvéglise C, Oeser B, Pearson M, Poulain J, Poussereau N, Quesneville H, Rascle C, Schumacher J, Ségurens B, Sexton A, Silva E, Sirven C, Soanes DM, Talbot NJ, Templeton M, Yandava C, Yarden O, Zeng Q, Rollins JA, Lebrun M-H, Dickman M (2011) Genomic analysis of the necrotrophic fungal pathogens *Sclerotinia sclerotiorum* and *Botrytis cinerea*. PLoS Genet 7(8):e1002230. doi:10.1371/journal.pgen.1002230

Amyotte SG, Tan X, Pennerman K, del Mar Jimenez-Gasco M, Klosterman SJ, Ma L-J, Dobinson KF, Veronese P (2012) Transposable elements in phytopathogenic *Verticillium* spp.: insights into genome evolution and inter- and intra-specific diversification. BMC Genomics 13:314. doi:10.1186/1471-2164-13-314

Anderson JP, Gleason CA, Foley RC, Thrall PH, Burdon JB, Singh KB (2010) Plants versus pathogens: an evolutionary arms race. Funct Plant Biol 37(6):499–512. doi:10.1071/FP09304

Anonymous (2011) Best is yet to come. Nature 470(7333):140. doi:10.1038/470140a

Atallah ZK, Hayes RJ, Subbarao KV (2011a) Fifteen years of Verticillium wilt of lettuce in America's salad bowl: a tale of immigration, subjugation, and abatement. Plant Dis 95(7):784–792. doi:10.1094/PDIS-01-11-0075

Atallah ZK, Maruthachalam K, du Toit L, Koike ST, Davis RM, Klosterman SJ, Hayes RJ, Subbarao KV (2010) Population analyses of the vascular plant pathogen *Verticillium dahliae* detect recombination and transcontinental gene flow. Fungal Genet Biol 47(5):416–422. doi:10.1016/j.fgb.2010.02.003

Atallah ZK, Maruthachalam K, Subbarao KV (2012) Sources of *Verticillium dahliae* affecting lettuce. Phytopathology 102(11):1071–1078. doi:10.1094/phyto-04-12-0067-r

Atallah ZK, Maruthachalam K, Vallad GE, Davis RM, Klosterman SJ, Subbarao KV (2011b) Analysis of *Verticillium dahliae* suggests a lack of correlation between genotypic diversity and virulence phenotypes. Plant Dis 95(10):1224–1232. doi:10.1094/pdis-02-11-0110

Atallah ZK, Subbarao KV (2012) Population biology of fungal plant pathogens. In: Bolton MD, Thomma BPHJ (eds) Plant fungal pathogens: methods and protocols, vol 835. Methods in Molecular Biology. Humana Press, Totowa, pp 333–363. doi:10.1007/978-1-61779-501-5_20

Babadoost M, Chen W, Bratsch AD, Eastman CE (2004) *Verticillium longisporum* and *Fusarium solani*: two new species in the complex of internal discoloration of horseradish roots. Plant Pathol 53(5):669–676. doi:10.1111/j.1365-3059.2004.01070.x

Baird NA, Etter PD, Atwood TS, Currey MC, Shiver AL, Lewis ZA, Selker EU, Cresko WA, Johnson EA (2008) Rapid SNP discovery and genetic mapping using sequenced RAD markers. PLoS ONE 3(10):e3376. doi:10.1371/journal.pone.0003376

Barasubiye T, Parent J-G, Hamelin RC, Laberge S, Richard C, Dostaler D (1995) Discrimination between alfalfa and potato isolates of *Verticillium albo-atrum* using RAPD markers. Mycol Res 99(12):1507–1512. doi:10.1016/S0953-7562(09)80800-9

Barbara DJ, Clewes E (2003) Plant pathogenic *Verticillium* species: How many of them are there? Mol Plant Pathol 4(4):297–305. doi:10.1046/J.1364-3703.2003.00172.X

Bartnicki-Garcia S (1968) Cell wall chemistry, morphogenesis, and taxonomy of fungi. Annu Rev Microbiol 22:87–108. doi:10.1146/annurev.mi.22.100168.000511

Berbegal M, Garzón CD, Ortega A, Armengol J, Jiménez-Díaz RM, Jiménez-Gasco MM (2011) Development and application of new molecular markers for analysis of genetic diversity in *Verticillium dahliae* populations. Plant Pathol 60(5):866–877. doi:10.1111/j.1365-3059.2011.02432.x

Bishop CD, Cooper RM (1983) An ultrastructural study of root invasion in three vascular wilt diseases. Physiol Plant Pathol 22(1):15–27. doi:10.1016/S0048-4059(83)81034-0

Bohin JP (2000) Osmoregulated periplasmic glucans in Proteobacteria. FEMS Microbiol Lett 186(1):11–19. doi:10.1111/j.1574-6968.2000.tb09075.x

Brasier CM (1991) *Ophiostoma novo-ulmi* sp. nov., causative agent of current Dutch elm disease pandemics. Mycopathologia 115(3):151–161. doi:10.1007/bf00462219

Cambareri EB, Jensen BC, Schabtach E, Selker EU (1989) Repeat-induced G–C to A–T mutations in *Neurospora*. Science 244(4912):1571–1575. doi:10.1126/science.2544994

Cambareri EB, Singer MJ, Selker EU (1991) Recurrence of repeat-induced point mutation (RIP) in *Neurospora crassa*. Genetics 127(4):699–710. http://www.genetics.org/content/127/4/699.long

Cannon PF, Buddie AG, Bridge PD, de Neergaard E, Lübeck M, Askar MM (2012) *Lectera*, a new genus of the *Plectosphaerellaceae* for the legume pathogen *Volutella colletotrichoides*. MycoKeys 3:23–36. doi:10.3897/mycokeys.3.3065

Cantarel BL, Coutinho PM, Rancurel C, Bernard T, Lombard V, Henrissat B (2009) The Carbohydrate-Active EnZymes database (CAZy): an expert resource for glycogenomics. Nucleic Acids Res 37(suppl 1):D233–D238. doi:10.1093/nar/gkn663

Carder JH, Barbara DJ (1991) Molecular variation and restriction fragment length polymorphisms (RFLPs) within and between six species of *Verticillium*. Mycol Res 95(8):935–942. doi:10.1016/S0953-7562(09)80090-7

Caten CE, Jinks JL (1966) Heterokaryosis: its significance in wild homothallic ascomycetes and fungi imperfecti. Trans Br Mycol Soc 49(1):81–93. doi:10.1016/S0007-1536(66)80038-4

Chen P, Lee B, Robb J (2004) Tolerance to a non-host isolate of *Verticillium dahliae* in tomato. Physiol Mol Plant Pathol 64(6):283–291. doi:10.1016/j.pmpp.2004.10.002

Chen RE, Thorner J (2007) Function and regulation in MAPK signaling pathways: lessons learned from the yeast *Saccharomyces cerevisiae*. Biochim Biophys Acta 1773(8):1311–1340. doi:10.1016/j.bbamcr.2007.05.003

Chuma I, Isobe C, Hotta Y, Ibaragi K, Futamata N, Kusaba M, Yoshida K, Terauchi R, Fujita Y, Nakayashiki H, Valent B, Tosa Y (2011) Multiple translocation of the *AVR-Pita* effector gene among chromosomes of the rice blast fungus *Magnaporthe oryzae* and related species. PLoS Path 7(7):e1002147. doi:10.1371/journal.ppat.1002147

Coleman JJ, Rounsley SD, Rodriguez-Carres M, Kuo A, Wasmann CC, Grimwood J, Schmutz J, Taga M, White GJ, Zhou S, Schwartz DC, Freitag M, Ma L-j, Danchin EGJ, Henrissat B, Coutinho PM, Nelson DR, Straney D, Napoli CA, Barker BM, Gribskov M, Rep

M, Kroken S, Molnár I, Rensing C, Kennell JC, Zamora J, Farman ML, Selker EU, Salamov A, Shapiro H, Pangilinan J, Lindquist E, Lamers C, Grigoriev IV, Geiser DM, Covert SF, Temporini E, VanEtten HD (2009) The genome of *Nectria haematococca*: contribution of supernumerary chromosomes to gene expansion. PLoS Genet 5(8):e1000618. doi:10.1371/journal.pgen.1000618

Collins A, Okoli CA, Morton A, Parry D, Edwards SG, Barbara DJ (2003) Isolates of *Verticillium dahliae* pathogenic to crucifers are of at least three distinct molecular types. Phytopathology 93(3):364–376. doi:10.1094/PHYTO.2003.93.3.364

Cooper RM, Wood RKS (1980) Cell wall degrading enzymes of vascular wilt fungi. III. Possible involvement of endo-pectin lyase in Verticillium wilt of tomato. Physiol Plant Pathol 16(2):285–300. doi:10.1016/0048-4059(80)90043-0

Corsini DL, Davis JR, Pavek JJ (1985) Stability of resistance of potato to strains of *Verticillium dahliae* from different vegetative compatibility groups. Plant Dis 69(11):980–982. doi:10.1094/PD-69-980

Cuomo CA, Güldener U, Xu J-R, Trail F, Turgeon BG, Di Pietro A, Walton JD, Ma L-J, Baker SE, Rep M, Adam G, Antoniw J, Baldwin T, Calvo S, Chang Y-L, DeCaprio D, Gale LR, Gnerre S, Goswami RS, Hammond-Kosack K, Harris LJ, Hilburn K, Kennell JC, Kroken S, Magnuson JK, Mannhaupt G, Mauceli E, Mewes H-W, Mitterbauer R, Muehlbauer G, Münsterkötter M, Nelson D, O'Donnell K, Ouellet T, Qi W, Quesneville H, Roncero MIG, Seong K-Y, Tetko IV, Urban M, Waalwijk C, Ward TJ, Yao J, Birren BW, Kistler HC (2007) The *Fusarium graminearum* genome reveals a link between localized polymorphism and pathogen specialization. Science 317(5843):1400–1402. doi:10.1126/science.1143708

de Jonge R, Bolton MD, Kombrink A, van den Berg GCM, Yadeta KA, Thomma BPHJ (2013) Extensive chromosomal reshuffling drives evolution of virulence in an asexual pathogen. Genome Res 23(8):1271–1282. doi:10.1101/gr.152660.112

de Jonge R, Bolton MD, Thomma BPHJ (2011) How filamentous pathogens co-opt plants: the ins and outs of fungal effectors. Curr Opin Plant Biol 14(4):400–406. doi:10.1016/j.pbi.2011.03.005

de Jonge R, Thomma BPHJ (2009) Fungal LysM effectors: extinguishers of host immunity? Trends Microbiol 17(4):151–157. doi:10.1016/j.tim.2009.01.002

de Jonge R, van Esse HP, Kombrink A, Shinya T, Desaki Y, Bours R, van der Krol S, Shibuya N, Joosten MH, Thomma BPHJ (2010) Conserved fungal LysM effector Ecp6 prevents chitin-triggered immunity in plants. Science 329(5994):953–955. doi:10.1126/science.1190859

de Jonge R, van Esse PH, Maruthachalam K, Bolton MD, Santhanam P, Saber MK, Zhang Z, Usami T, Lievens B, Subbarao KV, Thomma BPHJ (2012) Tomato immune receptor Ve1 recognizes effector of multiple fungal pathogens uncovered by genome and RNA sequencing. Proc Natl Acad Sci U S A 109(13):5110–5115. doi:10.1073/pnas.1119623109

Dean RA, Talbot NJ, Ebbole DJ, Farman ML, Mitchell TK, Orbach MJ, Thon M, Kulkarni R, Xu J-R, Pan H, Read ND, Lee Y-H, Carbone I, Brown D, Oh YY, Donofrio N, Jeong JS, Soanes DM, Djonovic S, Kolomiets E, Rehmeyer C, Li W, Harding M, Kim S, Lebrun M-H, Bohnert H, Coughlan S, Butler J, Calvo S, Ma L-J, Nicol R, Purcell S, Nusbaum C, Galagan JE, Birren BW (2005) The genome sequence of the rice blast fungus *Magnaporthe grisea*. Nature 434(7036):980–986. doi:10.1038/nature03449

Debets AJM (1998) Parasexuality in fungi: mechanisms and significance in wild populations. In: Bridge P, Couteaudier Y, Clarkson J (eds) Molecular variability of fungal pathogens. CAB International, Wallingford, pp 41–52

Debuchy R, Turgeon BG (2006) Mating-type structure, evolution, and function in Euascomycetes. In: Kües U, Fischer R (eds) The Mycota I: Growth, differentation and sexuality. Springer, Berlin, pp 293–323. doi:10.1007/3 540 28135 5_15

Dobinson K, Grant S, Kang S (2004) Cloning and targeted disruption, via *Agrobacterium tumefaciens*-mediated transformation, of a trypsin protease gene from the vascular wilt fungus *Verticillium dahliae*. Curr Genet 45(2):104–110. doi:10.1007/s00294-003-0464-6

Dobinson KF (1995) Genetic transformation of the vascular wilt fungus *Verticillium dahliae*. Can J Bot 73(5):710–715. doi:10.1139/b95-076

Dong S, Kong G, Qutob D, Yu X, Tang J, Kang J, Dai T, Wang H, Gijzen M, Wang Y (2012) The NLP toxin family in *Phytophthora sojae* includes rapidly evolving groups that lack necrosis-inducing activity. Mol Plant-Microbe Interact 25(7):896–909. doi:10.1094/mpmi-01-12-0023-r

Down G, Barbara D, Radišek S (2007) *Verticillium albo-atrum* and *V. dahliae* on hop. EPPO Bull 37(3):528–535. doi:10.1111/j.1365-2338.2007.01160.x

Dunker S, Keunecke H, Steinbach P, von Tiedemann A (2008) Impact of *Verticillium longisporum* on yield and morphology of winter oilseed rape (*Brassica napus*) in relation to systemic spread in the plant. J Phytopathol 156(11–12):698–707. doi:10.1111/j.1439-0434.2008.01429.x

Durrands PK, Cooper RM (1988a) The role of pectinases in vascular wilt disease as determined by defined mutants of *Verticillium albo-atrum*. Physiol Mol Plant Pathol 32(3):363–371. doi:10.1016/S0885-5765(88)80030-4

Durrands PK, Cooper RM (1988b) Selection and characterization of pectinase-deficient mutants of the vascular wilt pathogen *Verticillium albo-atrum*. Physiol Mol Plant Pathol 32(3):343–362. doi:10.1016/S0885-5765(88)80029-8

Ellis EA, McEachern GR, Clark S, Cobb BG (2010) Ultrastructure of pit membrane dissolution and movement of *Xylella fastidiosa* through pit membranes in petioles of *Vitis vinifera*. Botany 88(6):596–600. doi:10.1139/B10-025

Elshire RJ, Glaubitz JC, Sun Q, Poland JA, Kawamoto K, Buckler ES, Mitchell SE (2011) A robust, simple genotyping-by-sequencing (GBS) approach for high diversity species. PLoS ONE 6(5):e19379. doi:10.1371/journal.pone.0019379

Enebak SA (2012) Soil fumigation: the critical use exemption, quarantine pre-shipment rules, re-registration decision and their effect on the 2012 growing season. In: Haase DL, Pinto JR, Riley LE (eds) National proceedings: forest and conservation nursery associations—2011, Fort Collins, CO. USDA Forest Service, Rocky Mountain Research Station, 2012, pp 26–30. http://www.fs.fed.us/rm/pubs/rmrs_p068.html

Fahleson J, Lagercrantz U, Hu Q, Steventon LA, Dixelius C (2003) Estimation of genetic variation among *Verticillium* isolates using AFLP analysis. Eur J Plant Pathol 109(4):361–371. doi:10.1023/a:1023534005538

Faino L, de Jonge R, Thomma BPHJ (2012) The transcriptome of *Verticillium dahliae*-infected *Nicotiana benthamiana* determined by deep RNA sequencing. Plant Signal Behav 7(9):1065–1069. doi:10.4161/psb.21014

Fitzell R, Evans G, Fahy P (1980) Studies on the colonization of plant roots by *Verticillium dahliae* Klebahn with use of immunofluorescent staining. Aust J Bot 28(3):357–368. doi:10.1071/BT9800357

Fradin EF, Abd-El-Haliem A, Masini L, van den Berg GCM, Joosten MHAJ, Thomma BPHJ (2011) Interfamily transfer of tomato *Ve1* mediates *Verticillium* resistance in Arabidopsis. Plant Physiol 156(4):2255–2265. doi:10.1104/pp.111.180067

Fradin EF, Thomma BPHJ (2006) Physiology and molecular aspects of Verticillium wilt diseases caused by *V. dahliae* and *V. albo-atrum*. Mol. Plant Pathol 7(2):71–86. doi:10.1111/J.1364-3703.2006.00323.X

Fradin EF, Zhang Z, Juarez Ayala JC, Castroverde CDM, Nazar RN, Robb J, Liu C-M, Thomma BPHJ (2009) Genetic dissection of Verticillium Wilt resistance mediated by tomato Ve1. Plant Physiol 150(1):320–332. doi:10.1104/pp.109.136762

Freitag M, Williams RL, Kothe GO, Selker EU (2002) A cytosine methyltransferase homologue is essential for repeat-induced point mutation in *Neurospora crassa*. Proc Natl Acad Sci U S A 99(13):8802–8807. doi:10.1073/pnas.132212899

Friebertshauser GE, DeVay JE (1982) Differential effects of the defoliating and nondefoliating pathotypes of *Verticillium dahliae* upon the growth and development of *Gossypium hirsutum*. Phytopathology 72(7):872–877. doi:10.1094/Phyto-72-872

Friesen TL, Stukenbrock EH, Liu Z, Meinhardt S, Ling H, Faris JD, Rasmussen JB, Solomon PS, McDonald BA, Oliver RP (2006) Emergence of a new disease as a result of interspecific virulence gene transfer. Nat Genet 38(8):953–956. doi:10.1038/ng1839

Galagan JE, Calvo SE, Borkovich KA, Selker EU, Read ND, Jaffe D, FitzHugh W, Ma L-J, Smirnov S, Purcell S, Rehman B, Elkins T, Engels R, Wang S, Nielsen CB, Butler J, Endrizzi M, Qui D, Ianakiev P, Bell-Pedersen D, Nelson MA, Werner-Washburne M, Selitrennikoff CP, Kinsey JA, Braun EL, Zelter A, Schulte U, Kothe GO, Jedd G, Mewes W, Staben C, Marcotte E, Greenberg D, Roy A, Foley K, Naylor J, Stange-Thomann N, Barrett R, Gnerre S, Kamal M, Kamvysselis M, Mauceli E, Bielke C, Rudd S, Frishman D, Krystofova S, Rasmussen C, Metzenberg RL, Perkins DD, Kroken S, Cogoni C, Macino G, Catcheside D, Li W, Pratt RJ, Osmani SA, DeSouza CPC, Glass L, Orbach MJ, Berglund JA, Voelker R, Yarden O, Plamann M, Seiler S, Dunlap J, Radford A, Aramayo R, Natvig DO, Alex LA, Mannhaupt G, Ebbole DJ, Freitag M, Paulsen I, Sachs MS, Lander ES, Nusbaum C, Birren B (2003) The genome sequence of the filamentous fungus *Neurospora crassa*. Nature 422(6934):859–868. doi:10.1038/nature01554

Gan P, Ikeda K, Irieda H, Narusaka M, O'Connell RJ, Narusaka Y, Takano Y, Kubo Y, Shirasu K (2013) Comparative genomic and transcriptomic analyses reveal the hemibiotrophic stage shift of *Colletotrichum* fungi. New Phytol 197(4):1236–1249. doi:10.1111/nph.12085

Gao F, Zhou B-J, Li G-Y, Jia P-S, Li H, Zhao Y-L, Zhao P, Xia G-X, Guo H-S (2010) A glutamic acid-rich protein identified in *Verticillium dahliae* from an insertional mutagenesis affects microsclerotial formation and pathogenicity. PLoS ONE 5(12):e15319. doi:10.1371/journal.pone.0015319

Garber RH, Houston BR (1966) Penetration and development of *Verticillium albo-atrum* in the cotton plant. Phytopathology 56(10):1121–1126

García-Pedrajas MD, Paz Z, Andrews DL, Baeza-Montañez L, Gold SE (2013) Rapid deletion plasmid construction methods for protoplast and *Agrobacterium*-based fungal transformation systems. In: Gupta VK, Tuohy MG, Ayyachamy M, Turner KM, O'Donovan A (eds) Laboratory protocols in fungal biology. Fungal Biology. Springer, New York, pp 375–393. doi:10.1007/978-1-4614-2356-0_34

Gardiner DM, McDonald MC, Covarelli L, Solomon PS, Rusu AG, Marshall M, Kazan K, Chakraborty S, McDonald BA, Manners JM (2012) Comparative pathogenomics reveals horizontally acquired novel virulence genes in fungi infecting cereal hosts. PLoS Path 8(9):e1002952. doi:10.1371/journal.ppat.1002952

Garron M-L, Cygler M (2010) Structural and mechanistic classification of uronic acid-containing polysaccharide lyases. Glycobiology 20(12):1547–1573. doi:10.1093/glycob/cwq122

Gehring CA, Irving HR (2003) Natriuretic peptides—a class of heterologous molecules in plants. Int J Biochem Cell Biol 35(9):1318–1322. doi:10.1016/S1357-2725(03)00032-3

Gijzen M, Nürnberger T (2006) Nep1-like proteins from plant pathogens: recruitment and diversification of the NPP1 domain across taxa. Phytochemistry 67(16):1800–1807. doi:10.1016/j.phytochem.2005.12.008

Goodwin SB, Ben M'Barek S, Dhillon B, Wittenberg AHJ, Crane CF, Hane JK, Foster AJ, Van der Lee TAJ, Grimwood J, Aerts A, Antoniw J, Bailey A, Bluhm B, Bowler J, Bristow J, van der Burgt A, Canto-Canché B, Churchill ACL, Conde-Ferràez L, Cools HJ, Coutinho PM, Csukai M, Dehal P, De Wit P, Donzelli B, van de Geest HC, van Ham RCHJ, Hammond-Kosack KE, Henrissat B, Kilian A, Kobayashi AK, Koopmann E, Kourmpetis Y, Kuzniar A, Lindquist E, Lombard V, Maliepaard C, Martins N, Mehrabi R, Nap JPH, Ponomarenko A, Rudd JJ, Salamov A, Schmutz J, Schouten HJ, Shapiro H, Stergiopoulos I, Torriani SFF, Tu H, de Vries RP, Waalwijk C, Ware SB, Wiebenga A, Zwiers L-H, Oliver RP, Grigoriev IV, Kema GHJ (2011) Finished genome of the fungal wheat pathogen *Mycosphaerella graminicola* reveals dispensome structure, chromosome plasticity, and stealth pathogenesis. PLoS Genet 7(6):e1002070. doi:10.1371/journal.pgen.1002070

Gottig N, Garavaglia BS, Daurelio LD, Valentine A, Gehring C, Orellano EG, Ottado J (2008) *Xanthomonas axonopodis* pv. *citri* uses a plant natriuretic peptide-like protein to modify host homeostasis. Proc Natl Acad Sci U S A 105(47):18631–18636. doi:10.1073/pnas.0810107105

Gray YHM (2000) It takes two transposons to tango: Transposable-element-mediated chromosomal rearrangements. Trends Genet 16(10):461–468. doi:10.1016/S0168-9525(00)02104-1

Green ED, Guyer MS (2011) Charting a course for genomic medicine from base pairs to bedside. Nature 470(7333):204–213. doi:10.1038/nature09764

Guillén D, Sánchez S, Rodríguez-Sanoja R (2010) Carbohydrate-binding domains: multiplicity of biological roles. Appl Microbiol Biotechnol 85(5):1241–1249. doi:10.1007/s00253-009-2331-y

Guo M, Chen Y, Du Y, Dong Y, Guo W, Zhai S, Zhang H, Dong S, Zhang Z, Wang Y, Wang P, Zheng X (2011) The bZIP transcription factor MoAP1 mediates the oxidative stress response and is critical for pathogenicity of the rice blast fungus *Magnaporthe oryzae*. PLoS Path 7(2):e1001302. doi:10.1371/journal.ppat.1001302

Haas BJ, Zeng Q, Pearson MD, Cuomo CA, Wortman JR (2011) Approaches to fungal genome annotation. Mycology 2(3):118–141. doi:10.1080/21501203.2011.606851

Hamel L-P, Nicole M-C, Duplessis S, Ellis BE (2012) Mitogen-activated protein kinase signaling in plant-interacting fungi: distinct messages from conserved messengers. Plant Cell 24(4):1327–1351. doi:10.1105/tpc.112.096156

Harrison JAC, Isaac I (1969) Survival of the causal agents of 'early-dying disease' (Verticillium wilt) of potatoes. Ann Appl Biol 63(2):277–288. doi:10.1111/j.1744-7348.1969.tb05489.x

Hastie AC (1962) Genetic recombination in the hop-wilt fungus *Verticillium albo-atrum*. J Gen Microbiol 27(3):373–382. doi:10.1099/00221287-27-3-373

Hawksworth DL (1970a) *Verticillium nigrescens*. CMI Descr Pathog Fungi Bact 26:257. http://www.cabi.org/dfb/

Hawksworth DL (1970b) *Verticillium nubilum*. CMI Descr Pathog Fungi Bact 26:258. http://www.cabi.org/dfb/

Hawksworth DL (1970c) *Verticillium tricorpus*. CMI Descr Pathog Fungi Bact 26:260. http://www.cabi.org/dfb/

Hawksworth DL, Holliday P (1970) *Verticillium theobromae*. CMI Descr Pathog Fungi Bact 26:259. http://www.cabi.org/dfb/

Hawksworth DL, Talboys PW (1970a) *Verticillium albo-atrum*. CMI Descr Pathog Fungi Bact 26:255. http://www.cabi.org/dfb/

Hawksworth DL, Talboys PW (1970b) *Verticillium dahliae*. CMI Descr Pathog Fungi Bact 26:256. http://www.cabi.org/dfb/

Heale JB (2000) Diversification and speciation in *Verticillium*—an overview. In: Tjamos EC, Rowe RC, Heale JB, Fravel DR (eds) Advances in *Verticillium* research and disease management. APS Press, St. Paul, pp 1–14

Heale JB, Isaac I (1963) Wilt of lucerne caused by species of *Verticillium*. IV. Pathogenicity of *V. albo-atrum* and *V. dahliae* to lucerne and other crops; spread and survival of *V. albo-atrum* in soil and weeds; effect upon lucerne production. Ann Appl Biol 52(3):439–451. doi:10.1111/j.1744-7348.1963.tb03768.x

Heale JB, Karapapa VK (1999) The *Verticillium* threat to Canada's major oilseed crop: canola. Can J Plant Pathol 21(1):1–7. doi:10.1080/07060661.1999.10600114

Herskowitz I (1989) A regulatory hierarchy for cell specialization in yeast. Nature 342(6251):749–757. doi:10.1038/342749a0

Hirschberg HJHB, Simons J-WFA, Dekker N, Egmond MR (2001) Cloning, expression, purification and characterization of patatin, a novel phospholipase A. Eur J Biochem 268(19):5037–5044. doi:10.1046/j.0014-2956.2001.02411.x

Inderbitzin P, Asvarak T, Turgeon BG (2010) Six new genes required for production of T-Toxin, a polyketide determinant of high virulence of *Cochliobolus heterostrophus* to maize. Mol Plant-Microbe Interact 23(4):458–472. doi:10.1094/MPMI-23-4-0458

Inderbitzin P, Bostock RM, Davis RM, Usami T, Platt HW, Subbarao KV (2011a) Phylogenetics and taxonomy of the fungal vascular wilt pathogen *Verticillium*, with the descriptions of five new species. PLoS ONE 6(12):e28341. doi:10.1371/journal.pone.0028341

Inderbitzin P, Davis RM, Bostock RM, Subbarao KV (2011b) The ascomycete *Verticillium longisporum* is a hybrid and a plant pathogen with an expanded host range. PLoS ONE 6(3):e18260. doi:10.1371/journal.pone.0018260

Inderbitzin P, Davis RM, Bostock RM, Subbarao KV (2013) Identification and differentiation of *Verticillium* species and *V. longisporum* lineages by simplex and multiplex PCR assays. PLoS ONE 8(6):e65990. doi:10.1371/journal.pone.0065990

Inderbitzin P, Subbarao KV (2014) *Verticillium* systematics and evolution: Implications of information confusion on Verticillium wilt management and potential solutions. Phytopathology 104(6):564–574. doi: 10.1094/PHYTO-11-13-0315-IA

Inoue Y, Kimura A (1996) Identification of the structural gene for glyoxalase I from *Saccharomyces cerevisiae*. J Biol Chem 271(42):25958–25965. doi:10.1074/jbc.271.42.25958

Isaac I (1953) A further comparative study of pathogenic isolates of *Verticillium*: *V. nubilum* Pethybr. and *V. tricorpus* sp. nov. Trans Br Mycol Soc 36(3):180–195. doi:10.1016/S0007-1536(53)80002-1

Isaac I (1967) Speciation in *Verticillium*. Annu Rev Phytopathol 5(1):201–222. doi:10.1146/annurev.py.05.090167.001221

Isaac I, Harrison JAC (1968) The symptoms and causal agents of early-dying disease (Verticillium wilt) of potatoes. Ann Appl Biol 61(2):231–244. doi:10.1111/j.1744-7348.1968.tb04528.x

Joaquim TR, Rowe RC (1990) Reassessment of vegetative compatibility relationships among strains of *Verticillium dahliae* using nitrate-nonutilizing mutants. Phytopathology 80:1160–1166. doi:10.1094/Phyto-80-1160

Jonkers W, Dong Y, Broz K, Corby Kistler H (2012) The Wor1-like protein Fgp1 regulates pathogenicity, toxin synthesis and reproduction in the phytopathogenic fungus *Fusarium graminearum*. PLoS Path 8(5):e1002724. doi:10.1371/journal.ppat.1002724

Juzwik J, Appel DN, MacDonald WL, Burks S (2011) Challenges and successes in managing oak wilt in the United States. Plant Dis 95(8):888–900. doi:10.1094/pdis-12-10-0944

Kämper J, Kahmann R, Bölker M, Ma L-J, Brefort T, Saville BJ, Banuett F, Kronstad JW, Gold SE, Muller O, Perlin MH, Wosten HAB, de Vries R, Ruiz-Herrera J, Reynaga-Pena CG, Snetselaar K, McCann M, Perez-Martin J, Feldbrugge M, Basse CW, Steinberg G, Ibeas JI, Holloman W, Guzman P, Farman M, Stajich JE, Sentandreu R, Gonzalez-Prieto JM, Kennell JC, Molina L, Schirawski J, Mendoza-Mendoza A, Greilinger D, Munch K, Rossel N, Scherer M, Vranes M, Ladendorf O, Vincon V, Fuchs U, Sandrock B, Meng S, Ho ECH, Cahill MJ, Boyce KJ, Klose J, Klosterman SJ, Deelstra HJ, Ortiz-Castellanos L, Li W, Sanchez-Alonso P, Schreier PH, Hauser-Hahn I, Vaupel M, Koopmann E, Friedrich G, Voss H, Schluter T, Margolis J, Platt D, Swimmer C, Gnirke A, Chen F, Vysotskaia V, Mannhaupt G, Guldener U, Munsterkotter M, Haase D, Oesterheld M, Mewes H-W, Mauceli EW, DeCaprio D, Wade CM, Butler J, Young S, Jaffe DB, Calvo S, Nusbaum C, Galagan J, Birren BW (2006) Insights from the genome of the biotrophic fungal plant pathogen *Ustilago maydis*. Nature 444(7115):97–101. doi:10.1038/nature05248xxx

Kapitonov VV, Jurka J (2008) A universal classification of eukaryotic transposable elements implemented in Repbase. Nat Rev Genet 9(5):411–412. doi:10.1038/nrg2165-c1

Kawchuk LM, Hachey J, Lynch DR, Kulcsar F, van Rooijen G, Waterer DR, Robertson A, Kokko E, Byers R, Howard RJ, Fischer R, Prüfer D (2001) Tomato *Ve* disease resistance genes encode cell surface-like receptors. Proc Natl Acad Sci U S A 98(11):6511–6515. doi:10.1073/pnas.091114198

Khoshraftar S, Hung S, Khan S, Gong Y, Tyagi V, Parkinson J, Sain M, Moses A, Christendat D (2013) Sequencing and annotation of the *Ophiostoma ulmi* genome. BMC Genom 14:162. doi:10.1186/1471-2164-14-162

Kistler HC, Miao VPW (1992) New modes of genetic change in filamentous fungi. Annu Rev Phytopathol 30:131–153. doi:10.1146/annurev.py.30.090192.001023

Klimes A, Dobinson KF (2006) A hydrophobin gene, *VDH1*, is involved in microsclerotial development and spore viability in the plant pathogen *Verticillium dahliae*. Fungal Genet Biol 43(4):283–294. doi:10.1016/j.fgb.2005.12.006

Klimes A, Neumann MJ, Grant SJ, Dobinson KF (2006) Characterization of the glyoxalase I gene from the vascular wilt fungus *Verticillium dahliae*. Can J Microbiol 52(9):816–822. doi:10.1139/w06-033

Klosterman SJ, Atallah ZK, Vallad GE, Subbarao KV (2009) Diversity, pathogenicity, and management of *Verticillium* species. Annu Rev Phytopathol 47:39–62. doi:10.1146/annurev-phyto-080508-081748

Klosterman SJ, Perlin MH, Garcia-Pedrajas M, Covert SF, Gold SE (2007) Genetics of morphogenesis and pathogenic development of *Ustilago maydis*. Adv Genet 57:1–47. doi:10.1016/S0065-2660(06)57001-4

Klosterman SJ, Subbarao KV, Kang S, Veronese P, Gold SE, Thomma BPHJ, Chen Z, Henrissat B, Lee Y-H, Park J, Garcia-Pedrajas MD, Barbara DJ, Anchieta A, de Jonge R, Santhanam P, Maruthachalam K, Atallah Z, Amyotte SG, Paz Z, Inderbitzin P, Hayes RJ, Heiman DI, Young S, Zeng Q, Engels R, Galagan J, Cuomo CA, Dobinson KF, Ma L-J (2011) Comparative genomics yields insights into niche adaptation of plant vascular wilt pathogens. PLoS Path 7(7):e1002137. doi:10.1371/journal.ppat.1002137

Knight CJ, Bailey AM, Foster GD (2009) *Agrobacterium*-mediated transformation of the plant pathogenic fungus, *Verticillium albo-atrum*. J Plant Pathol 91(3):745–750. doi:10.4454/jpp.v91i3.573

Köhler GA, Brenot A, Haas-Stapleton E, Agabian N, Deva R, Nigam S (2006) Phospholipase A2 and phospholipase B activities in fungi. Biochim Biophys Acta 1761(11):1391–1399. doi:10.1016/j.bbalip.2006.09.011

Koike M, Fujita M, Nagao H, Ohshima S (1996) Random amplified polymorphic DNA analysis of Japanese isolates of *Verticillium dahliae* and *V. albo-atrum*. Plant Dis 80(11):1224–1227. doi:10.1094/PD-80-1224

Kronstad J, De Maria A, Funnell D, Laidlaw RD, Lee N, Moniz de Sá M, Ramesh M (1998) Signaling via cAMP in fungi: Interconnections with mitogen-activated protein kinase pathways. Arch Microbiol 170(6):395–404. doi:10.1007/s002030050659

Latunde-Dada AO (2001) *Colletotrichum*: tales of forcible entry, stealth, transient confinement and breakout. Mol Plant Pathol 2(4):187–198. doi:10.1046/j.1464-6722.2001.00069.x

Laurie JD, Ali S, Linning R, Mannhaupt G, Wong P, Güldener U, Münsterkötter M, Moore R, Kahmann R, Bakkeren G, Schirawski J (2012) Genome comparison of barley and maize smut fungi reveals targeted loss of RNA silencing components and species-specific presence of transposable elements. Plant Cell 24(5):1733–1745. doi:10.1105/tpc.112.097261

Leuthner B, Aichinger C, Oehmen E, Koopmann E, Müller O, Müller P, Kahmann R, Bölker M, Schreier PH (2005) A $H_2O_2$-producing glyoxal oxidase is required for filamentous growth and pathogenicity in *Ustilago maydis*. Mol Genet Genomics 272(6):639–650. doi:10.1007/s00438-004-1085-6

López-Berges MS, Capilla J, Turrà D, Schafferer L, Matthijs S, Jöchl C, Cornelis P, Guarro J, Haas H, Di Pietro A (2012) HapX-mediated iron homeostasis is essential for rhizosphere competence and virulence of the soilborne pathogen *Fusarium oxysporum*. Plant Cell 24(9):3805–3822. doi:10.1105/tpc.112.098624

Ma L-J, van der Does HC, Borkovich KA, Coleman JJ, Daboussi M-J, Di Pietro A, Dufresne M, Freitag M, Grabherr M, Henrissat B, Houterman PM, Kang S, Shim W-B, Woloshuk C, Xie X, Xu J-R, Antoniw J, Baker SE, Bluhm BH, Breakspear A, Brown DW, Butchko RAE, Chapman S, Coulson R, Coutinho PM, Danchin EGJ, Diener A, Gale LR, Gardiner DM, Goff S, Hammond-Kosack KE, Hilburn K, Hua-Van A, Jonkers W, Kazan K, Kodira CD, Koehrsen M, Kumar L, Lee Y-H, Li L, Manners JM, Miranda-Saavedra D, Mukherjee M, Park G, Park J, Park S-Y, Proctor RH, Regev A, Ruiz-Roldan MC, Sain D, Sakthikumar S, Sykes S, Schwartz DC, Turgeon BG, Wapinski I, Yoder O, Young S, Zeng Q, Zhou S, Galagan J, Cuomo CA, Kistler HC, Rep M (2010) Comparative genomics reveals mobile pathogenicity chromosomes in *Fusarium*. Nature 464(7287): 367–373. doi:10.1038/nature08850

Marshall R, Kombrink A, Motteram J, Loza-Reyes E, Lucas J, Hammond-Kosack KE, Thomma BPHJ, Rudd JJ (2011) Analysis of two in planta expressed LysM effector homologs from the fungus *Mycosphaerella graminicola* reveals novel functional properties and varying contributions to virulence on wheat. Plant Physiol 156(2):756–769. doi:10.1104/pp.111.176347

Martinez D, Berka RM, Henrissat B, Saloheimo M, Arvas M, Baker SE, Chapman J, Chertkov O, Coutinho PM, Cullen D, Danchin EGJ, Grigoriev IV, Harris P, Jackson M, Kubicek CP, Han CS, Ho I, Larrondo LF, de Leon AL, Magnuson JK, Merino S, Misra M, Nelson B, Putnam N, Robbertse B, Salamov AA, Schmoll M, Terry A, Thayer N, Westerholm-Parvinen A, Schoch CL, Yao J, Barabote R, Nelson MA, Detter C, Bruce D, Kuske CR, Xie G, Richardson P, Rokhsar DS, Lucas SM, Rubin EM, Dunn-Coleman N, Ward M, Brettin TS (2008) Genome sequencing and analysis of the biomass-degrading fungus *Trichoderma reesei* (syn. *Hypocrea jecorina*). Nat Biotechnol 26(5):553–560. doi:10.1038/nbt1403

Martinez D, Larrondo LF, Putnam N, Gelpke MDS, Huang K, Chapman J, Helfenbein KG, Ramaiya P, Detter JC, Larimer F, Coutinho PM, Henrissat B, Berka R, Cullen D, Rokhsar D (2004) Genome sequence of the lignocellulose degrading fungus *Phanerochaete chrysosporium* strain RP78. Nat Biotechnol 22(6):695–700. doi:10.1038/nbt967

Maruthachalam K, Atallah ZK, Vallad GE, Klosterman SJ, Hayes RJ, Davis RM, Subbarao KV (2010) Molecular variation among isolates of *Verticillium dahliae* and polymerase chain reaction-based differentiation of races. Phytopathology 100(11):1222–1230. doi:10.1094/phyto-04-10-0122

Maruthachalam K, Klosterman SJ, Kang S, Hayes RJ, Subbarao KV (2011) Identification of pathogenicity-related genes in the vascular wilt fungus *Verticillium dahliae* by *Agrobacterium tumefaciens*-mediated T-DNA insertional mutagenesis. Mol Biotechnol 49(3):209–221. doi:10.1007/s12033-011-9392-8

McDonald BA, Linde C (2002) Pathogen population genetics, evolutionary potential, and durable resistance. Annu Rev Phytopathol 40:349–379. doi:10.1146/annurev.phyto.40.120501.101443

Mehrabi R, Bahkali AH, Abd-Elsalam KA, Moslem M, Ben M'Barek S, Gohari AM, Jashni MK, Stergiopoulos I, Kema GHJ, de Wit PJGM (2011) Horizontal gene and chromosome transfer in plant pathogenic fungi affecting host range. FEMS Microbiol Rev 35(3):542–554. doi:10.1111/j.1574-6976.2010.00263.x

Mehrabi R, Zhao X, Kim Y, Xu J-R (2009) The cAMP signaling and MAP kinase pathways in plant pathogenic fungi. In: Deising HB (ed) Plant relationships. The mycota, vol 5. Springer, Berlin, pp 157–172. doi:10.1007/978-3-540-87407-2_8

Mentlak TA, Kombrink A, Shinya T, Ryder LS, Otomo I, Saitoh H, Terauchi R, Nishizawa Y, Shibuya N, Thomma BPHJ, Talbot NJ (2012) Effector-mediated suppression of chitin-triggered immunity by *Magnaporthe oryzae* is necessary for rice blast disease. Plant Cell 24(1):322–335. doi:10.1105/tpc.111.092957

Michielse CB, Becker M, Heller J, Moraga J, Collado IG, Tudzynski P (2011) The *Botrytis cinerea* Reg1 protein, a putative transcriptional regulator, is required for pathogenicity, conidiogenesis, and the production of secondary metabolites. Mol Plant-Microbe Interact 24(9):1074–1085. doi:10.1094/mpmi-01-11-0007

Michielse CB, Rep M (2009) Pathogen profile update: *Fusarium oxysporum*. Mol Plant Pathol 10(3): 311–324. doi:10.1111/j.1364-3703.2009.00538.x

Michielse CB, van Wijk R, Reijnen L, Manders EMM, Boas S, Olivain C, Alabouvette C, Rep M (2009) The nuclear protein Sge1 of *Fusarium oxysporum* is required for parasitic growth. PLoS Path 5(10):e1000637. doi:10.1371/journal.ppat.1000637

Mieczkowski PA, Lemoine FJ, Petes TD (2006) Recombination between retrotransposons as a source of chromosome rearrangements in the yeast *Saccharomyces cerevisiae*. DNA Repair 5(9–10):1010–1020. doi:10.1016/j.dnarep.2006.05.027

Milgroom MG (1996) Recombination and the multilocus structure of fungal populations. Annu Rev Phytopathol 34:457–477. doi:10.1146/annurev.phyto.34.1.457

Nazar RN, Hu X, Schmidt J, Culham D, Robb J (1991) Potential use of PCR-amplified ribosomal intergenic sequences in detection and differentiation of Verticillium wilt pathogens. Physiol Mol Plant Pathol 39(1):1–11. doi:10.1016/0885-5765(91)90027-F

Nguyen QB, Kadotani N, Kasahara S, Tosa Y, Mayama S, Nakayashiki H (2008) Systematic functional analysis of calcium-signalling proteins in the genome of the rice-blast fungus, *Magnaporthe oryzae*, using a high-throughput RNA-silencing system. Mol Microbiol 68(6):1348–1365. doi:10.1111/j.1365-2958.2008.06242.x

O'Connell RJ, Thon MR, Hacquard S, Amyotte SG, Kleemann J, Torres MF, Damm U, Buiate EA, Epstein L, Alkan N, Altmuller J, Alvarado-Balderrama L, Bauser CA, Becker C, Birren BW, Chen Z, Choi J, Crouch JA, Duvick JP, Farman MA, Gan P, Heiman D, Henrissat B, Howard RJ, Kabbage M, Koch C, Kracher B, Kubo Y, Law AD, Lebrun M-H, Lee Y-H, Miyara I, Moore N, Neumann U, Nordstrom K, Panaccione DG, Panstruga R, Place M, Proctor RH, Prusky D, Rech G, Reinhardt R, Rollins JA, Rounsley S, Schardl CL, Schwartz DC, Shenoy N, Shirasu K, Sikhakolli UR, Stuber K, Sukno SA, Sweigard JA, Takano Y, Takahara H, Trail F, van der Does HC, Voll LM, Will I, Young S, Zeng Q, Zhang J, Zhou S, Dickman MB, Schulze-Lefert P, Ver Loren van Themaat E, Ma L-J, Vaillancourt LJ (2012) Lifestyle transitions in plant pathogenic *Colletotrichum* fungi deciphered by genome and transcriptome analyses. Nat Genet 44(9):1060–1065. doi:10.1038/ng.2372

Oliver RP, Solomon PS (2008) Recent fungal diseases of crop plants: Is lateral gene transfer a common theme? Mol Plant-Microbe Interact 21(3):287–293. doi:10.1094/MPMI-21-3-0287

Olmedo-Monfil V, Cortés-Penagos C, Herrera-Estrella A (2004) Three decades of fungal transformation. In: Balbás P, Lorence A (eds) Recombinant gene expression. Methods in Molecular Biology, vol 267. Humana Press, Totowa, pp 297–313. doi:10.1385/1-59259-774-2:297

Ottmann C, Luberacki B, Küfner I, Koch W, Brunner F, Weyand M, Mattinen L, Pirhonen M, Anderluh G, Seitz HU, Nürnberger T, Oecking C (2009) A common toxin fold mediates microbial attack and plant defense. Proc Natl Acad Sci U S A 106(25):10359–10364. doi:10.1073/pnas.0902362106

Palmer JM, Keller NP (2010) Secondary metabolism in fungi: does chromosomal location matter? Curr Opin Microbiol 13(4):431–436. doi:10.1016/j.mib.2010.04.008

Pantou MP, Typas MA (2005) Electrophoretic karyotype and gene mapping of the vascular wilt fungus *Verticillium dahliae*. FEMS Microbiol Lett 245(2):213–220. doi:10.1016/j.femsle.2005.03.011

Parra G, Blanco E, Guigó R (2000) GeneID in *Drosophila*. Genome Res 10(4):511–515. doi:10.1101/gr.10.4.511

Paz Z, García-Pedrajas MD, Andrews DL, Klosterman SJ, Baeza-Montañez L, Gold SE (2011) One step construction of *Agrobacterium*-recombination-ready-plasmids (OSCAR), an efficient and robust tool for ATMT based gene deletion construction in fungi. Fungal Genet Biol 48(7):677–684. doi:10.1016/j.fgb.2011.02.003

Pegg GF, Brady BL (2002) Verticillium wilts. CABI Publishing, Wallingford

Pemberton CL, Salmond GPC (2004) The Nep1-like proteins—a growing family of microbial elicitors of plant necrosis. Mol Plant Pathol 5(4):353–359. doi:10.1111/j.1364-3703.2004.00235.x

Pietro A, Roncero MIG, Ruiz-Roldán MC (2009) From tools of survival to weapons of destruction: The role of cell wall-degrading enzymes in plant infection. In: Deising HB (ed) Plant relationships. The Mycota, vol 5. Springer, Berlin, pp 181–200. doi:10.1007/978-3-540-87407-2_10

Platt HW, MacLean V, Mahuku G, Maxwell P (2000) Verticillium wilt of potatoes caused by three *Verticillium* species. In: Tjamos EC, Rowe RC, Heale JB, Fravel DR (eds) Advances in *Verticillium*: research and disease management. APS Press, St. Paul, pp 59–62

Puhalla JE (1979) Classification of isolates of *Verticillium dahliae* based on heterokaryon incompatibility. Phytopathology 69(11):1186–1189. doi:10.1094/Phyto-69-118

Puhalla JE, Hummel M (1983) Vegetative compatibility groups within *Verticillium dahliae*. Phytopathology 73(9):1305–1308. doi:10.1094/Phyto-73-1305

Qin Q-M, Vallad GE, Subbarao KV (2008) Characterization of *Verticillium dahliae* and *V. tricorpus* isolates from lettuce and artichoke. Plant Dis 92(1):69–77. doi:10.1094/PDIS-92-1-0069

Qin QM, Vallad GE, Wu BM, Subbarao KV (2006) Phylogenetic analyses of phytopathogenic isolates of *Verticillium* spp. Phytopathology 96(6):582–592. doi:10.1094/PHYTO-96-0582

Qutob D, Kemmerling B, Brunner F, Küfner I, Engelhardt S, Gust AA, Luberacki B, Seitz HU, Stahl D, Rauhut T, Glawischnig E, Schween G, Lacombe B, Watanabe N, Lam E, Schlichting R, Scheel D, Nau K, Dodt G, Hubert D, Gijzen M, Nürnberger T (2006) Phytotoxicity and innate immune responses induced by Nep1-like proteins. Plant Cell 18(12):3721–3744. doi:10.1105/tpc.106.044180

Radišek S, Jakše J, Javornik B (2006) Genetic variability and virulence among *Verticillium albo-atrum* isolates from hop. Eur J Plant Pathol 116(4):301–314. doi:10.1007/s10658-006-9061-0

Raffaele S, Farrer RA, Cano LM, Studholme DJ, MacLean D, Thines M, Jiang RHY, Zody MC, Kunjeti SG, Donofrio NM, Meyers BC, Nusbaum C, Kamoun S (2010) Genome evolution following host jumps in the Irish potato famine pathogen lineage. Science 330(6010):1540–1543. doi:10.1126/science.1193070

Ramos B, González-Melendi P, Sánchez-Vallet A, Sánchez-Rodríguez C, López G, Molina A (2013) Functional genomics tools to decipher the pathogenicity mechanisms of the necrotrophic fungus *Plectosphaerella cucumerina* in *Arabidopsis thaliana*. Mol Plant Pathol 14(1):44–57. doi:10.1111/j.1364-3703.2012.00826.x

Rauyaree P, Ospina-Giraldo M, Kang S, Bhat R, Subbarao K, Grant S, Dobinson K (2005) Mutations in *VMK1*, a mitogen-activated protein kinase gene, affect microsclerotia formation and pathogenicity in *Verticillium dahliae*. Curr Genet 48(2):109–116. doi:10.1007/s00294-005-0586-0

Rep M, Kistler HC (2010) The genomic organization of plant pathogenicity in *Fusarium* species. Curr Opin Plant Biol 13(4):420–426. doi:10.1016/j.pbi.2010.04.004

Roman DG, Dancis A, Anderson GJ, Klausner RD (1993) The fission yeast ferric reductase gene *frp1*+ is required for ferric iron uptake and encodes a protein that is homologous to the gp91-*phox* subunit of the human NADPH phagocyte oxidoreductase. Mol Cell Biol 13(7):4342–4350. doi:10.1128/mcb.13.7.4342

Roper MC, Greve LC, Warren JG, Labavitch JM, Kirkpatrick BC (2007) *Xylella fastidiosa* requires polygalacturonase for colonization and pathogenicity in *Vitis vinifera* grapevines. Mol Plant-Microbe Interact 20(4):411–419. doi:10.1094/MPMI-20-4-0411

Rosewich UL, Kistler HC (2000) Role of horizontal gene transfer in the evolution of fungi. Annu Rev Phytopathol 38:325–363. doi:10.1146/annurev.phyto.38.1.325

Rouxel T, Grandaubert J, Hane JK, Hoede C, van de Wouw AP, Couloux A, Dominguez V, Anthouard V, Bally P, Bourras S, Cozijnsen AJ, Ciuffetti LM, Degrave A, Dilmaghani A, Duret L, Fudal I, Goodwin SB, Gout L, Glaser N, Linglin J, Kema GHJ, Lapalu N, Lawrence CB, May K, Meyer M, Ollivier B, Poulain J, Schoch CL, Simon A, Spatafora JW, Stachowiak A, Turgeon BG, Tyler BM, Vincent D, Weissenbach J, Amselem J, Quesneville H, Oliver RP, Wincker P, Balesdent M-H, Howlett BJ (2011) Effector diversification within compartments of the *Leptosphaeria maculans* genome affected by Repeat-Induced Point mutations. Nat Commun 2(202):1–10. doi:10.1038/ncomms1189

Rowe RC (1995) Recent progress in understanding relationships between *Verticillium* species and subspecific groups. Phytoparasitica 23(1):31–38. doi:10.1007/BF02980394

Rowe RC, Powelson ML (2002) Potato early dying: Management challenges in a changing production environment. Plant Dis 86(11):1184–1193. doi:10.1094/pdis.2002.86.11.1184

Rydholm C, Dyer PS, Lutzoni F (2007) DNA sequence characterization and molecular evolution of *MAT1* and *MAT2* mating-type loci of the self-compatible ascomycete mold *Neosartorya fischeri*. Eukaryot Cell 6(5):868–874. doi:10.1128/EC.00319-06

Salamov AA, Solovyev VV (2000) Ab initio gene finding in *Drosophila* genomic DNA. Genome Res 10(4):516–522. doi:10.1101/gr.10.4.516

Santhanam P (2012) Random insertional mutagenesis in fungal genomes to identify virulence factors. In: Bolton MD, Thomma BPHJ (eds) Plant fungal pathogens. Methods in molecular biology, vol 835. Humana Press, Totowa, pp 509–517. doi:10.1007/978-1-61779-501-5_31

Santhanam P, Thomma BPHJ (2013) *Verticillium dahliae Sge1* differentially regulates expression of candidate effector genes. Mol Plant-Microbe Interact 26(2):249–256. doi:10.1094/mpmi-08-12-0198-r

Santhanam P, van Esse HP, Albert I, Faino L, Nürnberger T, Thomma BPHJ (2013) Evidence for functional diversification within a fungal NEP1-like protein family. Mol Plant-Microbe Interact 26(3):278–286. doi:10.1094/mpmi-09-12-0222-r

Schaible L, Cannon OS, Waddoups V (1951) Inheritance of resistance to Verticillium wilt in a tomato cross. Phytopathology 41(11):986–990

Schwartz DC, Li X, Hernandez LI, Ramnarain SP, Huff EJ, Wang YK (1993) Ordered restriction maps of *Saccharomyces cerevisiae* chromosomes constructed by optical mapping. Science 262(5130):110–114. doi:10.1126/science.8211116

Selkoe KA, Toonen RJ (2006) Microsatellites for ecologists: a practical guide to using and evaluating microsatellite markers. Ecol Lett 9(5):615–629. doi:10.1111/j.1461-0248.2006.00889.x

Soanes DM, Alam I, Cornell M, Wong HM, Hedeler C, Paton NW, Rattray M, Hubbard SJ, Oliver SG, Talbot NJ (2008) Comparative genome analysis of filamentous fungi reveals gene family expansions associated with fungal pathogenesis. PLoS ONE 3(6):e2300. doi:10.1371/journal.pone.0002300

Stahl DJ, Schäfer W (1992) Cutinase is not required for fungal pathogenicity on pea. Plant Cell 4(6):621–629. doi:10.1105/tpc.4.6.621

Stukenbrock EH, Christiansen FB, Hansen TT, Dutheil JY, Schierup MH (2012) Fusion of two divergent fungal individuals led to the recent emergence of a unique widespread pathogen species. Proc Natl Acad Sci U S A 109(27):10954–10959. doi:10.1073/pnas.1201403109

Stukenbrock EH, Jørgensen FG, Zala M, Hansen TT, McDonald BA, Schierup MH (2010) Whole-genome and chromosome evolution associated with host adaptation and speciation of the wheat pathogen *Mycosphaerella graminicola*. PLoS Genet 6(12):e1001189. doi:10.1371/journal.pgen.1001189

Subbarao KV, Chassot A, Gordon TR, Hubbard JC, Bonello P, Mullin R, Okamoto D, Davis RM, Koike ST (1995) Genetic relationships and cross pathogenicities of *Verticillium dahliae* isolates from cauliflower and other crops. Phytopathology 85(10):1105–1112. doi:10.1094/Phyto-85-1105

Subbarao KV, Hubbard JC, Greathead AS, Spencer GA (1997) Verticillium wilt. In: Davis RM, Subbarao

KV, Raid RN, Kurtz EA (eds) Compendium of lettuce diseases. The American Phytopathological Society, St. Paul, pp 26–27

Sunnucks P (2000) Efficient genetic markers for population biology. Trends Ecol Evol 15(5):199–203. doi:10.1016/S0169-5347(00)01825-5

Ter-Hovhannisyan V, Lomsadze A, Chernoff YO, Borodovsky M (2008) Gene prediction in novel fungal genomes using an ab initio algorithm with unsupervised training. Genome Res 18(12):1979–1990. doi:10.1101/gr.081612.108

Thomma BPHJ, Nürnberger T, Joosten MHAJ (2011) Of PAMPs and effectors: the blurred PTI-ETI dichotomy. Plant Cell 23(1):4–15. doi:10.1105/tpc.110.082602

Tran VT, Braus-Stromeyer S, Timpner C, Braus G (2013) Molecular diagnosis to discriminate pathogen and apathogen species of the hybrid *Verticillium longisporum* on the oilseed crop *Brassica napus*. Appl Microbiol Biotechnol 97(10):4467–4483. doi:10.1007/s00253-012-4530-1

Turgeon BG, Christiansen SK, Yoder OC (1993) Mating type genes in ascomycetes and their imperfect relatives. In: Reynolds DR, Taylor JW (eds) The fungal holomorph: mitotic, meiotic and pleomorphic speciation in fungal systematics. CAB International, Wallingford, pp 199–215

Turgeon BG, Lu S-W (2000) Evolution of host specific virulence in *Cochliobolus heterostrophus*. In: Kronstad JW (ed) Fungal pathology. Kluwer Academic Publishers, Boston, pp 93–126. doi:10.1007/978-94-015-9546-9_4

Typas MA, Heale JB (1976) Heterokaryosis and role of cytoplasmic inheritance in dark resting structure formation in *Verticillium* spp. Mol Gen Genet 146(1):17–26. doi:10.1007/BF00267978

Tzima A, Paplomatas EJ, Rauyaree P, Kang S (2010) Roles of the catalytic subunit of cAMP-dependent protein kinase A in virulence and development of the soilborne plant pathogen *Verticillium dahliae*. Fungal Genet Biol 47(5):406–415. doi:10.1016/j.fgb.2010.01.007

Tzima AK, Paplomatas EJ, Rauyaree P, Ospina-Giraldo MD, Kang S (2011) *VdSNF1*, the sucrose nonfermenting protein kinase gene of *Verticillium dahliae*, is required for virulence and expression of genes involved in cell-wall degradation. Mol Plant-Microbe Interact 24(1):129–142. doi:10.1094/mpmi-09-09-0217

Tzima AK, Paplomatas EJ, Tsitsigiannis DI, Kang S (2012) The G protein beta subunit controls virulence and multiple growth- and development-related traits in *Verticillium dahliae*. Fungal Genet Biol 49(4):271–283. doi:10.1016/j.fgb.2012.02.005

Usami T, Amemiya Y, Shishido M (2001) Analyses of transcriptional region found in a tomato pathotype-specific DNA fragment of *Verticillium dahliae*. Soil Microorgan 55(1):11–20

Usami T, Ishigaki S, Takashina H, Matsubara Y, Amemiya Y (2007) Cloning of DNA fragments specific to the pathotype and race of *Verticillium dahliae*. J Gen Plant Pathol 73(2):89–95. doi:10.1007/s10327-006-0334-4

Usami T, Itoh M, Amemiya Y (2008) Mating type gene *MAT1-2-1* is common among Japanese isolates of *Verticillium dahliae*. Physiol Mol Plant Pathol 73(6):133–137. doi:10.1016/j.pmpp.2009.04.002

Usami T, Itoh M, Amemiya Y (2009) Asexual fungus *Verticillium dahliae* is potentially heterothallic. J Gen Plant Pathol 75(6):422–427. doi:10.1007/s10327-009-0197-6

Usami T, Kanto T, Inderbitzin P, Itoh M, Kisaki G, Ebihara Y, Suda W, Amemiya Y, Subbarao KV (2011) *Verticillium tricorpus* causing lettuce wilt in Japan differs genetically from California lettuce isolates. J Gen Plant Pathol 77(1):17–23. doi:10.1007/s10327-010-0282-x

Vallad GE, Qin Q-M, Grube R, Hayes RJ, Subbarao KV (2006) Characterization of race-specific interactions among isolates of *Verticillium dahliae* pathogenic on lettuce. Phytopathology 96(12):1380–1387. doi:10.1094/PHYTO-96-1380

Vallad GE, Subbarao KV (2008) Colonization of resistant and susceptible lettuce cultivars by a green fluorescent protein-tagged isolate of *Verticillium dahliae*. Phytopathology 98(8):871–885. doi:10.1094/PHYTO-98-8-0871

Van de Wouw AP, Cozijnsen AJ, Hane JK, Brunner PC, McDonald BA, Oliver RP, Howlett BJ (2010) Evolution of linked avirulence effectors in *Leptosphaeria maculans* is affected by genomic environment and exposure to resistance genes in host plants. PLoS Path 6(11):e1001180. doi:10.1371/journal.ppat.1001180

van den Brink J, de Vries RP (2011) Fungal enzyme sets for plant polysaccharide degradation. Appl Microbiol Biotechnol 91(6):1477–1492. doi:10.1007/s00253-011-3473-2

van den Burg HA, Harrison SJ, Joosten MHAJ, Vervoort J, de Wit PJGM (2006) *Cladosporium fulvum* Avr4 protects fungal cell walls against hydrolysis by plant chitinases accumulating during infection. Mol Plant-Microbe Interact 19(12):1420–1430. doi:10.1094/mpmi-19-1420

van Esse HP, Bolton MD, Stergiopoulos I, de Wit PJGM, Thomma BPHJ (2007) The chitin-binding *Cladosporium fulvum* effector protein Avr4 is a virulence factor. Mol Plant-Microbe Interact 20(9):1092–1101. doi:10.1094/mpmi-20-9-1092

Walker J (1990) *Verticillium albo-atrum* in Australia: A case study of information confusion in plant pathology. Australas Plant Path 19(3):57–69. doi:10.1071/APP9900057

Wang J-Y, Cai Y, Gou J-Y, Mao Y-B, Xu Y-H, Jiang W-H, Chen X-Y (2004) VdNEP, an elicitor from *Verticillium dahliae*, induces cotton plant wilting. Appl Environ Microbiol 70(8):4989–4995. doi:10.1128/aem.70.8.4989-4995.2004

Weld RJ, Plummer KM, Carpenter MA, Ridgway HJ (2006) Approaches to functional genomics in filamentous fungi. Cell Res 16:31–44. doi:10.1038/sj.cr.7310006

Wicker T, Sabot F, Hua-Van A, Bennetzen JL, Capy P, Chalhoub B, Flavell A, Leroy P, Morgante M, Panaud O, Paux E, SanMiguel P, Schulman AH (2007) A unified classification system for eukaryotic transposable elements. Nat Rev Genet 8(12):973–982. doi:10.1038/nrg2165

Wilhelm S (1955) Longevity of the Verticillium wilt fungus in the laboratory and field. Phytopathology 45(3):180–181

Wilhelm S, Paulus AO (1980) How soil fumigation benefits the California strawberry industry. Plant Dis 64(3):264–270. doi:10.1094/PD-64-264

Xu JR (2000) MAP kinases in fungal pathogens. Fungal Genet Biol 31(3):137–152. doi:10.1006/fgbi.2000.1237

Xu JR, Hamer JE (1996) MAP kinase and cAMP signaling regulate infection structure formation and pathogenic growth in the rice blast fungus *Magnaporthe grisea*. Genes Dev 10(21):2696–2706. doi:10.1101/gad.10.21.2696

Zare R, Gams W, Starink-Willemse M, Summerbell RC (2007) *Gibellulopsis*, a suitable genus for *Verticillium nigrescens*, and *Musicillium*, a new genus for *V. theobromae*. Nova Hedwigia 85(3–4):463–489. doi:10.1127/0029-5035/2007/0085-0463

Zeise K, von Tiedemann A (2002) Application of RAPD-PCR for virulence type analysis within *Verticillium dahliae* and *V. longisporum*. J Phytopathol 150(10): 557–563. doi:10.1046/j.1439-0434.2002.00799.x

Zhang J (2003) Evolution by gene duplication: an update. Trends Ecol Evol 18(6):292–298. doi:10.1016/S0169-5347(03)00033-8

Zhang N, Castlebury LA, Miller AN, Huhndorf SM, Schoch CL, Seifert KA, Rossman AY, Rogers JD, Kohlmeyer J, Volkmann-Kohlmeyer B, Sung G-H (2006) An overview of the systematics of the *Sordariomycetes* based on a four-gene phylogeny. Mycologia 98(6):1076–1087. doi:10.3852/mycologia.98.6.1076

Zhou B-J, Jia P-S, Gao F, Guo H-S (2012) Molecular characterization and functional analysis of a necrosis- and ethylene-inducing, protein-encoding gene family from *Verticillium dahliae*. Mol Plant-Microbe Interact 25(7):964–975. doi:10.1094/mpmi-12-11-0319

# Fusarium oxysporum

5

Seogchan Kang, Jill Demers, Maria del Mar Jimenez-Gasco, and Martijn Rep

## 5.1 Overview of Biology

*Fusarium oxysporum* (Fo) is a species complex that encompasses genetically and phenotypically diverse strains and has been found in a wide range of ecosystems. Most studies on the Fo species complex (FOSC) have focused on plant pathogenic strains because they cause diseases in >120 agriculturally and horticulturally important plants (Armstrong and Armstrong 1981; Michielse and Rep 2009). In a recent survey in the international community of fungal pathologists, Fo was ranked fifth in a list of top 10 fungal plant pathogens based on scientific and economic importance (Dean et al. 2012). In contrast to the broad collective host range of the FOSC, individual strains are typically host specific, only causing disease in one or a few related plant species (Armstrong and Armstrong 1981). Plant pathogenic isolates are classified based on their host range into *formae speciales* (ff. spp., plural; *forma specialis*, f. sp., singular). Some members of the FOSC also infect animals including humans and insects (Dignani and Anaissie 2004; O'Donnell et al. 2004, 2007, 2009, 2012; Ortoneda et al. 2004). The fungus carries a haploid genome that exhibits a high degree of variation in karyotype and can harbor segmental duplications (Boehm et al. 1994; Ma et al. 2010). Horizontal chromosome transfer within the species complex and perhaps across species boundaries seems to have potentially contributed to the generation of variants, particularly new pathogens (Ma et al. 2010). Movement of transposable elements (TEs) and chromosomal rearrangements caused by recombination among homologous TEs scattered around the genome are other mechanisms that appear to have generated new variants.

### 5.1.1 Taxonomy

A well-resolved phylogeny is fundamental for studying the evolution, population biology, and reproductive mode of members of the FOSC. Clear answers to questions such as if FOSC strains can reproduce sexually and whether host specificity has evolved vertically or horizontally should help in the development of effective disease management strategies and quarantine regulations. Although the phylogenetic relationship of the FOSC with other fusaria is well established (Geiser et al. 2013), species boundaries within the

S. Kang (✉) · M. del Mar Jimenez-Gasco
Department of Plant Pathology & Environmental Microbiology, The Pennsylvania State University, University Park, PA 16802, USA
e-mail: sxk55@psu.edu

J. Demers
USDA-ARS Systematic Mycology & Microbiology Laboratory, 10300 Baltimore Ave., Beltsville, MD 20705, USA

M. Rep (✉)
Swammerdam Institute for Life Sciences, University of Amsterdam, Science Park, 1098 XH, Amsterdam, The Netherlands
e-mail: m.rep@uva.nl

R. A. Dean et al. (eds.), *Genomics of Plant-Associated Fungi and Oomycetes: Dicot Pathogens*,
DOI: 10.1007/978-3-662-44056-8_5, © Springer-Verlag Berlin Heidelberg 2014

FOSC have yet to be clearly resolved—if the concept of "species" is applicable at all. Markers such as the IGS (inter-genic spacer) region of ribosomal RNA encoding genes (rDNA) and genes encoding EF-1$\alpha$, polygalacturonases, mitochondrial small subunit rDNA, phosphate permease, $\beta$-tubulin, nitrate reductase, and mating type have been evaluated for phylogenetic resolution of the FOSC. Among these markers, only the *EF-1$\alpha$* gene and IGS rDNA appear to provide sufficient phylogenetic signal (O'Donnell et al. 2009). These two markers were employed to analyze 850 isolates that represent the phylogenetic breadth and ecological diversity of the FOSC, toward the development of a multilocus sequence typing (MLST) database that can support the identification of *formae speciales* and pathogens of humans (O'Donnell et al. 2009). This analysis revealed 101 *EF-1$\alpha$*, 203 IGS rDNA, and 256 two-locus sequence types (STs), confirming a high degree of genetic diversity within the FOSC. In addition, the potential for producing mycotoxins such as moniliformin, fumonisin, and enniatin was assessed in the resulting phylogenetic context. However, the phylogenetic resolution conferred by this MLST dataset was insufficient to distinguish all of the *formae speciales* and VCGs (vegetative compatibility groups). In addition, the homoplastic evolutionary history of the IGS rDNA locus (Fourie et al. 2009; Mbofung et al. 2007; O'Donnell et al. 2009) obscured accurate phylogenetic relationships, underscoring the need for identifying a new set of markers for more robust phylogenetic inference. However, given recent findings that *formae speciales* may be determined by the accessory genome (Ma et al. 2010), irrespective of the core genome, it is highly doubtful that distinguishing *formae speciales* can be done based on sequences of genes conserved across the FOSC.

Rapid increase in fungal genome sequencing offers a potential solution for this need. The Broad Institute of MIT and Harvard (http://www.broadinstitute.org/annotation/genome/fusarium_group/) sequenced the genomes of 11 phylogenetically diverse Fo strains. Thousands of orthologous loci in these genomes were evaluated to identify markers that provide good phylogenetic signals, resulting in a large number of potential new markers (Park 2013). Some of these markers will help build a more robust phylogenetic framework that will guide inquires concerning the evolution of important traits in the FOSC and also support management of pathogenic Fo.

### 5.1.2 Reproduction

*Fusarium oxysporum* is generally considered to reproduce asexually, because a teleomorph has never been observed in nature or induced in the laboratory. However, the possibility of a cryptic sexual cycle cannot be completely ruled out. Both mating-type genes have been found in members of the FOSC (Yun et al. 2000). Although some studies have supported the clonality of Fo based on association of alleles (Bentley et al. 1998; Koenig et al. 1997), re-analysis of the data (Taylor et al. 1999) showed that the possibility of recombination could not be excluded. Three types of asexual spores are produced. Microconidia are oval or elliptical and consist of one or two cells. They are produced under diverse conditions, such as liquid and solid culture media, in the rhizosphere, and within the vascular system of infected plants. Macroconidia are three- to five-celled, gradually pointed and curved toward the ends. Macroconidia are commonly found on the surface of plants killed by Fo (Katan et al. 1997) as well as in sporodochia. Thick-walled chlamydospores ensure long-term survival of Fo. Chlamydospores are produced either terminally or intercalary on older mycelium or in macroconidia and may remain viable in the soil for many years, making them a long-term constraint on crop production in previously infested fields. The molecular mechanism underpinning the production of these spores in Fo is poorly understood. However, this deficiency can be quickly addressed via a combination of genome-enabled approaches (e.g., profiling of expressed genes and proteins, systematic mutagenesis of candidate genes). Given the significance of these spores for Fo pathology, new insights into asexual reproduction may offer novel means to manage Fo diseases.

### 5.1.3 Ecological Niches

Members of the FOSC are ubiquitous in soil and have been found in diverse ecosystems, including grasslands (Opperman and Wehner 1994), forests (Cabello and Arambarri 2002), and deserts (Mandeel et al. 2005) and ranging from the tropics (Sangalang et al. 1995) to the Arctic (Kommedahl et al. 1988). Although they are commonly known as plant pathogens, they can survive as saprophytes or as endophytes colonizing asymptomatic plants, and most isolates are presumed nonpathogenic. A review by Kuldau and Yates (2000) found reports of Fo endophytes in nearly 100 plant species from a limited number of surveys, suggesting that Fo is intimately associated with many more plant species. Some putatively nonpathogenic strains have been employed successfully as biocontrol agents to suppress soilborne pathogens including Fo itself (Bao et al. 2004; Blok et al. 1997; Fravel et al. 2003; Fuchs et al. 1999). However, the null hypothesis that saprophytic, endophytic, or biocontrol strains are nonpathogenic to plants is difficult to test due to the large number of potential hosts.

### 5.1.4 Plant Pathogens

Plant pathogenic Fo strains mainly cause vascular wilts but also have been shown to cause root rots (Jarvis and Shoemaker 1978; Vakalounakis 1996), crown rots (Van Bakel and Kerstens 1970), bulb rots (Linderman 1981), and damping-off (Bloomberg 1979). The collective host range of the FOSC is very broad, with over 120 plant species reported as hosts (Armstrong and Armstrong 1981; Michielse and Rep 2009), including important crops such as banana, cotton, palm, tomato, melon, and many other vegetables and ornamentals. Given that minor diseases in crop plants and diseases in wild plants are often overlooked, the number of plant species in which the FOSC can cause symptoms is likely to be much greater.

#### 5.1.4.1 Origins of Pathogenicity

How pathogenicity originally evolved within the FOSC is still a matter of speculation. However, genomics-enabled inquiries into this question have begun to provide new insights (Ma et al. 2010). New *formae speciales* are frequently reported; recent examples include f. sp. *palmarum* (pathogenic to queen palm and Mexican fan palm) in 2010 (Elliott et al. 2010), f. sp. *loti* (pathogenic to birdsfoot trefoil) in 2009 (Wunsch et al. 2009), and f. sp. *rapae* (pathogenic to *Brassica rapa*) in 2008 (Enya et al. 2008). It is not clear whether these reports of novel hosts are due to better surveillance, movement of the host or pathogen, or the evolution of new *formae speciales* from nonpathogenic strains or other *formae speciales*. Increasing agricultural trade likely plays a major role in the spread of pathogenic strains by facilitating their movement via infected plant material. Many *formae speciales* are also known to be seed-transmitted, such as *basilici* (Vannacci et al. 1999), *erythroxyli* (Gracia-Garza et al. 1999), *lactucae* (Garibaldi et al. 2004), and over twenty other *formae speciales* (Agarwal and Sinclair 1997).

The majority of *formae speciales* appear to be polyphyletic (e.g., Fig. 5.1), meaning that pathogenicity to a given host seems to have evolved multiple times independently (Baayen et al. 2000; O'Donnell et al. 1998, 2009). O'Donnell et al. first demonstrated polyphyletic origins of f. sp. *cubense* isolates, pathogenic to banana, through a phylogenetic analysis with isolates from other *formae speciales* (O'Donnell et al. 1998). This analysis showed that f. sp. *cubense* isolates belong to multiple distinct clades within the FOSC and that isolates of f. sp. *cubense* are therefore commonly more closely related to isolates of other *formae speciales* than to other f. sp. *cubense* isolates. The finding that *formae speciales* are often polyphyletic raised the question of how pathogenicity has evolved in the FOSC. One hypothesis is that pathogenicity to a particular host has evolved independently in each clade of a polyphyletic *forma specialis*.

According to this hypothesis, strains within a *forma specialis* are similar only by convergent evolution and pathogenicity to a given host may have arisen from different ancestral traits. An alternative hypothesis is that polyphyletic *formae speciales* have emerged by the horizontal transfer of pathogenicity genes to unrelated isolates. In this scenario, pathogenicity to a given host may have evolved only once, but then has been transferred to other strains, so that all strains within a *forma specialis* share the same basic mechanism for causing disease in that host. This seems to be the case at least for f. sp. *lycopersici*, pathogenic to tomato (Ma et al. 2010). Sequence analysis of candidate virulence (effector) genes in other polyphyletic *formae speciales* led to a similar conclusion (Rep unpublished observations), however, independent parallel emergence of pathogenicity toward some plant species remains a distinct possibility. Although polyphyly seems to be frequent in Fo *formae speciales*, there are also several monophyletic *formae speciales*.

#### 5.1.4.2 Pathogenic Races

*Formae speciales* frequently comprise several pathogenic races, defined by virulence patterns on differentially resistant varieties of the host plant species. Mutations in virulence-related genes can lead to the emergence of new races within a *forma specialis*. The best evidence of how this process could occur comes from the polyphyletic f. sp. *lycopersici*. Fourteen effector genes encoding small proteins secreted in tomato have been identified so far in f. sp. *lycopersici* (Houterman et al. 2007; Ma et al. 2010; Schmidt et al. 2013). Three of these are also avirulence (*AVR*) genes, corresponding to the three known pathogenic races of f. sp. *lycopersici* (races 1, 2, and 3) (Takken and Rep 2010). The effectors Avr2 and Avr3 are needed for full virulence on tomato and are recognized by the resistance (R) proteins I-2 and I-3, respectively. Avr1 suppresses host defenses mediated by *I-2* and *I-3* but is recognized by the R proteins *I* and *I-1*. New races arose through mutation in several clonal lines and were apparently selected for by the extensive use of resistant cultivars, carrying *I* and/or *I-2*. Race 2 appears to have emerged several times from race 1 by loss of *AVR1* through deletion of a genomic region (Chellappan and Cornelissen personal communication). Subsequently, point mutations in *AVR2* have led to a loss of recognition by *I-2*, leading to emergence of race 3 (Takken and Rep 2010). Race 3 also appears to have evolved independently several times, given that three different point mutations have been found in the *AVR2* gene (Takken and Rep 2010) and that race 3 isolates can be closely related to race 2 isolates in the same area (Cai et al. 2003). *AVR* genes have not yet been identified in other *formae speciales*, and whether pathogenic races for other *formae speciales* fit into the gene-for-gene model of host resistance has not yet been definitely shown.

#### 5.1.4.3 Model Plants for Studying the Mechanisms Underpinning Fo Pathogenicity and Host Specificity

Considering the intimate and dynamic interactions between Fo and its hosts throughout the disease cycle, the capability of genetically manipulating the host is critical for understanding molecular mechanisms underlying host colonization and disease progression. *Arabidopsis thaliana* and tomato have been widely utilized as experimental model hosts due to their rich genetic resources and tractability. The use of *A. thaliana* as a model has greatly advanced our understanding of the mechanisms underpinning defense, susceptibility and disease progression with a diverse array of pathogenic organisms ranging from viruses to fungi to nematodes and insects (Nishimura and Dangl 2010). A number of studies have employed *A. thaliana* to study plant responses to Fo infection and the genetic requirements of colonization by Fo in the host and the pathogen (Berrocal-Lobo and Molina 2008; Czymmek et al. 2007; Diener and Ausubel 2005; Kidd et al. 2011; Kim et al. 2011; Ospina-Giraldo et al. 2003; Shen and Diener 2013). Several ecotypes of *A. thaliana* have been shown to differentially interact with different strains of the FOSC (Diener and Ausubel 2005), offering materials for identifying and mapping putative resistance (*R*) genes. The first such gene identified, *RFO1*, which confers non-race specific resistance against Fo, encodes a receptor-like kinase (RLK) (Diener and Ausubel 2005; Ospina-Giraldo et al. 2003). *RFO2* and *RFO3* both confer resistance to Fo f.sp. *matthioli* and encode, respectively, a receptor-like protein (RLP) (Shen and Diener 2013) and a RLK (Cole and Diener 2013). Roles of specific *A. thaliana* genes and signaling pathways in defense or susceptibility to Fo have been studied by taking advantage of its vast mutant resources (Berrocal-Lobo and Molina 2008; Diener and Ausubel 2005; Kidd et al. 2011). As described below (Sect. 5.3.4.1), the small size of *A. thaliana* made it possible to grow and infect it with Fo in chambers with a cover-glass bottom, which enabled time-lapse confocal microscopic imaging of the colonization and penetration of *A. thaliana* roots by Fo without physically disrupting roots (Czymmek et al. 2007; Kim et al. 2011).

The tomato-Fo f. sp. *lycopersici* (Fol) interaction has mainly been used to identify Fo pathogenicity factors (Di Pietro et al. 2003; Michielse and Rep 2009; Michielse et al. 2009a) and to analyze the molecular basis of tomato resistance toward Fo. All three avirulence genes corresponding to tomato resistance genes *I*, *I-2*, and *I-3* have been identified (Takken and Rep 2010). The *I-2* resistance gene, which encodes a protein of the NB-LRR class (Simons et al. 1998), is mainly expressed in tissue surrounding the xylem vessels (Mes et al. 2000b). In accordance with NB-LRR proteins being intracellular receptors, the corresponding avirulence protein from Fol, AVR2, is recognized in the cytoplasm, even though it was identified in xylem sap of Fol-infected tomato plants (Houterman et al. 2007). Apart from the observation that never-ripe (ethylene insensitive) tomato plants exhibit reduced symptoms upon Fol infection (Lund et al. 1998), the role of hormones and various signaling pathways under their influence in the tomato–Fol interaction has hardly been explored.

### 5.1.5 Human/Animal Pathogens

Besides infecting diverse plants, certain members of the FOSC also cause localized or deeply invasive infections in humans, causing very high mortality in immunocompromised patients (Dignani and Anaissie 2004; O'Donnell et al. 2004, 2007, 2009; Ortoneda et al. 2004). Fo can also infect the corneas of immune-competent humans, frequently causing blindness, and was associated with outbreaks of Fusarium keratitis in Asia and the United States (Chang et al. 2006; O'Donnell et al. 2007). A comprehensive phylogenetic analysis with 850 isolates that represent diverse *formae speciales* and pathogens of humans showed that isolates associated with opportunistic human infection are genetically diverse, corresponding to many distinct sequence types (STs), and are nested within the three major clades that comprise the phylogenetic breadth of the FOSC (O'Donnell et al. 2004, 2009). Fusaria that are associated with insects are nested within 10 species complexes, including the FOSC, spanning the phylogenetic breadth of the genus *Fusarium* (O'Donnell et al. 2012). Their relationships to human or plant pathogenic species and strains were studied via multilocus phylogenetic analysis to investigate the utility of insecticolous fusaria as biological control agents of insect pests and to minimize the risk of inadvertently employing plant or animal pathogens as biological control agents by providing robust marker sequences for identifying these fusaria (O'Donnell et al. 2012).

## 5.2 Genome Structure and Evolution

### 5.2.1 Genome Structure and Notable Features

Annotated genome sequences of 11 Fo strains are publicly available through the Broad Institute Fusarium Comparative Genomics platform (http://www.broadinstitute.org/annotation/genome/fusarium_group/GenomesIndex.htm), and additional strains are currently being sequenced. The genome of 4287, a strain pathogenic to tomato, was the first to be decoded, which was accomplished via whole genome shotgun sequencing (6X coverage) with Sanger's method (Ma et al. 2010). Optical maps of its chromosomes provided physical scaffolds that guided the genome assembly by helping anchor sequence contigs. The Broad platform provides a description of sequencing and assembly strategies and genome- and gene-associated statistics of this and other Fo strains.

#### 5.2.1.1 Genome Structure

Compared to the *F. graminearum* (Fg) and *F. verticillioides* (Fv) genomes, the most notable feature of Fo genomes is the presence of large "accessory" chromosomes and chromosomal regions, also called lineage-specific (LS) chromosomes and regions. These regions do not display synteny to sequences in the Fg and Fv genomes and appear to have evolutionary histories that are different from the core genome (Ma et al. 2010). The core genome of Fo is similar to the genome of Fv with regard to chromosome number (11), gene number (~14,200), and sequence (91 % average identity between orthologs) (Ma et al. 2010). The high TE content in Fo, especially compared to Fg and Fv, has long been recognized (Daboussi and Capy 2003) and is largely attributable to the high TE content of accessory genomic regions (Ma et al. 2010). Fo is not unique in this sense; transposon-rich accessory genomes have been identified, for instance, in *Fusarium solani* (=*Nectria haematococca*) (Coleman et al. 2009) and *Alternaria arborescens* (Hu et al. 2012), and TE-rich chromosomal subregions have been observed, for example, in *Leptosphaeria maculans* (Rouxel et al. 2011) and *Verticillium dahliae* (Klosterman et al. 2011).

#### 5.2.1.2 Gene Content

We take Fol4287 here as representative for the gene content and features for the FOSC, even though exact gene numbers differ between strains. In Fol4287, about 20,900 genes were predicted in the most recent annotation (http://www.broadinstitute.org/annotation/genome/fusarium_group/GenomesIndex.htm). This number is among the highest in fungi, which is mostly due to the large number of predicted genes in LS regions (~3,500). Fg and Fv were predicted to have ~13,300 and 15,900 genes, respectively, with the three species sharing around 9,000 conserved, syntenic orthologs. Compared to other filamentous ascomycetes, including *Magnaporthe oryzae*, a rice-infecting pathogen within the same class as *Fusarium* (*Sordariomycetes*), the genomes of the three *Fusarium* species encode larger numbers of transcription factors, carbohydrate-active enzymes, and ABC and PDR transmembrane transporters (Ma et al. 2010). Around 80 % of the predicted gene products encoded in LS regions could not be classified. Classification of the remainder revealed enrichment for secreted effectors and virulence factors, transcription factors and proteins involved in signal transduction, and a notable deficiency in house-keeping functions (Ma et al. 2010). A more detailed, manual analysis of LS chromosome 14, which contains the genes conferring virulence toward tomato (Ma et al. 2010), revealed a predicted 245 non-transposon protein-coding genes, occupying only 13 % of the ~2.2 Mb of this chromosome (Schmidt et al. 2013). Again, most genes (140) could not be classified. The largest classes of the remaining ~100 genes encode secreted proteins (29), proteins involved in secondary metabolism (35) or transcription factors (11). Among the secreted proteins are nine secreted enzymes and all but one of the small proteins (named "Six" for *S*ecreted *i*n *x*ylem) found in xylem sap of infected tomato plants (Schmidt et al. 2013). Expression of most genes for secreted proteins and a secondary metabolite gene cluster on this "pathogenicity" chromosome is highly induced upon infection of tomato (Schmidt et al. 2013; Schmidt and Rep unpublished observations).

#### 5.2.1.3 Transposable Elements

Early work, mostly of the Daboussi lab, has shown that the Fo genome is a particularly rich treasure trove of transposable elements (TEs) (Daboussi and Capy 2003). Compared to Fv, which has 0.14 % of its genome corresponding to TEs, TEs make up about 4 % of the Fo genome (Ma et al. 2010). The Fo TEs include comparable numbers of LTR (long terminal repeat) and non-LTR retrotransposons (class I TEs) and DNA transposons (class II) (Daboussi and Capy 2003). Some of these TEs have been demonstrated to be active by TE trapping in genes such as the nitrate reductase gene (Daboussi and Capy 2003). Analysis of the Fol4287 genome sequence has confirmed and further expanded our appreciation of this treasure (Bergemann et al. 2008; Dufresne et al. 2011; Schmidt et al. 2013). Most of these TEs, especially class II TEs, reside in LS regions. The core genome of Fo is in fact rather TE-poor (Ma et al. 2010) with the 0.6 kb LINE element Foxy (Mes et al. 2000a) being the most abundant with a few hundred copies (Rep unpublished observations). About 24 % of the sequence of the Fol4287 "pathogenicity" chromosome (chromosome 14) was annotated to TEs or TE remnants, almost twice the space of (other) protein-coding sequences (Schmidt et al. 2013). On this chromosome alone, over 500 TEs and TE remnants were identified, roughly equally divided between class I and class II TEs. Class II TEs appear mostly aggregated in large subregions of the chromosome, which also harbor the known effector (*SIX*) genes. Remarkably, a miniature impala (mimp), a class II non-autonomous TE of about 200 bp, consistently resides within 1.5 kb upstream of the start codon of effector genes and several other in planta induced genes with potential virulence functions (Schmidt et al. 2013).

### 5.2.2 Genome Dynamics

#### 5.2.2.1 Translocation and Duplication

The FOSC is known for its extreme karyotope variability between strains, even within a vegetative compatibility group (Boehm et al. 1994; Kistler et al. 1995; Ma et al. 2010). From the optical map of Fol4287 (Ma et al. 2010) and other strains (Ma personal communication), it appears that most of this variation is due to variability in size of LS chromosomes and large subtelomeric LS regions—the 11 core chromosomes appear largely similar between strains. Presumably, LS regions are prone to recombination, perhaps enhanced by their large TE content. In Fol4287, the two largest LS chromosomes (3 and 6) harbor large (>1 Mb) inter- and intra-chromosomal segmental duplications, revealed by the optical map, resulting in threefold or fourfold duplications of many genes. These duplications appear recent (99 % sequence identity) and together comprise ~7 Mb in this strain (Ma et al. 2010). In addition, ~1 Mb of chromosome 15 (supercontig 24) is almost identical to a subtelomeric LS region of chromosome 1 (supercontig 27). These particular duplications appear to be unique to this strain and it is unknown how frequent such duplications are in other strains—their high sequence identity requires optical mapping to resolve.

#### 5.2.2.2 TE Activity

There is experimental evidence, such as TE trapping, for activity of many DNA TEs in Fo (Daboussi and Capy 2003). The LINE Foxy, a retroelement, is also active (Mes et al. 2000a); even two strains within the same clonal line (Fol4287 and Fol007) share only a part (~70 %) of Foxy insertion sites in their core genome, and a strain from a different clonal line (Fo47) does not share any Foxy insertion site with Fol4287 (Ma et al. 2010). TE activity can lead to phenotypic novelty. A dramatic example of this is the loss of *AVR1*—hence the overcoming of disease resistance conferred by the *I* gene in tomato—due to the insertion of the class II TE Hormin (Inami et al. 2012). Another example is the apparent past inactivation of an effector gene homolog (*SIX1-H*) by the class II TE Drifter (Rep et al. 2005). There are also many examples of nested TEs on chromosome 14 of Fol4287 (Schmidt et al. 2013). In addition to transposition, recombination between (almost) identical TEs with the same orientation on the same chromosome can lead to deletion of chromosomal regions of considerable size. An example of this phenomenon is the loss of a ~30- or ~70 kb region containing *AVR1* in race 2 and race 3 strains of Fol due to recombination between Helitrons (Chellappan and Cornelissen personal communication).

##### 5.2.2.3 Chromosome Transfer Between Strains as a Mechanism Driving Genome Innovation

Besides translocations, duplications, and deletions within a genome, horizontal chromosome transfer is a driver of genome dynamics in Fo. Although the sexual cycle of Fo has not been observed in nature or in the lab, the mixing of two strains on a Petri dish can lead to transfer of one or more LS chromosomes from one strain to the other. With the chromosomes, the recipient strain can gain virulence toward a particular host (Ma et al. 2010). Although in the lab, selection for drug resistance markers is required for identification of these rare events and analysis of the resulting strains, this process likely contributes significantly to the enormous genetic diversity of Fo strains in soil and plant tissue, and to the emergence of novel clonal lines pathogenic to a particular plant species (Ma et al. 2010). The mechanism is unclear. Hyphal fusion is common in filamentous fungi including Fo, and chromosome transfer after hyphal fusion may either involve nuclear fusion followed by loss of most of one "parental" genome or uptake of one or more chromosomes by a nucleus of one of the "parents." For progeny to develop from such an event, the nucleus with the new chromosome(s) must undergo mitosis and populate new hyphae and, eventually, conidia. Whether only LS chromosomes and/or small chromosomes can undergo horizontal transfer is currently under investigation.

## 5.3 Toolbox and Other Resources

To advance our knowledge of the biology, ecology, and evolution of the FOSC by taking advantage of a rapidly increasing number of available genome sequences, a diverse array of experimental tools and resources are needed. We here mention some important resources that support *Fusarium* genomics and gene functional studies.

### 5.3.1 Culture Collections

Availability of well-curated and well-characterized cultures is critical to guide comparative genomics of the FOSC. Three facilities offer well-curated *Fusarium* cultures including diverse members of the FOSC. First, the *Fusarium* Research Center (FRC) at Penn State University houses ~20,000 cultures, which were isolated from over 100 countries and every continent except Antarctica and include >1,000 Fo strains. The FRC collection emphasizes plant pathogens and toxin producers, but also contains environmental isolates from a variety of substrates and isolates from human and animal infections. Historically, the FRC has offered the following services: (a) isolate storage and distribution, (b) identification, using both morphological and DNA-sequence-based methods, (c) training in basic laboratory protocols to work with *Fusarium*, and (d) providing irradiated carnation leaf pieces, a favored growth medium for the fungus. Unfortunately, the information associated with its cultures has not yet been converted into a format that can be searched by outside users. Second, the ARS Culture Collection (also known as the NRRL Collection) in Peoria, IL also houses approximately 20,000 *Fusarium* isolates in addition to other fungal and bacterial cultures. Its online catalog (http://nrrl.ncaur.usda.gov/cgi-bin/usda/) allows users to search and request strains. However, many *Fusarium* accessions have yet to be archived in the online catalog. The Centraalbureau voor Schimmelcultures (CBS) Fungal Biodiversity Centre (http://www.cbs.knaw.nl/) in the Netherlands maintains a smaller collection of *Fusarium* (1,129 accessions) as well as other filamentous fungi, yeasts, and bacteria. Most *Fusarium* isolates that have been used for recent phylogenetic analyses, including those belonging to the FOSC (O'Donnell et al. 2009, 2012, 2013), are archived in one or more of these three collection facilities, and resulting sequence data are available via the two online platforms described below.

### 5.3.2 Web Platforms Supporting Strain Identification, Phylogenetic Analysis, or Comparative Genomics

#### 5.3.2.1 Strain Identification and Phylogenetic Analysis

A significant impediment to the study of *Fusarium* has been the incorrect and confused application of species names. Without a robust phylogenetic framework that guides species and strain identification, documenting the global *Fusarium* diversity and identifying both old and new problems caused by *Fusarium* continues to be fragmented, creating confusion instead of the order that taxonomy should provide. Extensive molecular phylogenetic studies based on the publicly available culture collections mentioned above have been performed and will continue to clarify relationships at various taxon levels. Web-based community platforms have been developed to enable researchers worldwide to quickly and accurately identify fusaria based on the resulting phylogenetic frameworks (Geiser et al. 2004; O'Donnell et al. 2009; Park et al. 2011).

In the early 2000s, Fusarium-ID v.1.0, a first-generation online database for identifying *Fusarium* using *EF-1α*, was publicly opened (Geiser et al. 2004). It has evolved to offer multiple markers (5,535 sequences from 14 loci; 1,847 isolates) (Park et al. 2011). Fusarium-ID v.2.0 (http://isolate.fusariumdb.org/) allows users to use sequences from an isolate in order to determine its identity using BLAST, tree-building, and other functions. It averages >2,000 uses per month. The primary identification markers are portions of the *EF-1α* and second largest RNA polymerase II subunit (*RPB2*) genes, with several additional markers employed depending on the species complex. A parallel site, Fusarium MLST, exists at the CBS. Both platforms archive multilocus phylogenetic data derived from analyzing the FOSC (O'Donnell et al. 2009, 2012) and other fusaria. While one can (and should) conduct BLAST searches against GenBank, sequences in GenBank are frequently not attached to vouchered isolates and have a number of issues (Kang et al. 2010). By contrast, every sequence in Fusarium-ID and Fusarium MLST is connected to a culture available from FRC, CBS, or NRRL. A number of improvements are planned for Fusarium-ID: (a) tutorials for interpreting the results and connecting it to useful information about an isolate or taxon; (b) GIS tools for visualizing the geospatial and temporal contexts of archived isolates; and (c) a comprehensive electronic monographic resource for the genus *Fusarium*, which will critically assess the boundaries of the genus and define phylogenetically robust subgeneric taxa and evolutionary units, to make inferences about evolutionary patterns and mechanisms underlying its considerable phenotypic and genotypic diversity.

#### 5.3.2.2 Comparative Genomics Platforms

Genome sequences of diverse fungal species and multiple isolates within species have been rapidly accumulating (e.g., 1,000 fungal genome project, which aims to sequence fungal genomes across the Fungal Tree of Life), presenting vast opportunities for comparative genomic analyses at various taxon levels. Well-annotated, published genomes for Fg, Fo, Fv, and *F. solani* (Fs) are available (Coleman et al. 2009; Cuomo et al. 2007; Ma et al. 2010). In addition, mostly next-generation sequence data are in various states of production for many other *Fusarium* species and additional strains of the FOSC. The rapid accumulation of sequence data necessitates a robust mechanism and tools for data curation and utilization to enable the effective management and analysis of huge quantities of genomics data. Genome sequences of the FOSC and other *Fusarium* species can be accessed through multiple online platforms.

The *Fusarium* Comparative Genomics platform at the Broad Institute (http://www.broadinstitute.org/annotation/genome/fusarium_

group/GenomesIndex.htm) provides annotated genome sequences of Fg, Fv, and Fo and associated statistics. MycoCosm, the genome portal of the Department of Energy Joint Genome Institute (http://jgi.doe.gov/fungi), offers annotated genome sequences of diverse fungi, including some *Fusarium* species, generated by JGI and others and provides an array of interactive web-based genome analysis tools (Grigoriev et al. 2012). Results from the 1,000 fungal genome projects will be made available via MycoCosm, offering a comprehensive comparative genomics reference that will guide research on fungi. Cyber-infrastructure for Fusarium (CiF; http://www.fusariumdb.org/), in which Fusarium-ID is one of the three main components, offers a genomics platform specialized for *Fusarium*, entitled the *Fusarium* Comparative Genomics Platform (http://genomics.fusariumdb.org/). This platform provides an interactive genome browser, in which sequence data, contig information, and annotation information in a chosen region can be displayed and also presents computed characteristics of multiple gene families and functional groups to support quick comparison and analysis across species (Park et al. 2011). The PEDANT genome database archives annotated genome sequences from viruses to eukaryotes including Fo and supports analyses of genomic sequences through a large variety of bioinformatics tools (Walter et al. 2009). PhylomeDB provides collections of gene phylogenies (phylomes), including data for Fo, to facilitate interactive exploration of the evolutionary history of individual genes via phylogenetic trees and multiple sequence alignments (Huerta-Cepas et al. 2011). It also provides genome-wide orthology and paralogy predictions and allows downloading of all stored data.

### 5.3.3 Tools for Functional Studies of Genes

Transformation-mediated mutagenesis techniques have driven gene functional studies in Fo. Integration of DNA into the fungal genome occurs via either ectopic insertion or homologous recombination. The frequencies of these events appear to depend on the genetic background of individual species and strains, types of transforming DNA, chromosomal location of the targeted locus, and methods of introducing DNA. Ectopic integration has been widely utilized to randomly mutagenize the genome of Fo and other filamentous fungi (Michielse et al. 2009a), mainly because the inserted DNA provides a convenient tag for subsequent isolation of the mutated gene. Homologous recombination between a target gene and its in vitro engineered mutant allele including selection marker results in gene disruption and has been routinely performed to study gene function in Fo. Two methods mainly have been employed for delivering DNA into fungal cells. One is mixing fungal protoplasts with DNA. This involves generation of protoplasts, which is critical for high transformation efficiency and requires the digestion of the fungal cell wall using a mixture of hydrolytic enzymes. *Agrobacterium tumefaciens*-mediated transformation (ATMT) offers an alternative means for introducing DNA (Khang et al. 2006; Mullins et al. 2001). This method has been successfully applied to transform phylogenetically diverse fungi, including Fo (Khang et al. 2006; Michielse et al. 2009a). Advantages of ATMT over protoplast-based transformation include: (a) no dependence on hydrolytic enzymes, (b) higher and more reliable transformation efficiency, (c) increased frequency of homologous recombination (Michielse et al. 2005), and (d) low-copy number of inserted DNA per genome (Michielse et al. 2009a). A number of binary vectors for ATMT with different features and utilities have been developed to support gene functional studies in Fo and other fungi (Frandsen 2011; Khang et al. 2005, 2006; Michielse et al. 2009a; Mullins et al. 2001; Paz et al. 2011).

### 5.3.4 Imaging Tools

Rapid advances in the development of microscopic and spectroscopic tools present a vast opportunity for studying biological structure–function relationships at levels ranging from individual molecules to cells and tissues. In combination with exponentially increasing genomics data and various fluorescent proteins (FPs) and FP-based sensors, these tools are indispensable for understanding biological processes at the cellular level and also help study the function and dynamics of individual gene products in relation to organism–organism and organism–environment interactions. Here, we introduce a number of such tools that have been utilized to study Fo.

#### 5.3.4.1 Fluorescence Microscopy

A large collection of natural and engineered FPs with diverse biochemical and physical properties is available for labeling molecules and organisms (Stepanenko et al. 2011). Since FPs do not require substrates or co-factors for fluorescence, they are excellent markers for visualizing the dynamics of individual proteins, organelles, and organisms through multiphoton/confocal imaging (Czymmek et al. 2007; Wu et al. 2011). Multiple molecules or organisms can be simultaneously imaged when they are tagged with FPs with non-overlapping spectral properties. In combination with certain FPs or organic fluorophores, recently developed super-resolution fluorescence microscopy techniques even allow imaging of single proteins (Huang et al. 2009).

Genetic transformation of Fo and other fungi with FP genes has vastly facilitated in vivo imaging of pathogenesis with unequivocal determination of the locality of labeled fungus within the host tissue (Czymmek et al. 2007; Duyvesteijn et al. 2005; Kim et al. 2011; van der Does et al. 2008). A main technical challenge to imaging pathogenic processes by soilborne root pathogens like Fo is that, unlike foliar pathogens that follow readily visible pathogenesis processes, disease progression in roots is concealed within the soil. To image progression of plant root infection, different plants were harvested at specific time points during the disease cycle for observation, which eliminates the possibility for following individual infection sites at multiple time points. Besides, uprooting from the soil or other media and mounting in an optically clear medium likely cause stress to plant roots. Furthermore, unless the samples were chemically or cryogenically fixed, the pathogen and host will actively grow and respond to their altered environment, likely causing artifacts that may significantly impact some experiments. These problems necessitate the creation of a system in which the chance for artifacts due to root manipulation is minimized. The use of cover-glass chambers with an agar medium or soil to grow and infect *A. thaliana* helped apply time-lapse confocal microscopy to analyze how FP-labeled Fo colonizes roots and how roots respond to infection at high resolution without disrupting roots (Czymmek et al. 2007; Kim et al. 2011). Using this system, images from individual encounter sites were acquired repeatedly over a several day period (Czymmek et al. 2007), which not only revealed fungal structures associated with infection, preferred locations of fungal colonization and penetration, and host cellular responses as an indicator of defense and pathogenesis, but also enabled the documentation of the exact timing and velocity of colonization of the root surface and within the vascular tissue (e.g., Fig. 5.2). This system is not suitable for plants bigger than *A. thaliana*. For tomato and other plants, infection by FP-labeled Fo has been monitored in intact plant seedlings by fixing the root system in agarose under a cover slip in a Petri dish, after prior exposure to fungal spores for 1–4 days (Rep unpublished observations; Fig. 5.3). Nevertheless, development of a miniature plant growth chamber with precise control of environmental conditions will further improve microscopic rhizosphere imaging of Fo infection of a variety of plants. Apart from monitoring infection, fluorescence microscopy also has been used in Fo to determine when and where specific promoters are activated (van der Does et al. 2008), to localize proteins in cells (Jonkers et al. 2009; Michielse et al. 2009b) and to monitor behavior of nuclei using histone-FP (Ruiz-Roldan et al. 2010).

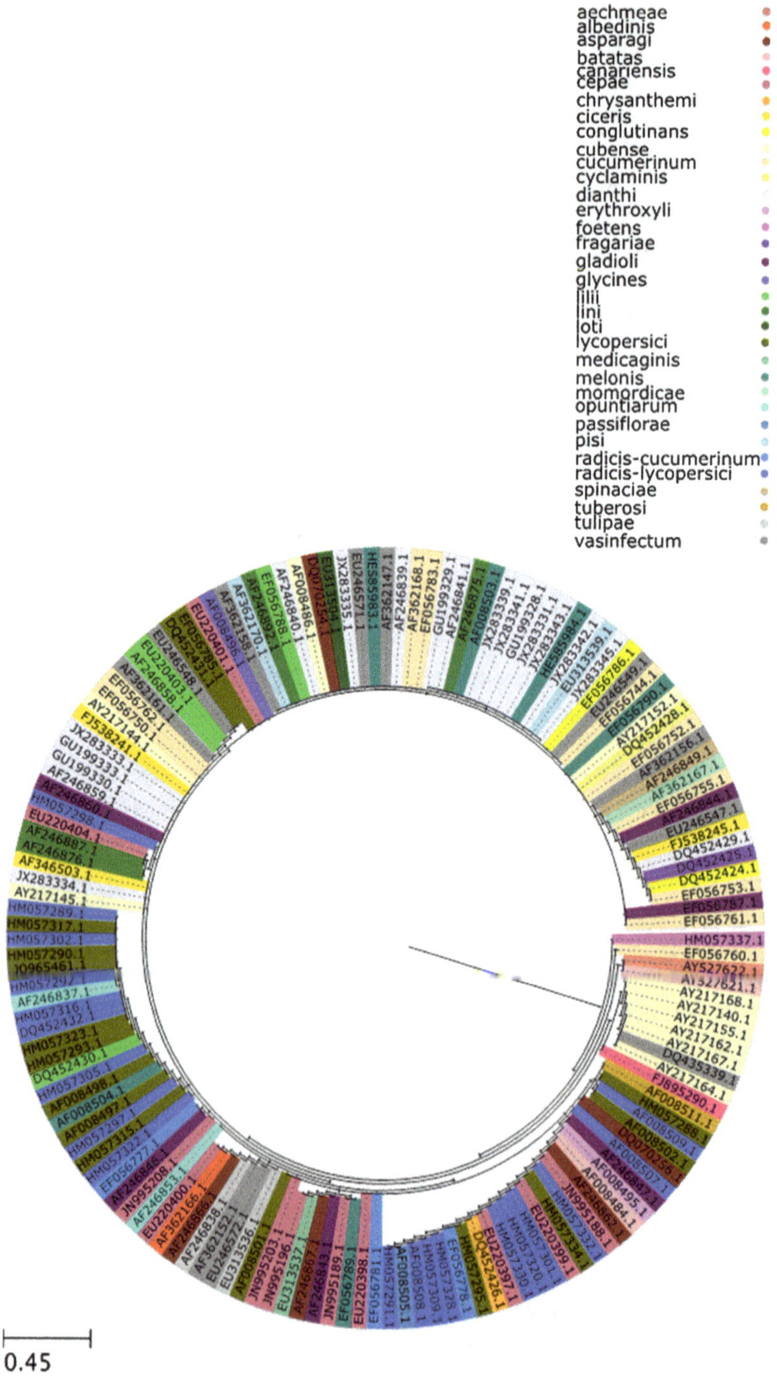


**Fig. 5.1** *Polyphyletic origins of many formae speciales*. Sequences of *F. oxysporum* for *EF-1α* (at least 648 nt) were retrieved from GenBank. Whenever two sequences were identical and both strains belonged to the same *forma speciales*, one of the two (chosen randomly) was removed. The 188 remaining sequences were aligned using Muscle with default settings. Subsequent to trimming of the alignment using GBlocks (again with default settings), pairwise distances were calculated with Protdist and the tree topology inferred with BioNJ with a 1,000 bootstraps. Clades that only contain sequences that belong to the same *forma speciales* were collapsed: a single sequence was randomly picked to represent the clade. Nodes with less than 500/1,000 bootstrap support were collapsed as well. The *F. foetens EF-1α* sequence was used as outgroup (*arrow*). GenBank accession numbers are noted. Each *color* represents a *forma specialis*, as indicated

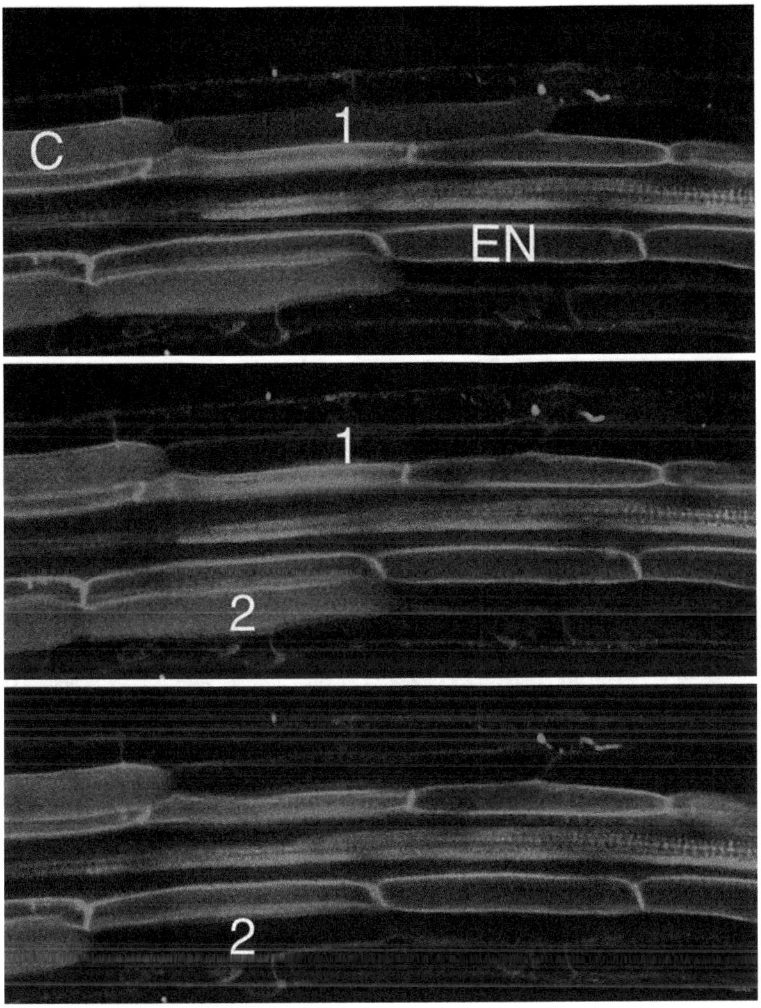


**Fig. 5.2** *Time-lapse microscopy of Fo growth within vascular bundle of A. thaliana (Col-0) and host response.* Fungal progression was documented using confocal microscopy for a Fo strain expressing ZsGreen and multiphoton microscopy for plant walls and cortex vacuolar contents. As hyphae grew from *right* to *left*, vacuolar autofluorescence significantly decreased, probably caused by the loss of cellular integrity and the subsequent loss of vacuolar content, typically within ~6 min in cortex cells adjacent to the leading hyphal tips. This change suggests that Fo phytotoxin(s) could effectively move through the endodermal layer to negatively affect root cortex cells. Numbers (*1 2*) in consecutive frames (6-min interval) indicate the cortex cells before and after loss of vacuolar fluorescence. *EN* endodermis; *C* cortex. Leading hypha growth rate 2.8 μm/min. This *black* and *white* figure is a condensed version of a *figure* and a movie in Czymmek et al. (2007)

#### 5.3.4.2 Engineered FP-Based in Vivo Sensors for Specific Signaling Molecules

Like other organisms, Fo employs elaborate signaling pathways for translating external stimuli to appropriate responses so as to preserve cellular homeostasis in the short term, carry out orderly developmental changes in the long term, and control interactions with other organisms. These signaling mechanisms involve many gene products as well as a number of secondary messengers such as $Ca^{2+}$, diacylglycerol, phosphatidylinositols, inositol triphosphate, and cyclic nucleotides to execute and coordinate these processes. Characterization of the function and mode of action of fungal genes involved in

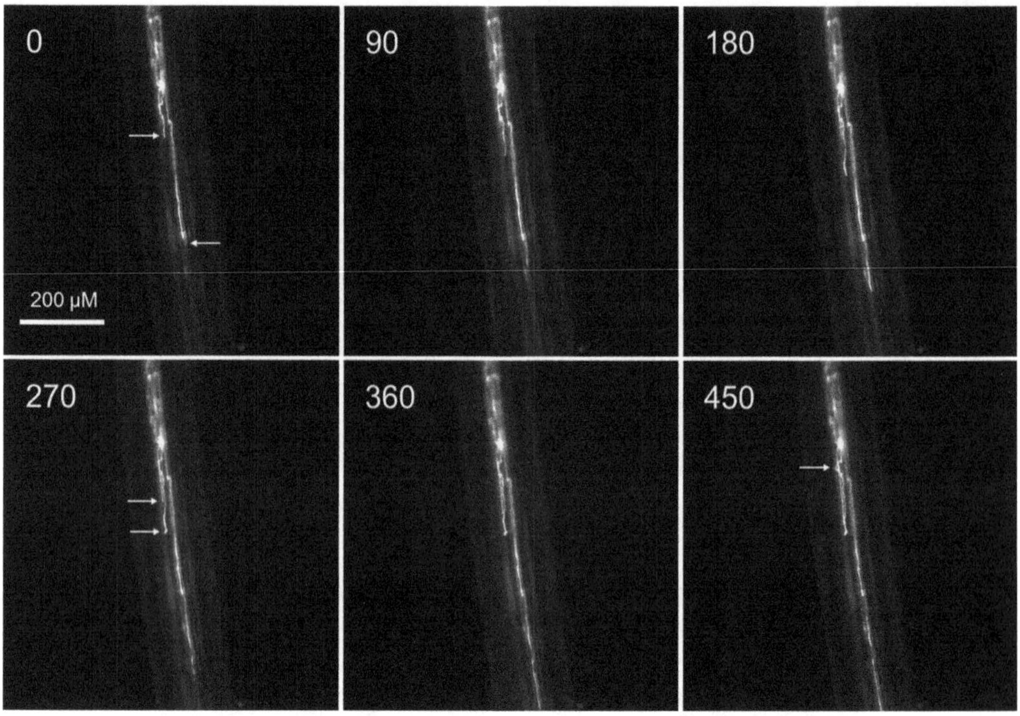


**Fig. 5.3** *Time-lapse microscopy of Fol growth within the vascular bundle of a tomato root*. Hyphal growth of a Fol strain expressing eGFP (*green*) through vascular tissue was documented using fluorescence microscopy (EVOS® FL, Life Technologies). Minutes after the first frame are indicated at the *top left* of each frame. At 0 min, one hyphal tip is next to a xylem vessel (*top arrow*) and another inside a xylem vessel (*bottom arrow*). At 270 min, a third hyphal tip is visible (*top arrow*) and the hypha that is next to a vessel (*bottom arrow*) apparently encountered a barrier and the hyphal tip has swollen (later it will break through the barrier—not shown). At 450 min, a fourth hyphal tip has emerged (*arrow*)

signaling pathways has been greatly facilitated by a rapid increase in available genome sequences and gene manipulation tools, some of which are described above. To connect the activity and cellular location of signaling proteins to dynamic changes of specific secondary messengers, tools for visualizing and quantifying the spatial and temporal dynamics of secondary messengers are needed. Many types of FP-based, in vivo sensors have been developed for such applications (Okumoto et al. 2012; Plaxco and Soh 2011; Zhang and Keasling 2011). However, visualization of the spatial and temporal dynamics of secondary messengers in fungi using such sensors has not yet been widely employed.

Expression of a FP-based sensor for calcium ions ($Ca^{2+}$), CaMeleon, in three fungi, including Fo, Fg, and *M. oryzae*, demonstrated the utility of such sensors for studying cell signaling in fungi (Kim et al. 2012). Calcium ions are involved in controlling many disparate cellular responses in organisms ranging from microbes to humans. A prevailing model of $Ca^{2+}$-mediated signaling is that the influx and efflux of $Ca^{2+}$ through several types of channels and transporters in plasma and organellar membranes in response to external stimuli create spatial and temporal patterns in cytoplasmic $Ca^{2+}$ ($[Ca^{2+}]_c$) known as "$Ca^{2+}$ signatures" (Dodd et al. 2010). Such signatures are believed to carry messages that are decoded by combined actions of several types of $Ca^{2+}$-binding proteins (CBPs) and downstream proteins that interact with these CBPs (Dodd et al. 2010). Expression of CaMeleon YC3.60, a FRET (Förster resonance energy transfer)-based $Ca^{2+}$ sensor, enabled for the first time time-lapse imaging of rapidly changing spatial and temporal patterns of $[Ca^{2+}]_c$ in fungi, both in vitro and in planta, at the subcellular level (Kim et al. 2012). Analyses of $Ca^{2+}$ signatures in Fo, Fg, and *M. oryzae* revealed that $[Ca^{2+}]_c$ change occurs in a pulsatile manner at hyphal tips with each species exhibiting distinct $Ca^{2+}$ signatures and that occurrence of pulsatile $Ca^{2+}$ signatures is age and development dependent. More recently, expression of the circularly permuted FP-based $Ca^{2+}$ sensor GCaMP5 (Akerboom et al. 2012) made high-sensitivity/high-speed imaging of $Ca^{2+}$ dynamics possible (Kang unpublished data). These sensors enabled characterization of the roles of known and predicted $Ca^{2+}$ signaling proteins in generating $Ca^{2+}$ signatures and how alterations of $Ca^{2+}$ signature caused by loss of individual $Ca^{2+}$ signaling proteins affect the organization and function of subcellular machineries that drive hyphal growth and macroscopic phenotypes.

#### 5.3.4.3 Atomic Force Microscopy

The cell surface protects the cell and participates in regulating cellular and developmental activities by generating signals in response to interactions with various substrates and other cells and organisms. Elucidation of physical and chemical properties of the cell surface and how these properties change in response to various stimuli and genetic changes, therefore, will provide important clues to understanding how organisms function. Diverse scanning probe microscopes (SPMs), which utilize a sharp physical probe that scans the specimen to image the surface and measure its physicochemical properties down to the nanoscale, have been developed for many different applications. Among SPMs, Atomic Force Microscopy (AFM) has been most widely applied to analyze biological materials, including fungi (Adams et al. 2012; Alsteens et al. 2010; Dague et al. 2008; Dufrene 2000; Kaminskyj and Dahms 2008; Kim et al. 2011), and offers many different ways for analyzing biological surfaces (Adams et al. 2012; Webb et al. 2011). AFM allows quantification of forces in solutions that mimic physiological conditions. AFM probes can also be functionalized by covalently attaching a range of materials from individual molecules to whole cells. This enables the measurement of binding forces between the functionalized molecule and other molecules and various substrates, as well as imaging of the distribution pattern of receptors of the functionalized molecule across the scanned surface (Webb et al. 2011).

Fungal spore attachment to host surfaces initiates infection, and their interaction is governed by factors such as the chemical and physical properties of both the fungal cell and the host surface and environmental conditions. Kim et al. (2011) used AFM to study factors governing the interaction between Fo spores and the plant root surface (Kim et al. 2011). Using AFM probes functionalized with spores of a wild-type Fo strain and its mutant in *FoCPKA*, encoding a cAMP-dependent protein kinase, the binding affinities of the spores to the root surface of *A. thaliana* were quantified and compared, which revealed that the binding affinity of mutant spores was significantly reduced (Kim et al. 2011). Imaging the surface topography of these spores via AFM also revealed that the surface of mutant spores was noticeably smoother than that of wild-type spores, suggesting that alteration of the spore surface by the loss of *FoCPKA* may be responsible for the reduced attachment of mutant spores to the root surface.

### 5.3.5 Mass Spectrometry for Protein Identification

Proteomic tools have been especially useful for the identification of proteins secreted by Fol inside tomato xylem vessels. With Mass Spectrometry (MS), to date 20 small proteins without apparent enzymatic activity and 30 enzymes secreted by Fol have been identified in xylem sap of infected tomato plants (Houterman et al. 2007; Schmidt et al. 2013). The genes for 14 of the small proteins are located in LS regions of Fol4287, of which 13 are located on chromosome 14, the "pathogenicity chromosome." The genes for these 14 proteins (SIX1–14) have no ortholog in other *Fusarium* species (except *SIX2* in Fv). Initially, 1-D or 2-D gels were used to separate and identify xylem sap proteins (Houterman et al. 2007; Rep et al. 2004). More recently, tryptic digestion of total xylem sap protein combined with the availability of a Fol genome sequence has greatly expanded the number of fungal proteins identified in xylem sap (Schmidt et al. 2013). Identification of fungal proteins in xylem sap of infected plants depends on sufficient accumulation of fungal biomass before severe wilting symptoms prevent retrieval of xylem sap. This was the case for tomato but may not be true for all Fo–host interactions (Rep unpublished observations). An alternative method could be proteomics of infected roots, although here, too, fungal proteins would need to be sufficiently abundant for detection among a majority of plant proteins. To circumvent these problems, growth conditions can potentially be manipulated to mimic conditions that cause host-induced changes in gene expression and protein secretion in an axenic culture. For Fol, this has turned out not to be straightforward (van der Does et al. 2008; Rep et al. unpublished observations).

Besides secreted proteins, cell wall proteins are of interest for understanding pathogenicity of Fo because these can mediate host-attachment, host sensing, immune suppression or elicitation, morphological adaptations, protection against host enzymes or chemicals or help secure nutrients. An initial survey of Fo cell wall proteins has been published (Prados Rosales and Di Pietro 2008). A next step could be the quantification of changes in cell wall protein composition in relation to Fo–host interactions, for which MS is ideally suited.

## 5.4 Synthesis and Future Direction

Soilborne fungal pathogens like Fo pose a serious threat to many crop plants, yet few control strategies are available compared with those causing diseases on aboveground parts of plants. One strategy that has successfully protected certain high value crops, such as strawberries, from soilborne diseases is soil fumigation, mainly using methyl bromide. However, the mandatory phase-out of this ozone-depleting chemical and increasing organic farming call for urgent development of new environmentally friendly and economically viable control strategies. These might be more forthcoming with a better understanding of the complex molecular interactions that determine disease resistance and susceptibility in encounters between fungi and plant roots. The available knowledge and experimental tools summarized here currently available for the FOSC make it an excellent "soilborne fungal disease" model. In particular, rapidly increasing availability of genome sequences from many Fo strains presents unprecedented opportunities for systematic analysis of the evolution, ecology, and pathology of the FOSC.

**Acknowledgments** We would like to thank Like Fokkens at University of Amsterdam for creating Fig. 5.1.

## References

Adams E, Emerson D, Croker S, Kim H-S, Modla S, Kang S, Czymmek K (2012) Atomic force microscopy: a tool for studying biophysical surface properties underpinning fungal interactions with plants and substrates. In: Bolton MD, Thomma BPHJ (eds) Plant fungal pathogens. Humana Press Inc., Totowa, pp 151–164

Agarwal VK, Sinclair JB (1997) Principles of seed pathology, 2nd edn. CRC Press Inc, Boca Raton

Akerboom J, Chen T-W, Wardill TJ, Tian L, Marvin JS, Mutlu S, Calderón NC, Esposti F, Borghuis BG, Sun XR, Gordus A, Orger MB, Portugues R, Engert F, Macklin JJ, Filosa A, Aggarwal A, Kerr RA, Takagi R, Kracun S, Shigetomi E, Khakh BS, Baier H, Lagnado L, Wang SSH, Bargmann CI, Kimmel BE, Jayaraman V, Svoboda K, Kim DS, Schreiter ER, Looger LL (2012) Optimization of a gcamp calcium indicator for neural activity imaging. J Neurosci 32:13819–13840

Alsteens D, Garcia MC, Lipke PN, Dufrêne YF (2010) Force-induced formation and propagation of adhesion nanodomains in living fungal cells. Proc Natl Acad Sci U S A 107:20744–20749

Armstrong GM, Armstrong JK (1981) Formae speciales and races of *Fusarium oxysporum* causing wilt diseases. In: Nelson PE, Toussoun TA, Cook RJ (eds) *Fusarium*: diseases, biology, and taxonomy.

Pennsylvania State University, University Park, pp 391–399

Baayen RP, O'Donnell K, Bonants PJM, Cigelnik E, Kroon LPNM, Roebroeck EJA, Waalwijk C (2000) Gene genealogies and AFLP analyses in the *Fusarium oxysporum* complex identify monophyletic and non-monophyletic formae speciales causing wilt and rot disease. Phytopathology 90:891–900

Bao J, Fravel D, Lazarovits G, Chellemi D, van Berkum P, O'Neill N (2004) Biocontrol genotypes of *Fusarium oxysporum* from tomato fields in Florida. Phytoparasitica 32:9–20

Bentley S, Pegg KG, Moore NY, Davis RD, Buddenhagen IW (1998) Genetic variation among vegetative compatibility groups of *Fusarium oxysporum* f. sp. *cubense* analyzed by DNA fingerprinting. Phytopathology 88:1283–1293

Bergemann M, Lespinet O, Ben M'Barek S, Daboussi MJ, Dufresne M (2008) Genome-wide analysis of the *Fusarium oxysporum* mimp family of MITEs and mobilization of both native and de novo created mimps. J Mol Evol 67:631–642

Berrocal-Lobo M, Molina A (2008) *Arabidopsis* defense response against *Fusarium oxysporum*. Trends Plant Sci 13:145–150

Blok WJ, Zwankhuizen MJ, Bollen GJ (1997) Biological control of *Fusarium oxysporum* f.sp. *asparagi* by applying non-pathogenic isolates of *F. oxysporum*. Biocontrol Sci Technol 7:527–541

Bloomberg WJ (1979) A model of damping-off and root rot of douglas-fir seedlings caused by *Fusarium oxysporum*. Phytopathology 69:74–81

Boehm EWA, Ploetz RC, Kistler HC (1994) Statistical analysis of electrophoretic karyotype variation among vegetative compatibility groups of *Fusarium oxysporum* f. sp. *cubense*. Mol Plant-Microbe Interact 7:196–207

Cabello M, Arambarri AL (2002) Diversity in soil fungi from undisturbed and disturbed *Celtis tala* and *Scutia buxifolia* forests in the eastern Buenos Aires province (Argentina). Microbiol Res 157:115–125

Cai G, Rosewich Gale L, Schneider RW, Kistler HC, Davis RM, Elias KS, Miyao EM (2003) Origin of race 3 of *Fusarium oxysporum* f. sp. *lycopersici* at a single site in California. Phytopathology 93:1014–1022

Chang DC, Grant GB, O'Donnell K, Wannemuehler KA, Noble-Wang J, Rao CY, Jacobson LM, Crowell CS, Sneed RS, Lewis FMT, Schaffzin JK, Kainer MA, Genese CA, Alfonso EC, Jones DB, Srinivasan A, Fridkin SK, Park BJ (2006) Multistate outbreak of *Fusarium keratitis* associated with use of a contact lens solution. J Am Med Assoc 296:953–963

Cole SJ, Diener AC (2013) Diversity in receptor-like kinase genes is a major determinant of quantitative resistance to *Fusarium oxysporum* f.sp *matthioli*. New Phytol 200:172–184

Coleman JJ, Rounsley SD, Rodriguez-Carres M, Kuo A, Wasmann CC, Grimwood J, Schmutz J, Taga M, White GJ, Zhou S, Schwartz DC, Freitag M, Ma L-J, Danchin EGJ, Henrissat B, Coutinho PM, Nelson DR, Straney D, Napoli CA, Barker BM, Gribskov M, Rep M, Kroken S, Molnar I, Rensing C, Kennell JC, Zamora J, Farman ML, Selker EU, Salamov A, Shapiro H, Pangilinan J, Lindquist E, Lamers C, Grigoriev IV, Geiser DM, Covert SF, Temporini E, VanEtten HD (2009) The genome of *Nectria haematococca*: contribution of supernumerary chromosomes to gene expansion. PLoS Genet 5:e1000618

Cuomo CA, Guldener U, Xu J-R, Trail F, Turgeon BG, Di Pietro A, Walton JD, Ma L-J, Baker SE, Rep M, Adam G, Antoniw J, Baldwin T, Calvo S, Chang Y-L, DeCaprio D, Gale LR, Gnerre S, Goswami RS, Hammond-Kosack K, Harris LJ, Hilburn K, Kennell JC, Kroken S, Magnuson JK, Mannhaupt G, Mauceli E, Mewes H-W, Mitterbauer R, Muehlbauer G, Munsterkotter M, Nelson D, O'Donnell K, Ouellet T, Qi W, Quesneville H, Roncero MIG, Seong K-Y, Tetko IV, Urban M, Waalwijk C, Ward TJ, Yao J, Birren BW, Kistler HC (2007) The *Fusarium graminearum* genome reveals a link between localized polymorphism and pathogen specialization. Science 317:1400–1402

Czymmek K, Fogg M, Powell D, Sweigard J, Park S, Kang S (2007) In vivo time-lapse documentation using confocal and multi-photon microscopy reveals the mechanisms of invasion into the *Arabidopsis* root vascular system by *Fusarium oxysporum*. Fungal Genet Biol 44:1011–1023

Daboussi MJ, Capy P (2003) Transposable elements in filamentous fungi. Annu Rev Microbiol 57:275–299

Dague E, Alsteens D, Latge J-P, Dufrene YF (2008) High resolution cell surface dynamics of germinating *Aspergillus fumigatus* conidia. Biophys J 94:656–660

Dean R, Van Kan JAL, Pretorius ZA, Hammond-Kosack KE, Di Pietro A, Spanu PD, Rudd JJ, Dickman M, Kahmann R, Ellis J, Foster GD (2012) The top 10 fungal pathogens in molecular plant pathology. Mol Plant Pathol 13:414–430

Di Pietro A, Madrid MP, Caracuel Z, Delgado-Jarana J, Roncero MIG (2003) *Fusarium oxysporum*: exploring the molecular arsenal of a vascular wilt fungus. Mol Plant Pathol 4:315–325

Diener AC, Ausubel FM (2005) Resistance to *Fusarium oxysporum* 1, a dominant *Arabidopsis* disease-resistance gene, is not race specific. Genetics 171:305–321

Dignani MC, Anaissie E (2004) Human fusariosis. Clin Microbiol Infec 10:67–75

Dodd AN, Kudla J, Sanders D (2010) The language of calcium signaling. Annu Rev Plant Biol 61:593–620

Dufrene YF (2000) Direct characterization of the physicochemical properties of fungal spores using functionalized AFM probes. Biophys J 78:3286–3291

Dufresne M, Lespinet O, Daboussi MJ, Hua-Van A (2011) Genome-wide comparative analysis of *pogo*-like transposable elements in different *Fusarium* species. J Mol Evol 73:230–243

Duyvesteijn RG, van Wijk R, Boer Y, Rep M, Cornelissen BJ, Haring MA (2005) Frp1 is a *fusarium oxysporum* f-box protein required for pathogenicity on tomato. Mol Microbiol 57:1051–1063

Elliott ML, Des Jardin EA, O'Donnell K, Geiser DM, Harrison NA, Broschat TK (2010) *Fusarium oxysporum* f. sp. *palmarum*, a novel forma specialis causing a lethal disease of *Syagrus romanzoffiana* and *Washingtonia robusta* in Florida. Plant Dis 94:31–38

Enya J, Togawa M, Takeuchi T, Yoshida S, Tsushima S, Arie T, Sakai T (2008) Biological and phylogenetic characterization of *Fusarium oxysporum* complex, which causes yellows on *Brassica spp.*, and proposal of *F. oxysporum* f. sp. *rapae*, a novel forma specialis pathogenic on *B. rapa* in Japan. Phytopathology 98:475–483

Fourie G, Steenkamp ET, Gordon TR, Viljoen A (2009) Evolutionary relationships among the *Fusarium oxysporum* f. sp. *cubense* vegetative compatibility groups. Appl Environ Microbiol 75:4770–4781

Frandsen RJN (2011) A guide to binary vectors and strategies for targeted genome modification in fungi using *Agrobacterium tumefaciens*-mediated transformation. J Microbiol Meth 87:247–262

Fravel D, Olivain C, Alabouvette C (2003) *Fusarium oxysporum* and its biocontrol. New Phytol 157:493–502

Fuchs JG, Moenne-Loccoz Y, Defago G (1999) Ability of nonpathogenic *Fusarium oxysporum* Fo47 to protect tomato against Fusarium wilt. Biol Control 14:105–110

Garibaldi A, Gilardi G, Gullino M (2004) Seed transmission of *Fusarium oxysporum* f.sp. *lactucae*. Phytoparasitica 32:61–65

Geiser DM, Aoki T, Bacon CW, Baker SE, Bhattacharyya MK, Brandt ME, Brown DW, Burgess LW, Chulze S, Coleman JJ, Correll JC, Covert SF, Crous PW, Cuomo CA, De Hoog GS, Di Pietro A, Elmer WH, Epstein L, Frandsen RJN, Freeman S, Gagkaeva T, Glenn AE, Gordon TR, Gregory NF, Hammond-Kosack KE, Hanson LE, Jímenez-Gasco MM, Kang S, Kistler HC, Kuldau GA, Leslie JF, Logrieco A, Lu G, Lysøe E, Ma L-J, McCormick SP, Migheli Q, Moretti A, Munaut F, O'Donnell K, Pfenning L, Ploetz RC, Proctor RH, Rehner SA, Robert VARG, Rooney AP, bin Salleh B, Scandiani MM, Scauflaire J, Short DPG, Steenkamp E, Suga H, Summerell BA, Sutton DA, Thrane U, Trail F, Van Diepeningen A, VanEtten HD, Viljoen A, Waalwijk C, Ward TJ, Wingfield MJ, Xu J-R, Yang X-B, Yli-Mattila T, Zhang N (2013) One fungus, one name: defining the genus *Fusarium* in a scientifically robust way that preserves longstanding use. Phytopathology 103:400–408

Geiser DM, Jiménez-Gasco MM, Kang SC, Makalowska I, Veeraraghavan N, Ward TJ, Zhang N, Kuldau GA, O'Donnell K (2004) Fusarium-ID v. 1.0: a DNA sequence database for identifying *Fusarium*. Eur J Plant Pathol 110:473–479

Gracia-Garza JA, Fravel DR, Nelson AJ, Elias KS, Bailey BA, Arévalo Gardini E, Darlington LC (1999) Potential for dispersal of *Fusarium oxysporum* f. sp. *erythroxyli* by infested seed. Plant Dis 83:451–455

Grigoriev IV, Nordberg H, Shabalov I, Aerts A, Cantor M, Goodstein D, Kuo A, Minovitsky S, Nikitin R, Ohm RA, Otillar R, Poliakov A, Ratnere I, Riley R, Smirnova T, Rokhsar D, Dubchak I (2012) The genome portal of the department of energy joint genome institute. Nucleic Acids Res 40:D26–D32

Houterman PM, Speijer D, Dekker HL, De Koster CG, Cornelissen BJC, Rep M (2007) The mixed xylem sap proteome of *Fusarium oxysporum*-infected tomato plants. Mol Plant Pathol 8:215–221

Hu JN, Chen CX, Peever T, Dang H, Lawrence C, Mitchell T (2012) Genomic characterization of the conditionally dispensable chromosome in *Alternaria arborescens* provides evidence for horizontal gene transfer. BMC Genome 13:171

Huang B, Bates M, Zhuang X (2009) Super-resolution fluorescence microscopy. Annu Rev Biochem 78:993–1016

Huerta-Cepas J, Capella-Gutierrez S, Pryszcz LP, Denisov I, Kormes D, Marcet-Houben M, Gabaldón T (2011) PhylomeDB v3.0: an expanding repository of genome-wide collections of trees, alignments and phylogeny-based orthology and paralogy predictions. Nucleic Acids Res 39:D556–D560

Inami K, Yoshioka-Akiyama C, Morita Y, Yamasaki M, Teraoka T, Arie T (2012) A genetic mechanism for emergence of races in *Fusarium oxysporum* f. sp *lycopersici*: inactivation of avirulence gene *avr1* by transposon insertion. PLoS ONE 7:e44101

Jarvis WR, Shoemaker RA (1978) Taxonomic status of *Fusarium oxysporum* causing foot and root rot of tomato. Phytopathology 68:1679–1680

Jonkers W, Rodrigues CDA, Rep M (2009) Impaired colonization and infection of tomato roots by the Δ*frp1* mutant of *Fusarium oxysporum* correlates with reduced CWDE gene expression. Mol Plant Microbe Interact 22:507–518

Kaminskyj SGW, Dahms TES (2008) High spatial resolution surface imaging and analysis of fungal cells using SEM and AFM. Micron 39:349–361

Kang S, Mansfield MA, Park B, Geiser DM, Ivors KL, Coffey MD, Grunwald N, Martin FM, Levesque A, Blair JE (2010) The promise and pitfalls of sequence-based identification of plant pathogenic fungi and oomycetes. Phytopathology 100:732–737

Katan T, Shlevin E, Katan J (1997) Sporulation of *Fusarium oxysporum* f. sp. *lycopersici* on stem surfaces of tomato plants and aerial dissemination of inoculum. Phytopathology 87:712–719

Khang C, Park S, Lee Y, Kang S (2005) A dual selection based, targeted gene replacement tool for *Magnaporthe grisea* and *Fusarium oxysporum*. Fungal Genet Biol 42:483–492

Khang C, Park S, Rho H, Lee Y, Kang S (2006) Filamentous fungi (*Magnaporthe grisea* and *Fusarium oxysporum*). In: Wang K (ed) *Agrobacterium* protocols. Humana Press Inc., Totowa, pp 403–420

Kidd BN, Kadoo NY, Dombrecht B, Tekeoglu M, Gardiner DM, Thatcher LF, Aitken EAB, Schenk PM, Manners JM, Kazan K (2011) Auxin signaling and transport promote susceptibility to the root-infecting fungal pathogen *Fusarium oxysporum* in *Arabidopsis*. Mol Plant Microbe Interact 24:733–748

Kim H-S, Czymmek KJ, Patel A, Modla S, Nohe A, Duncan R, Gilroy S, Kang S (2012) Expression of the cameleon calcium biosensor in fungi reveals distinct $Ca^{2+}$ signatures associated with polarized growth, development, and pathogenesis. Fungal Genet Biol 49:598–601

Kim H-S, Park S-Y, Lee S, Adams EL, Czymmek K, Kang S (2011) Loss of cAMP-dependent protein kinase A affects multiple traits important for root pathogenesis by *Fusarium oxysporum*. Mol Plant Microbe Interact 24:719–732

Kistler HC, Benny U, Boehm EWA, Katan T (1995) Genetic duplication in *Fusarium oxysporum*. Curr Genet 28:173–176

Klosterman SJ, Subbarao KV, Kang S, Veronese P, Gold SE, Thomma BPHJ, Chen Z, Henrissat B, Lee Y-H, Park J, Garcia-Pedrajas MD, Barbara DJ, Anchieta A, de Jonge R, Santhanam P, Maruthachalam K, Atallah Z, Amyotte SG, Paz Z, Inderbitzin P, Hayes RJ, Heiman DI, Young S, Zeng Q, Engels R, Galagan J, Cuomo CA, Dobinson KF, Ma L-J (2011) Comparative genomics yields insights into niche adaptation of plant vascular wilt pathogens. PLoS Pathog 7:e1002137

Koenig RL, Ploetz RC, Kistler HC (1997) *Fusarium oxysporum* f. sp. *cubense* consists of a small number of divergent and globally distributed clonal lineages. Phytopathology 87:915–923

Kommedahl T, Abbas HK, Burnes PM, Mirocha CJ (1988) Prevalence and toxigenicity of *Fusarium* species from soils of Norway near the Arctic Circle. Mycologia 80:790–794

Kuldau GA, Yates IE (2000) Evidence for *Fusarium* endophytes in cultivated and wild plants. In: Bacon CW, White JF Jr (eds) Microbial endophytes. Marcel Dekker Inc, New York, pp 85–117

Linderman RG (1981) Fusarium diseases of flowering bulb crops. In: Nelson PE, Toussoun TA, Cook RJ (eds) *Fusarium*: diseases, biology, and taxonomy. Penn State University Press, University Park, pp 129–141

Lund ST, Stall RE, Klee HJ (1998) Ethylene regulates the susceptible response to pathogen infection in tomato. Plant Cell 10:371–382

Ma L-J, van der Does HC, Borkovich KA, Coleman JJ, Daboussi M-J, Di Pietro A, Dufresne M, Freitag M, Grabherr M, Henrissat B, Houterman PM, Kang S, Shim W-B, Woloshuk C, Xie X, Xu J-R, Antoniw J, Baker SE, Bluhm BH, Breakspear A, Brown DW, Butchko RAE, Chapman S, Coulson R, Coutinho PM, Danchin EGJ, Diener A, Gale LR, Gardiner DM, Goff S, Hammond-Kosack KE, Hilburn K, Hua-Van A, Jonkers W, Kazan K, Kodira CD, Koehrsen M, Kumar L, Lee Y-H, Li L, Manners JM, Miranda-Saavedra D, Mukherjee M, Park G, Park J, Park S-Y, Proctor RH, Regev A, Ruiz-Roldan MC, Sain D, Sakthikumar S, Sykes S, Schwartz DC, Turgeon BG, Wapinski I, Yoder O, Young S, Zeng Q, Zhou S, Galagan J, Cuomo CA, Kistler HC, Rep M (2010) Comparative genomics reveals mobile pathogenicity chromosomes in *Fusarium*. Nature 464:367–373

Mandeel Q, Ayub N, Gul J (2005) Survey of *Fusarium* species in an arid environment of Bahrain. VI: Biodiversity of the genus *Fusarium* in root-soil ecosystem of halophytic date palm (*Phoenix dactylifera*) community. Cryptogamie Mycol 26:365–404

Mbofung GY, Hong SG, Pryor BM (2007) Phylogeny of *Fusarium oxysporum* f. sp. *lactucae* inferred from mitochondrial small subunit, elongation factor 1-α, and nuclear ribosomal intergenic spacer sequence data. Phytopathology 97:87–98

Mes JJ, Haring MA, Cornelissen BJC (2000a) *Foxy*: an active family of short interspersed nuclear elements from *Fusarium oxysporum*. Mol Gen Genet 263:271–280

Mes JJ, Van Doorn AA, Wijbrandi J, Simons G, Cornelissen BJC, Haring MA (2000b) Expression of the *Fusarium* resistance gene *I-2* colocalizes with the site of fungal containment. Plant J 23:183–193

Michielse CB, Arentshorst M, Ram AFJ, van den Hondel CAMJJ (2005) *Agrobacterium* mediated transformation leads to improved gene replacement efficiency in *Aspergillus awamori*. Fungal Genet Biol 42:9–19

Michielse CB, Rep M (2009) Pathogen profile update: *Fusarium oxysporum*. Mol Plant Pathol 10:311–324

Michielse CB, van Wijk R, Reijnen L, Cornelissen BJC, Rep M (2009a) Insight into the molecular requirements for pathogenicity of *Fusarium oxysporum* f. sp *lycopersici* through large-scale insertional mutagenesis. Genome Biol 10:R4

Michielse CB, van Wijk R, Reijnen L, Manders EMM, Boas S, Olivain C, Alabouvette C, Rep M (2009b) The nuclear protein Sge1 of *Fusarium oxysporum* is required for parasitic growth. PLoS Pathog 5:e1000637

Mullins E, Romaine CP, Chen X, Geiser D, Raina R, Kang S (2001) *Agrobacterium tumefaciens*-mediated transformation of *Fusarium oxysporum*: An efficient tool for insertional mutagenesis and gene transfer. Phytopathology 91:173–180

Nishimura MT, Dangl JL (2010) Arabidopsis and the plant immune system. Plant J 61:1053–1066

O'Donnell K, Gueidan C, Sink S, Johnston PR, Crous PW, Glenn A, Riley R, Zitomer NC, Colyer P, Waalwijk C, Lee Tvd, Moretti A, Kang S, Kim H-S, Geiser DM, Juba JH, Baayen RP, Cromey MG, Bithel S, Sutton DA, Skovgaard K, Ploetz R, Corby Kistler H, Elliott M, Davis M, Sarver BAJ (2009) A two-locus DNA sequence database for typing plant and human pathogens within the *Fusarium oxysporum* species complex. Fungal Genet Biology 46:936–948

O'Donnell K, Humber RA, Geiser DM, Kang S, Park B, Robert VARG, Crous PW, Johnston PR, Aoki T, Rooney AP, Rehner SA (2012) Phylogenetic diversity of insecticolous fusaria inferred from multilocus DNA sequence data and their molecular identification via FUSARIUM-ID and *Fusarium MLST*. Mycologia 104:427–445

O'Donnell K, Kistler HC, Cigelnik E, Ploetz RC (1998) Multiple evolutionary origins of the fungus causing panama disease of banana: concordant evidence from nuclear and mitochondrial gene genealogies. Proc Natl Acad Sci U S A 95:2044–2049

O'Donnell K, Rooney AP, Proctor RH, Brown DW, McCormick SP, Ward TJ, Frandsen RJN, Lysøe E, Rehner SA, Aoki T, Robert VARG, Crous PW, Groenewald JZ, Kang S, Geiser DM (2013) Phylogenetic analyses of *RPB1* and *RPB2* support a middle cretaceous origin for a clade comprising all agriculturally and medically important fusaria. Fungal Genet Biol 52:20–31

O'Donnell K, Sarver B, Brandt M, Chang DC, Nobel-Wang J, Park BJ, Sutton D, Benjamin L, Lindsley M, Padhye A, Geiser DM, Ward TJ (2007) Phylogenetic diversity and microsphere array-based genotyping of human pathogenic fusaria, including isolates from the multistate contact lens-associated U.S. keratitis outbreaks of 2005 and 2006. J Clin Microbiol 45:2235–2248

O'Donnell K, Sutton DA, Rinaldi MG, Magnon KC, Cox PA, Revankar SG, Sanche S, Geiser DM, Juba JH, van Burik J-AH, Padhye A, Anaissie EJ, Francesconi A, Walsh TJ, Robinson JS (2004) Genetic diversity of human pathogenic members of the *Fusarium oxysporum* complex inferred from multilocus DNA sequence data and amplified fragment length polymorphism analyses: evidence for the recent dispersion of a geographically widespread clonal lineage and nosocomial origin. J Clin Microbiol 42:5109–5120

Okumoto S, Jones A, Frommer WB (2012) Quantitative imaging with fluorescent biosensors. Annu Rev Plant Biol 63:663–706

Opperman L, Wehner FC (1994) Survey of fungi associated with grass-roots in virgin soils on the springbok flats. S African J Bot 60:67–72

Ortoneda M, Guarro J, Madrid MP, Caracuel Z, Roncero MIG, Mayayo E, Di Pietro A (2004) *Fusarium oxysporum* as a multihost model for the genetic dissection of fungal virulence in plants and mammals. Infect Immun 72:1760–1766

Ospina-Giraldo M, Mullins E, Kang S (2003) Loss of function of the *Fusarium oxysporum SNF1* gene reduces virulence on cabbage and *Arabidopsis*. Curr Genet 44:49–57

Park B (2013) Cyber-infrastructure supporting fungal and oomycete phylogenetics and genomics. Ph.D. dissertation. Pennsylvania State University

Park B, Park J, Cheong K-C, Choi J, Jung K, Kim D, Lee Y-H, Ward TJ, O'Donnell K, Geiser DM, Kang S (2011) Cyber infrastructure for *Fusarium*: three integrated platforms supporting strain identification, phylogenetics, comparative genomics and knowledge sharing. Nucleic Acids Res 39:D640–D646

Paz Z, García-Pedrajas MD, Andrews DL, Klosterman SJ, Baeza-Montañez L, Gold SE (2011) One step construction of *Agrobacterium*-recombination-ready-plasmids (OSCAR), an efficient and robust tool for ATMT based gene deletion construction in fungi. Fungal Genet Biol 48:677–684

Plaxco KW, Soh HT (2011) Switch-based biosensors: a new approach towards real-time, *in vivo* molecular detection. Trends Biotechnol 29:1–5

Prados Rosales RC, Di Pietro A (2008) Vegetative hyphal fusion is not essential for plant infection by *Fusarium oxysporum*. Eukaryot Cell 7:162–171

Rep M, van der Does HC, Cornelissen BJC (2005) *Drifter*, a novel, low copy hat-like transposon in *Fusarium oxysporum* is activated during starvation. Fungal Genet Biol 42:546–553

Rep M, van der Does HC, Meijer M, van Wijk R, Houterman PM, Dekker HL (2004) A small, cysteine-rich protein secreted by *Fusarium oxysporum* during colonization of xylem vessels is required for I-3-mediated resistance in tomato. Mol Microbiol 53:1373–1383

Rouxel T, Grandaubert J, Hane JK, Hoede C, van de Wouw AP, Couloux A, Dominguez V, Anthouard V, Bally P, Bourras S, Cozijnsen AJ, Ciuffetti LM, Degrave A, Dilmaghani A, Duret L, Fudal I, Goodwin SB, Gout L, Glaser N, Linglin J, Kema GHJ, Lapalu N, Lawrence CB, May K, Meyer M, Ollivier B, Poulain J, Schoch CL, Simon A, Spatafora JW, Stachowiak A, Turgeon BG, Tyler BM, Vincent D, Weissenbach J, Amselem J, Quesneville H, Oliver RP, Wincker P, Balesdent MH, Howlett BJ (2011) Effector diversification within compartments of the *Leptosphaeria maculans* genome affected by repeat-induced point mutations. Nat Commun 2:202

Ruiz-Roldan MC, Kohli M, Roncero MIG, Philippsen P, Di Pietro A, Espeso EA (2010) Nuclear dynamics during germination, conidiation, and hyphal fusion of *Fusarium oxysporum*. Eukaryot Cell 9:1216–1224

Sangalang AE, Burgess LW, Backhouse D, Duff J, Wurst M (1995) Mycogeography of *Fusarium* species in soils from tropical, arid and mediterranean regions of Australia. Mycol Res 99:523–528

Schmidt SM, Houterman PM, Schreiver I, Ma LS, Amyotte S, Chellappan B, Boeren S, Takken FLW, Rep M (2013) MITEs in the promoters of effector genes allow prediction of novel virulence genes in *Fusarium oxysporum*. BMC Genom 14:119

Shen YP, Diener AC (2013) *Arabidopsis thaliana* resistance to *Fusarium oxysporum 2* implicates tyrosine-sulfated peptide signaling in susceptibility and resistance to root infection. PLoS Genet 9:e1003525

Simons G, Groenendijk J, Wijbrandi J, Reijans M, Groenen J, Diergaarde P, Van der Lee T, Bleeker M, Onstenk J, de Both M, Haring M, Mes J, Cornelissen B, Zabeau M, Vos P (1998) Dissection of the *Fusarium I2* gene cluster in tomato reveals six homologs and one active gene copy. Plant Cell 10:1055–1068

Stepanenko OV, Stepanenko OV, Shcherbakova DM, Kuznetsova IM, Turoverov KK, Verkhusha VV (2011) Modern fluorescent proteins: from chromophore formation to novel intracellular applications. Biotechniques 51:313–327

Takken F, Rep M (2010) The arms race between tomato and *Fusarium oxysporum*. Mol Plant Pathol 11:309–314

Taylor JW, Jacobson DJ, Fisher MC (1999) The evolution of asexual fungi: reproduction, speciation and classification. Annu Rev Phytopathol 37:197–246

Vakalounakis DJ (1996) Root and stem rot of cucumber caused by *Fusarium oxysporum* f. sp. *radicis-cucumerinum* f. sp. nov. Plant Dis 80:313–316

Van Bakel J, Kerstens J (1970) Footrot in asparagus caused by *Fusarium oxysporum* f. sp. *asparagi*. Eur J Plant Pathol 76:320–325

van der Does HC, Duyvesteijn RGE, Goltstein PM, van Schie CCN, Manders EMM, Cornelissen BJC, Rep M (2008) Expression of effector gene *SIX1* of *Fusarium oxysporum* requires living plant cells. Fungal Genet Biol 45:1257–1264

Vannacci G, Cristani C, Forti M, Kontoudakis G, Gambogi P (1999) Seed transmission of *Fusarium oxysporum* f. sp. *basilici* in sweet basil. J Plant Pathol 81:47–53

Walter MC, Rattei T, Arnold R, Güldener U, Münsterkötter M, Nenova K, Kastenmüller G, Tischler P, Wölling A, Volz A, Pongratz N, Jost R, Mewes H-W, Frishman D (2009) PEDANT covers all complete RefSeq genomes. Nucleic Acids Res 37:D408–D411

Webb HK, Truong VK, Hasan J, Crawford RJ, Ivanova EP (2011) Physico-mechanical characterisation of cells using atomic force microscopy—current research and methodologies. J Microbiol Meth 86:131–139

Wu B, Piatkevich KD, Lionnet T, Singer RH, Verkhusha VV (2011) Modern fluorescent proteins and imaging technologies to study gene expression, nuclear localization, and dynamics. Curr Opin Cell Biol 23:310–317

Wunsch MJ, Baker AH, Kalb DW, Bergstrom GC (2009) Characterization of *Fusarium oxysporum* f. sp. *loti* forma specialis nov., a monophyletic pathogen causing vascular wilt of birdsfoot trefoil. Plant Dis 93:58–66

Yun S-H, Arie T, Kaneko I, Yoder OC, Turgeon BG (2000) Molecular organization of mating type loci in heterothallic, homothallic, and asexual *Gibberella/Fusarium* species. Fungal Genet Biol 31:7–20

Zhang F, Keasling J (2011) Biosensors and their applications in microbial metabolic engineering. Trends Microbiol 19:323–329

# Illuminating the *Phytophthora capsici* Genome

6

Kurt Lamour, Jian Hu, Véronique Lefebvre, Joann Mudge, Andrew Howden, and Edgar Huitema

## 6.1 Introduction

The filamentous oomycete plant pathogen *Phytophthora capsici* causes root, crown, foliar, and fruit rot in many important vegetables. *P. capsici* was described in 1922 after being recovered from chili peppers at the New Mexico Agricultural Experiment Station in 1918. Originally described as host specific to pepper, it was soon reported on tomato, eggplant, cucurbits (cucumber, melon, pumpkin, and others), and more recently, it has become a problem on green and lima bean (Lamour et al. 2011). Under laboratory conditions, *P. capsici* infects a wide array of plants including the important model hosts *Arabidopsis* and *Nicotianae* (Erwin and Ribeiro 1996). Research efforts have focused on understanding the epidemiology, genetics, as well as the mechanisms of infection and virulence (Hausbeck and Lamour 2004). Many outcrossing *Phytophthora* species do not complete the sexual stage and *P. capsici* is unique because it routinely employs both sexual outcrossing and rapid asexual reproduction for propagation and survival (Lamour and Kamoun 2009). *P. capsici*-infested fields often feature high levels of inoculum, genetic diversity, and dormant spores that can persist for many years. At many locations, *P. capsici* appears to be exotic and once introduced it is difficult to manage and often impossible to eradicate. Fungicide-resistant strains are common and there is a critical lack of resistant cultivars for all hosts (Hausbeck and Lamour 2004).

Genetic and genomic resources have recently been developed for *P. capsici* including a high-quality reference genome, a dense genetic linkage map, and thousands of validated single nucleotide polymorphism (SNP) markers (Lamour et al. 2011). Although application of these tools is just beginning, progress has been made in the areas of genomics, genome stability, and the roles of secreted proteins contributing to *P. capsici* virulence (effectors). Here we highlight recent research advances and discuss future directions of investigation.

### 6.1.1 Life Cycle and Population Structure

*P. capsici* requires the interaction of an A1 and A2 mating type to initiate sexual recombination to produce thick-walled oospores (Erwin and

K. Lamour (✉) · J. Hu
University of Tennessee, Knoxville, TN, USA
e-mail: klamour@utk.edu

J. Hu
China Agricultural University, Beijing, China

V. Lefebvre
INRA, UR1052, Génétique et Amélioration des Fruits et Légumes, Montfavet, France

J. Mudge
National Center for Genome Resources, Santa Fe, NM, USA

A. Howden · E. Huitema
University of Dundee, Dundee, UK

R. A. Dean et al. (eds.), *Genomics of Plant-Associated Fungi and Oomycetes: Dicot Pathogens*,
DOI: 10.1007/978-3-662-44056-8_6, 

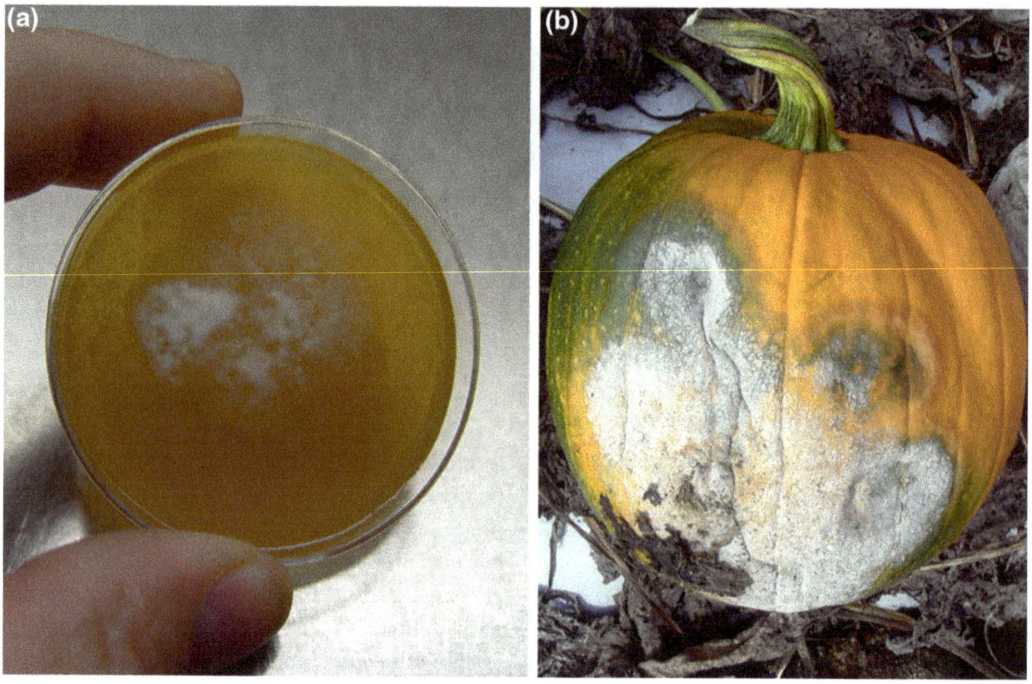

**Fig. 6.1** **a** Typical sectoring during mycelial growth of an isolate *P. capsici* on V8 agar. **b** Typical heavy sporangia production (white powder = millions of sporangia) on the surface of infected pumpkin in Tennessee

Ribeiro 1996). Oospores require a dormancy period of at least a month and the sexual stage is important in areas with cold winter conditions. Isolates recovered from sexually active populations are often fecund, whereas isolates recovered from regions dominated by long-lived clonal lineages often appear to be less fertile. Laboratory crosses have been used to develop extensive genetic resources including a dense genetic linkage map that includes more than 20,000 single nucleotide variant markers (discussed below).

The asexual phase of the *P. capsici* life cycle plays a very important role in the development of epidemics (Lamour 2013). Infection is initiated by germinating spores and includes growth as coenocytic (aseptate) mycelium within the plant tissue followed by production of deciduous sporangia on the surface of the infected plants (Fig. 6.1). Infected fruit (e.g., squash or pumpkin) can successively produce many millions of sporangia over a period of days, and large-scale epidemics are often initiated from a limited number of infected plants. The deciduous sporangia can directly germinate to cause infection or release 20–40 motile biflagellate swimming zoospores to cause infection indirectly. Aboveground waterways (streams, rivers, and ponds) draining infested fields are often contaminated and irrigation from these sources is problematic and may contribute to infestation of new sites.

Unlike many *Phytophthora* species which prefer cool conditions (e.g., *P. infestans*, *P. ramorum,* and *P. sojae*), *P. capsici* grows optimally at warm temperatures. All *P. capsici* hosts require warm conditions for growth and in areas where water cannot be strictly controlled (e.g., strict irrigation regimes in dry areas like the southwestern US or high tunnels in southern France), a single rain event can precipitate rapid spread throughout a field and 100 % crop loss. In the US, epidemics often peak when vegetable fruit are near harvest, which can dramatically exacerbate overall production losses. Due to the initial biotrophic phase of the infection process (discussed below), visible symptoms may not appear until the fruit are delivered to a processing plant or the market, leading to a scenario where the producer loses all the inputs for the crop and then has to pay for the return and disposal of the infected fruit (Hausbeck and Lamour 2004).

In the South American countries of Peru and Argentina and across much of China, long-lived clonal lineages are widely distributed and sexual recombination appears to play a minor role in maintaining populations or contributing to the overall population diversity (Gobena et al. 2012b; Hu et al. 2013; Hurtado-Gonzales et al. 2008). The diversity of clonal lineages is currently being explored in detail using whole genome approaches and it is becoming increasingly clear that a tremendous amount of genetic variation can occur as clonal lineages survive and spread. In the US, South Africa, and southern France, populations have a significant clonal component in conjunction with significant outcrossing and sexual recombination (Gobena et al. 2012a). This creates a highly dynamic scenario where genotypic diversity is very high and short-lived clonal lineages play an important role in epidemics but rarely persist (Lamour and Hausbeck 2001, 2002, 2003).

### 6.1.2 Genetic Tractability

Genetic modification of pathogens is important to understand virulence and infection mechanisms. In *Phytophthora*, genetic manipulation is limited to the introduction of gene cassettes to over-express specific genes or to induce gene silencing. *Phytophthora* transformation generally utilizes protoplasts or zoospores where the introduced DNA is integrated via heterologous recombination into the nuclear genome (Lamour and Kamoun 2009). The process typically occurs at low frequencies but the prolific growth and sporulation of *P. capsici* facilitates the relatively rapid development of transgenic strains. These features have led to the adoption of *P. capsici* as a heterologous model allowing the study of function *in vivo*. Examples include the development of translocation assays in which translational fusions between protein domains are used to test for translocation (Schornack et al. 2010; Kemen et al. 2011) and application of transgenic *P. capsici* strains that express fluorescent proteins, allowing cytological studies on living tissues by means of confocal microscopy (Schornack et al. 2010; Jupe et al. 2013). In addition to gene expression, delivery of specific cassettes could also allow the application of gene silencing. Both the introduction of hairpin and antisense constructs into *Phytophthora* can give rise to endogenous gene silencing. The recent emergence of genome editing tools, however, should allow the delivery and expression of sequence specific nucleases such as TALENs or CRISPRs to specifically mutate genes of interest. Given the importance of genetic manipulation in *Phytophthora* research, *P. capsici* is an attractive model for understanding *Phytophthora* biology.

## 6.2 Genome Resources

### 6.2.1 Reference Genome

The *P. capsici* reference genome is typical for the genus in size and content and provides a useful resource for exploring the underpinnings of a broad host range oomycete (Lamour et al. 2012). A partially inbred line of *P. capsici* was sequenced using 454 FLX and Titanium technology (30X paired and single reads) supplemented by Sanger reads (5X) (Lamour et al. 2012). The *P. capsici* assembly covered 64 Mb of the genome with 56 Mb (87.5 %) of non-gap sequence. Nearly all the genes were assembled as estimated by the capture of 97 % of *P. capsici* full-length cDNAs and ~93 % of highly conserved core eukaryotic orthologs, with an additional 4–5 % captured as partial genes. The assembly contains 917 scaffolds with half of the genome (N50) captured in 29 scaffolds of size 706 kb or greater. A dense SNP-based genetic linkage map (see below) further ordered and oriented 84 % of the assembly and 90 % of the predicted genes into 18 linkage groups. A total of 17,123 genes plus 2,682 transposable element-like genes were identified with 89 % having protein- or cDNA-based evidence. Approximately 86 % of the genome assembly was non-repetitive.

**Table 6.1** Sequenced oomycetes

| Organism | Genome size | Genes | Reference |
|---|---|---|---|
| *Albugo laibachii* | 37 Mb | 13,032 | Kemen et al. 2011 |
| *Pythium ultimum* | 43 Mb | 15,290 | Levesque et al. 2010 |
| *Phytophthora ramorum* | 65 Mb | 14,451 | Tyler et al. 2006 |
| *Phytophthora sojae* | 95 Mb | 16,988 | Tyler et al. 2006 |
| *Hyaloperonospora arabidopsidis* | 100 Mb | 14,543 | Baxter et al. 2010 |
| *Phytophthora infestans* | 240 Mb | 17,797 | Haas et al. 2009 |

The size of the *P. capsici* genome is intermediate among the sequenced oomycetes and comparable to *P. ramorum* (Table 6.1). Repeat composition was similar to other *Phytophthora* with 84 % of the repeats being long terminal-repeat retrotransposons, of which 57 % were Gypsy elements. Gene content was also comparable to other sequenced *Phytophthora* genomes and oomycetes (Table 6.1), with slightly more core eukaryotic genes reinforcing the assembly completeness. The non-repetitive gene density, at 268 genes per Mb, was higher in *P. capsici* compared to the other *Phytophthora* species, though it was lower than some of the other oomycetes (see Table 6.1). Genetic organization was also similar with two-thirds of the genes in conserved gene-rich blocks (Haas et al. 2009). Over three-quarters of the genes could be grouped into potential gene families (Lamour et al. 2012).

*P. capsici* and other available sequenced *Phytophthora* species have a high level of synteny. Scaffold-level synteny was nearly complete for *P. capsici* compared to *P. ramorum* and *P. sojae* with minimal duplications, rearrangements, and non-syntenic stretches. *P. capsici* and *P. infestans* also showed high levels of synteny though this comparison showed more duplications and translocations and >20 % of the *P. infestans* genome had no syntenic genes in *P. capsici* (Lamour et al. 2012).

Novel genes were found in *P. capsici.* At a protein level, 365 genes had no homolog in the genome assemblies of *P. infestans, P. sojae, and P. ramorum.* These were in gene poor regions, a trend seen in other *Phytophthora* species (Lamour et al. 2012; Haas et al. 2009).

### 6.2.2 SNPs and Linkage Map

*P. capsici* has one of the highest densities of SNP sites of any known eukaryote (at least one every 40 bp) (Lamour et al. 2012). To assist the assembly of a high-quality reference genome, a series of inbreeding crosses (an F1 and two successive backcrosses) were developed to produce an isolate with decreased heterozygosity. In addition, sexual progeny from the initial F1 cross were genotyped using restriction site-associated DNA (RAD) sequencing to produce a genetic linkage map that includes 20,568 single nucleotide variant sites. The map was built successively by initially analyzing the largest 108 scaffolds to determine if the markers contained on each scaffold were inherited consistent with their placement in the reference genome. This revealed a limited number of large-scale rearrangements where portions of individual scaffolds were either inverted or clearly not inherited as contiguous linked entities (Fig. 6.2). A final linkage analysis placed the largest 108 scaffolds into 18 linkage groups which include 84 % of the genome and 90 % of the genes. The mapping project provided novel insight into the surprisingly high level of mitotic, asexual genomic rearrangements that occurred over the 5 years of this study. The genomic changes, known as loss of heterozygosity (LOH), were common for the wild-type parental field isolates as well as the sexual progeny (discussed below).

### 6.2.3 Other Resources

To help with annotating the genome, transcriptome sequencing was also completed and made

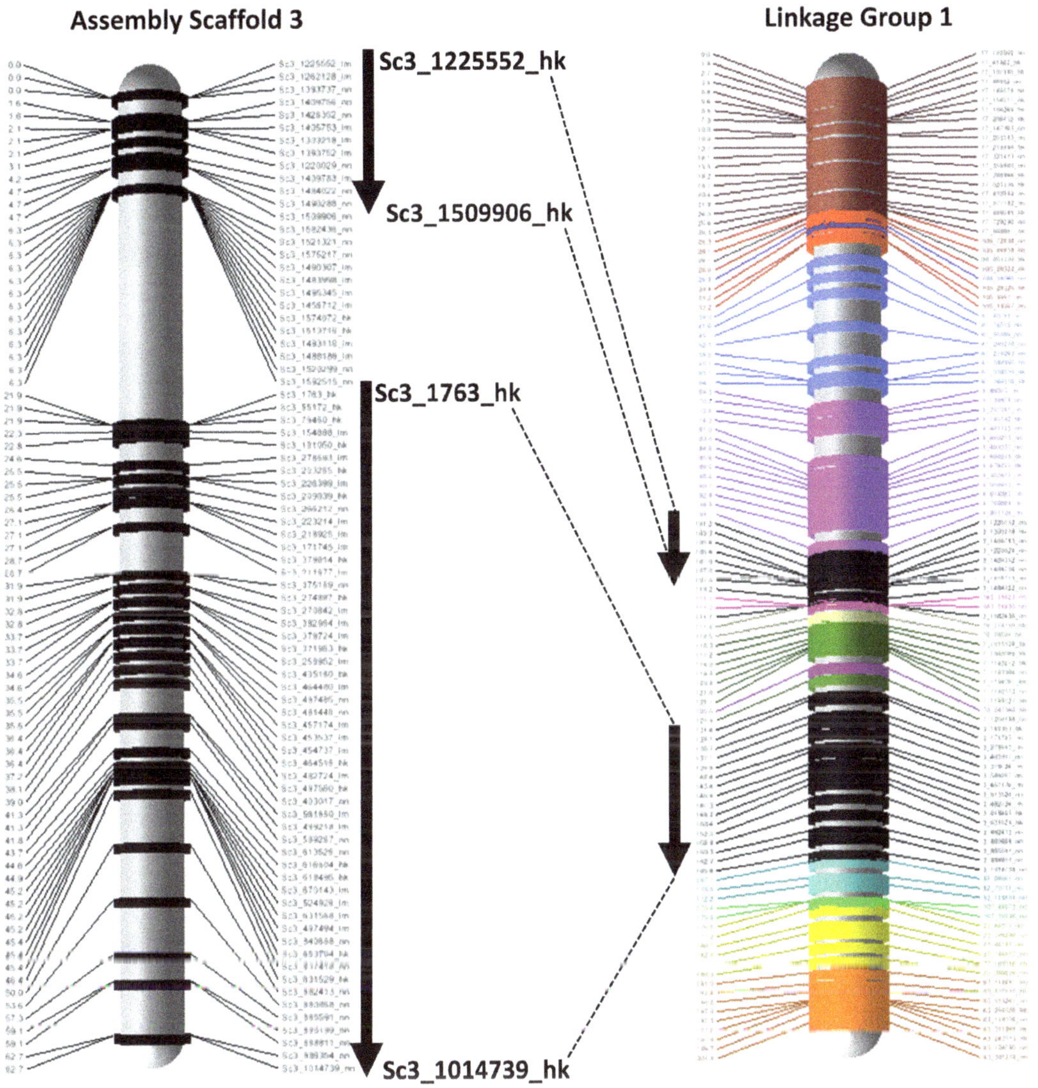


**Fig. 6.2** Example of a mis-assembled scaffold. Linkage mapping of the scaffold alone and in conjunction with markers from the largest 108 scaffolds indicates Scaffold 3 contains two blocks of correctly ordered markers on linkage group 1. SNP markers at the beginning and end of each block are labeled by scaffold (Sc3), nucleotide position on scaffold followed by the type of marker (hk = Aa × Aa). The marker at position 1763 on scaffold was the closest to the beginning of the scaffold, while the marker at position 1509906 was closest to the end. In the reference genome, these would be misrepresented as a single contiguous entity

available in Genbank. These include 1260 full-length cDNAs sequenced with Sanger technology (accessions BT031383.1-BT032642.1) as well as 56,448 Sanger ESTs (available under "DOE Joint Genome Institute *Phythophthora capsici* EST project") that have been assembled into 11,090 consensus sequences. All data are also available at http://genome.jgi-psf.org/Phyca11/Phyca11.home.html.

## 6.3 Genome-Enabled Discoveries

### 6.3.1 Mitotic Instability (Loss of Heterozygosity)

*Phytophthora* is infamously unstable. Sectoring (see Fig. 6.1 for example) is common for isolates grown on agar media and numerous studies indicate asexually produced propagules such as zoospores can vary significantly from their clonal progenitor (Fig. 6.1). Very little is known concerning the underlying genetics of asexual variation. The *P. capsici* genome project spanned a period of 5 years—from laboratory crosses in 2005 to the RAD sequencing and mapping in 2010. Over these 5 years, copies of the isolates were maintained in long-term storage and transferred on agar media roughly 10 times. Although not obvious initially, it was eventually revealed that variable length tracts of the *P. capsici* genome had spontaneously converted to either one or the other of the two possible haplotypes in the diploid genome. This occurred in sexual progeny as well as in clonal copies of the parental isolates and the converted tracts ranged in length from 299 bp to >1 Mbp (Lamour et al. 2012). The phenomenon is known as loss of heterozygosity (LOH) and has been described previously in the diploid phase of the yeast life cycle, in some cancerous tumor lines in humans, and for *Phytophthora* in population studies of highly clonal species *P. cinnamomi* and following chemical mutagenesis in *P. capsici* (Dobrowolski et al. 2003; Hulvey et al. 2010; Lamour et al. 2012). The mechanism for LOH in *Phytophthora* is unknown.

The overall extent of LOH in a relatively small cohort of *P. capsici* isolates was quite impressive. Of the 60 sexual progeny used for linkage mapping, 23 had at least one instance of LOH. Of the 20,568 markers showing normal Mendelian inheritance in the mapping cross, more than half (11,048) were impacted by LOH in at least one of the progeny isolates. LOH was found on 14 of the 18 linkage groups and in 59 of the 108 mapped scaffolds. The frequency and distribution of LOH tracts varied among isolates with some isolates having very little LOH and other isolates having as much as 10 % of their genomes impacted. In total, 36.5 % (19.7 Mbp) of the genome was affected by LOH in at least one isolate. Mating type was mapped to a region of about 300 Kbp and over the course of the study, mating type switched from the A2 to the A1 type for 5 isolates. For each of these isolates, LOH occurred across the mating type region and appears to have catalyzed the mating type switch. And finally, the two isolates with the most cumulative LOH entirely lost the ability to infect a highly susceptible variety of pepper; a variety they both infected easily in 2005 (Lamour 2013).

Genome re-sequencing and application of individual SNP markers are currently being used to analyze asexual variation in both laboratory and field scenarios. In the laboratory, single zoospore-derived isolates have been genotyped using known heterozygous markers and the rate of LOH measured. In field populations, clonal lineages have been identified and the extent of LOH-based variation is being analyzed. In both scenarios, it appears the overall extent of LOH-driven variation is extensive and ongoing studies aim to investigate the importance for evolution and adaptation.

### 6.3.2 Effectors

Plants are continuously challenged by potential pathogens. In most cases, infection is limited through the perception of pathogen (or microbe) associated molecular patterns (PAMPs or MAMPs) by pattern recognition receptors (PRRs) (Boller and Felix 2009). Recognition results in PAMP-triggered immunity (PTI) and features a shift in transcriptional activity toward defense. Obviously, in some cases, pathogens successfully infect plants, suggesting that host immune responses are suppressed or evaded (Dodds and Rathjen 2010; Jones and Dangl 2006).

Genome sequencing projects, combined with the development of computational pipelines and high-throughput functional assays, have revolutionized our thinking about plant pathogens

(Oliva et al. 2010). State-of-the-art models emanating from the field now describe pathogen molecules that are secreted and delivered into host tissues (effectors) to subvert host immunity. Both prokaryotic and filamentous (eukaryotic) pathogens form specialized structures that interface with the host. These structures are the sites where subsets of effectors (cytoplasmic effectors) are translocated into the host cell. Effector delivery and function have proven critical for disease and support the hypothesis that effectors are the principal determinants of virulence. Assessment of the *P. capsici* genome revealed a large effector repertoire including those active in the host apoplast (apoplastic effectors) and the host cytoplasm (intracellular effectors) (Lamour et al. 2012). Apoplastic effectors often act on pre-formed and induced biochemical barriers (degradative enzymes) or aim to inhibit the catabolic activities of host defense proteins. Among the secreted effector repertoire, *P. capsici* carries two distinct classes of proteins that act against cytoplasmic host targets. The RXLR protein family, named after a conserved N-terminal RXLR motif thought to be required for translocation, forms a class of small modular proteins with diverse C-termini that specify effector function. The highly diverse C-termini are thought to be due to adaptive evolution to changing host cell targets and environments. Interestingly, *P. capsici* has a significant number of RXLR proteins with homologs in other *Phytophthora* species (e.g., *P. infestans*). These homologs, especially in the face of the extreme levels of polymorphism observed among *P. capsici* isolates, suggests important roles for conserved effectors toward virulence. Studies aimed at understanding the roles of conserved effectors may help us understand basic functions required for infection.

Besides the RXLR protein family, *P. capsici* carries another class of effectors. The crinklers (CRN) were initially named after a crinkling and necrosis phenotype, observed in plants upon over-expression of secreted *P. infestans* proteins (Torto et al. 2003). CRN proteins carry a conserved N-terminal domain that includes the highly conserved LxLFLAK motif which was found to functionally complement the RXLR domain, suggesting functional analogy and translocation activity. In addition, similar to the RXLRs, CRN proteins also feature diverse C-terminal domains, although most domains are highly conserved in *Phytophthora*. In sharp contrast to the RXLRs, CRN-coding genes are widespread in divergent plant pathogenic oomycetes, including those that do not appear to form haustoria or biotrophic associations. In addition and unlike the RXLRs, all CRN effector domains localized thus far accumulate in the host nucleus, suggesting this ancient class of effectors targets nuclear signaling pathways during infection (Schornack et al. 2010). Recent studies carried out in *P. capsici* identified 84 full-length CRN candidates (Stam et al. 2013a). Detailed comparative analyses revealed significant diversity within the C-terminal effector domains, suggesting functional diversity (Stam et al. 2013a). Ectopic expression of *P. capsici* CRN effectors revealed diverse functions evidenced by variation in cell death, enhanced virulence in over-expression studies, and distinct subnuclear localization patterns (Stam et al. 2013a, b). Further detailed analyses of three necrosis inducing CRN effectors revealed that each of the three exhibited distinct activities in plant cells, leading to varying levels of cell death induction and differentially affecting other signaling pathways (Stam et al. 2013b).

Members of both RXLR and CRN effector classes are highly regulated during infection. Temporal analyses (using microarrays) of infection on tomato revealed dynamic changes in *P. capsici* RXLR gene expression with groups of genes highly expressed during initial infection, biotrophy, and the later necrotrophic phase (Jupe et al. 2013). A set of 73 RXLR effectors with dynamic expression patterns were divided into early, biotrophic, necrotrophic, and sporulation-induced gene classes.

Similar to the RXLRs, the CRN proteins can be divided in two classes based on expression (Class I and Class II). Class I genes are highly expressed in the early infection stages, but drop in expression as the pathogen invades. For Class II genes, expression is not detectable in the early phase of infection, but gradually increases in the

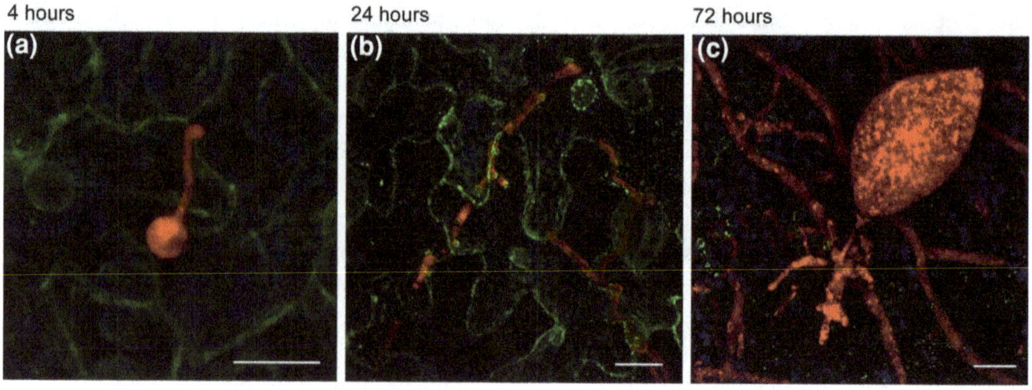


**Fig. 6.3** Asexual infection, growth, and sporulation of *P. capsici* on *Nicotiana benthamiana* expressing ER GFP. Zoospores suspensions of a transgenic *P. capsici* strain carrying the fluorescent protein TdTomato were inoculated onto transgenic *N. benthamiana* plants expressing ER-EGFP. Germination of cysts is evident 4 h after inoculation (**a**) with occasional penetration of the plant surface. Successful infection leads to the pathogen ingress, growth, and the formation of haustoria (**b**) that push against the host cell membrane to form an intimate cellular interface. After disease establishment, host tissues collapse and sporangia (**c**) emerge after 72 h under optimal laboratory conditions. Scale bars = 20 μM

later stages of infection (Stam et al. 2013a). Overall, it appears regulation of effectors is important for the progression of *P. capsic*i from invasion to biotrophy and necrotrophy and that they play a role in modulating host cellular signaling. Much work is still needed to determine the specific role(s) effectors play in the hemi-biotrophic infection process of *P. capsici*.

## 6.4 Infection and Epidemiology

Asexually produced sporangia and zoospores are the key spores driving *P. capsici*'s explosive epidemiology. It is likely most infections start after motile zoospores land on the plant surface and germinate to form infectious hyphae—although the deciduous sporangia and oospores can also produce hyphae capable of causing infection. *P. capsici* directly penetrates the plant cuticle to gain access to host tissues. Once inside, hyphae grow intercellularly and form haustorial structures that push against the host membrane to form a close membrane-to-membrane interface. Initial ingress and growth does not appear to affect cells as active cellular trafficking is commonly observed in haustoriated host cells (Fig. 6.3). Biotrophy is short-lived with tissue necrosis and collapse occurring around 48 h after infection. Tissue necrosis rapidly leads to the formation of deciduous sporangia on the surface of the host tissue, and the cycle of infection–sporulation–spread occurs within a few days under wet and warm field conditions. Host and pathogen transcripts, particularly those involving effectors and host immunity, are highly dynamic during the infection process (Jupe et al. 2013). During the biotrophic stage of infection genes important for protein metabolism, gene expression and biosynthetic processes are differentially expressed and during the transition to necrotrophy there is increased expression of genes involved in catabolic processes and degradation, including a number of peptidases and proteasomal subunits. The end of the infection cycle, which ultimately leads to the formation of sporangia, is marked by the expression of *P. capsici* genes implicated in signaling, regulation, and expression.

## 6.5 Host Range and Characteristics

*Phytophthora* spp. may be entirely host specific (e.g., *P. colocasiae* on taro or *P. sojae* on soybean) or have a seemingly unlimited host range (e.g., *P. cinnamomi*) (Lamour 2013). Evolutionary models describe a co-evolutionary arms race between host and pathogen—with specialized pathogen proteins (e.g., effectors) interacting with specialized plant defenses. Although highly co-evolved models are appropriate for host specific pathogens, the genetic factors important for broad-range pathogens such as *P. capsici* are much less clear.

Recently, *Arabidopsis thaliana* has been developed as a model for studying *P. capsici*–host interactions (Wang et al. 2013). A range of Arabidopsis accessions have been screened for susceptibility to *P. capsici* isolates revealing both compatible and incompatible interactions. Compatible interaction include successful completion of the infection cycle including initial zoospore attachment and germination on leaf tissue, the formation of appressoria and infection pegs, the penetration of hyphae into leaf tissue, and the eventual formation of sporangia. In contrast, incompatible interactions failed to progress beyond appressoria formation and triggered the production of reactive oxygen species, callose deposition, and localized cell death; all of which are key characteristics of a plant defense response (Wang et al. 2013). The variation in susceptibility between different *Arabidopsis* accessions suggests multiple resistance traits within *Arabidopsis,* and further studies of these traits may provide valuable information on the mechanisms of host and non-host resistance.

## 6.6 Host Resistance

Pepper is the only major economically important host of *P. capsici* where naturally occurring resistance has been studied extensively. Over the past decades, multiple pepper germplasm collections have been screened for resistance to *P. capsici*, and overall, resistance to *P. capsici* is rare (Candole et al. 2012). Among breeders, it is generally agreed that the landrace collected from Mexico known as CM 334 has a high level of resistance, although other sources, often derived from local or regional sources, are also used in developing breeding programs for pepper. All sources of resistance described thus far are polygenic and a recent meta-analysis of the published data on resistance to *P. capsici* indicates all of the loci with a major effect are located on the pepper chromosome P5 (Mallard et al. 2013). Further studies of the important regions of the P5 chromosome are underway with the goal of cloning the underlying genes of major effect and gaining a better understanding of the mechanisms underlying resistance.

Scoring resistance to *P. capsici* can be difficult. The overall diversity of *P. capsici* can make it difficult to choose isolates for screening and it is common for isolates to lose virulence over the course of a research program. Even when appropriate isolates have been identified and appear to retain their virulence, the kind and amount of inoculum (e.g., zoospores vs. mycelial suspension), the age of the plants, the plant part inoculated, whether it is wounded or non-wounded, and the environmental conditions for the screen can impact the host–pathogen interaction. It is increasingly recognized that the genetic background plays a crucial role in the overall strength and durability of resistance and it was recently shown that a novel disease resistance inhibitor gene found in a susceptible pepper accession can fully suppress resistance to *P. capsici* (Palloix et al. 2009; Reeves et al. 2013).

To overcome some of the challenges to breeding plants with resistance to *P. capsici*, there is work underway to determine if there are effector proteins that interact with the known sources of resistance in pepper. If this is the case, it may be possible to utilize cloned effectors in a transient assay (e.g., Agrobacterium to deliver the protein to the plant transiently) to assist with breeding and better understand how resistance is accomplished and how it may be overcome (Lamour and Kamoun 2009).

## 6.7 Future Perspectives

The high-quality well-ordered reference genome of *P. capsici* provides a valuable foundation for genetic and genomic investigations. The high level of genetic diversity within *P. capsici* is a challenge but also provides tremendous opportunity. For example, studies exploring the genome plasticity of *P. capsici* during asexual growth may provide novel insights into the asexual evolution of *Phytophthora*. This is important for field and laboratory research. In the field, a better understanding of the mechanisms allowing rapid adaptation to novel selection pressures is important as new chemicals and resistant plants are developed and deployed. In the laboratory, a better understanding of the genomic stability (and instability) of individual isolates is critical, especially as investigators share isolates and expect to replicate experiments.

The adaptation of techniques to transiently express candidate effectors in promising host germplasm holds promise for *P. capsici* and may prove useful to develop novel resistant hosts. In light of the asexual changes possible within single isolates, the use of cloned effectors to interrogate novel and known sources of resistance may help mitigate some of the problems resulting from genome plasticity and changes in overall virulence. With an increasing number of Solanaceous plant genomes available, some of which are resistant to *P. capsici*, it is increasingly feasible to identify the genes and/or molecular processes driving immunity (non-host resistance). This knowledge and resulting technologies can then be used to strengthen susceptible crops. And finally, *P. capsici* shares the host tomato with two other economically important *Phytophthora* species, *P. infestans* and *P. parasitica*, both with genomes available (Haas et al. 2009, http://www.broadinstitute.org/annotation/genome/Phytophthora_parasitica/MultiHome.html). Comparative studies on the similarities and differences of these three evolutionarily distinct species on tomato may be useful to identify species-specific as well as genus-wide factors important for plant infection.

## References

Baxter L, Tripathy S, Ishaque N, Boot N, Cabral A, Kemen E,Thines M, Ah-Fong A, Anderson R, Badejoko W, Bittner-Eddy P, Boore JL, Chibucos MC, Coates M, Dehal P, Delehaunty K, Dong S, Downton P, Dumas B, Fabro G, Fronick C, Fuerstenberg SI, Fulton L, Gaulin E, Govers F, Hughes L, Humphray S, Jiang RH, Judelson H, Kamoun S, Kyung K, Meijer H, Minx P, Morris P, Nelson J, Phuntumart V, Qutob D, Rehmany A, Rougon-Cardoso A, Ryden P, Torto-Alalibo T, Studholme D, Wang Y, Win J, Wood J, Clifton SW, Rogers J, Van den Ackerveken G, Jones JD, McDowell JM, Beynon J, Tyler B M (2010) Signatures of adaptation to obligate biotrophy in the *Hyaloperonospora arabidopsidis* genome. Science 330:1549–1551

Boller T, Felix G (2009) A renaissance of elicitors: perception of microbe-associated molecular patterns and danger signals by pattern-recognition receptors. Annu Rev Plant Biol 60:379–406

Candole B, Conner P, Ji P (2012) Evaluation of phytophthora root rot-resistant *Capsicum annuum* accessions for resistance to *phytophthora* foliar blight and *phytophthora* stem blight. Agric Sci 3:732–737

Dobrowolski MP, Tommerup IC, Shearer BL, O'Brien PA (2003) Three clonal lineages of *Phytophthora cinnamomi* in Australia revealed by microsatellites. Phytopathology 93:695–704

Dodds PN, Rathjen JP (2010) Plant immunity: towards an integrated view of plant-pathogen interactions. Nat Rev Genet 11:539–548

Erwin DC, Ribeiro OK (1996) *Phytophthora* diseases worldwide. The American Phytopathological Society, St Paul

Gobena D, McGrath M, Lamour K (2012a) Survival and spread of *Phytophthora capsici* on Long Island, New York. Mycol Prog 11:761–768

Gobena D, Roig J, Galmarini C, Hulvey J, Lamour K (2012b) Genetic diversity of *Phytophthora capsici* isolates from pepper and pumpkin in Argentina. Mycologia 104:102–107

Haas BJ, Kamoun S, Zody MC, Jiang RH, Handsaker RE, Cano LM, Grabherr M, Kodira CD, Raffaele S, Torto-Alalibo T, Bozkurt TO, Ah-Fong AM, Alvarado L, Anderson VL, Armstrong MR, Avrova A, Baxter L, Beynon J, Boevink PC, Bollmann SR, Bos JI, Bulone V, Cai G, Cakir C, Carrington JC, Chawner M, Conti L, Costanzo S, Ewan R, Fahlgren N, Fischbach MA, Fugelstad J, Gilroy EM, Gnerre S, Green PJ, Grenville-Briggs LJ, Griffith J, Grunwald NJ, Horn K, Horner NR, Hu CH, Huitema E, Jeong DH, Jones AM, Jones JD, Jones RW, Karlsson EK, Kunjeti SG, Lamour K, Liu Z, Ma L, Maclean D, Chibucos MC, McDonald H, McWalters J, Meijer HJ, Morgan W, Morris PF, Munro CA, O'Neill K, Ospina-Giraldo M, Pinzon A, Pritchard L, Ramsahoye B, Ren Q, Restrepo S, Roy S, Sadanandom A, Savidor A, Schornack S, Schwartz DC, Schumann UD, Schwessinger B, Seyer L, Sharpe T, Silvar C, Song J, Studholme DJ, Sykes S, Thines M, van de Vondervoort PJ, Phuntumart V, Wawra S, Weide R, Win J, Young C, Zhou S, Fry W, Meyers BC, van West P, Ristaino J, Govers F, Birch PR, Whisson SC, Judelson HS, Nusbaum C (2009) Genome sequence and analysis of the Irish potato famine pathogen *Phytophthora infestans*. Nature 461:393–398

Hausbeck MK, Lamour KH (2004) *Phytophthora capsici* on vegetable crops: research progress and management challenges. Plant Dis 88:1292–1303

Hu J, Pang Z, Bi Y, Shao J, Diao Y, Guo J, Liu Y, Lu H, Lamour K, Liu XL (2013) Genetically diverse long-lived clonal lineages of *Phytophthora capsici* from pepper in Gansu, China. Phytopathology 103:920–926

Hulvey J, Young J, Finley L, Lamour K (2010) Loss of heterozygosity in *Phytophthora capsici* after N-ethyl-nitrosourea mutagenesis. Mycologia 102:27–32

Hurtado-Gonzales O, Aragon-Caballero L, Apaza-Tapia W, Donahoo R, Lamour K (2008) Survival and spread of *Phytophthora capsici* in Coastal Peru. Phytopathology 98:688–694

Jones JD, Dangl JL (2006) The plant immune system. Nature 444:323–329

Jupe J, Stam R, Howden AJM, Morris JA, Hedley PE, Huitema E (2013) *Phytophthora capsici*-tomato host interaction features dramatic shifts in gene expression associated with a hemi-biotrophic lifestyle. Genome Biol 14:R63. doi:10.1186/gb-2013-14-6-r63

Kemen E, Gardiner A, Schultz-Larsen T, Kemen AC, Balmuth AL, Robert-Seilaniantz A, Bailey K, Holub E, Studholme DJ, MacLean D, Jones JDG (2011) Gene gain and loss during evolution of obligate parasitism in the white rust pathogen of *Arabidopsis thaliana*. PLoS Biol 9:e1001094

Lamour K (2013) *Phytophthora*: a global perspective. CABI plant protection series. CABI, wallingford, p 244

Lamour K, Mudge J, Gobena D, Hurtado-Gonzalez O, Shmutz J, Kuo A, Miller NA, Rice BJ, Raffaele S, Cano LM, Bharti AK, Donahoo RS, Finley S, Huitema E, Hulvey J, Platt D, Salamov A, Savidor A, Sharma R, Stam R, Storey D, Thines M, Win J, Haas BJ, Dinwiddie DL, Jenkins J, Knight JR, Affourtit JP, Han CS, Chertkov O, Lindquist EA, Detter C, Grigoriev IV, Kamoun S, Kingsmore SK (2012) Genome sequencing and mapping reveal loss of heterozygosity as a mechanism for rapid adaptation in the vegetable pathogen *Phytophthora capsici*. Mol. Plant Microbe Interact. 25:1350–1360

Lamour KH, Hausbeck MK (2001) The dynamics of mefenoxam insensitivity in a recombining population of *Phytophthora capsici* characterized with amplified fragment length polymorphism markers. Phytopathology 91:553–557

Lamour KH, Hausbeck MK (2002) The spatiotemporal genetic structure of *Phytophthora capsici* in Michigan and implications for disease management. Phytopathology 92:681–684

Lamour KH, Hausbeck MK (2003) Effect of crop rotation on the survival of *Phytophthora capsici* in Michigan. Plant Dis 87:841–845

Lamour KH, Kamoun S (2009) Oomycete genetics and genomics : diversity, interactions, and research tools. Wiley, Hoboken

Lamour KH, Stam R, Jupe J, Huitema E (2011) The oomycete broad-host-range pathogen *Phytophthora capsici*. Mol Plant Pathol 13:329–337

Levesque CA, Brouwer H, Cano L, Hamilton JP, Holt C, Huitema E, Raffaele S, Robideau GP, Thines M, Win J, Zerillo MM, Beakes GW, Boore JL, Busam D, Dumas B, Ferriera S, Fuerstenberg SI, Gachon CM, Gaulin E, Govers F,Grenville-Briggs L, Horner N, Hostetler J, Jiang RH, Johnson J, Krajaejun T, Lin H, Meijer HJ, Moore B, Morris P, Phuntmart V, Puiu D, Shetty J, Stajich JE, Tripathy S, Wawra S, van West P, Whitty BR, Coutinho PM, Henrissat B, Martin F, Thomas PD, Tyler BM, De Vries RP, Kamoun S, Yandell M, Tisserat N, Buell CR (2010) Genome sequence of the necrotrophic plant pathogen *Pythium ultimum* reveals original pathogenicity mechanisms and effector repertoire. Genome Biol 11:R73

Mallard S, Cantet M, Massire A, Bachellez A, Ewert S, Lefebvre V (2013) A key QTL cluster is conserved among accessions and exhibits broad-spectrum resistance to *Phytophthora capsici*: a valuable locus for pepper breeding. Mol Breeding 32:349–364

Oliva R, Win J, Raffaele S, Boutemy L, Bozkurt TO, Chaparro-Garcia A, Segretin ME, Stam R, Schornack S, Cano LM, van Damme M, Huitema E, Thines M, Banfield MJ, Kamoun S (2010) Recent developments in effector biology of filamentous plant pathogens. Cell Microbiol 12:705–715

Palloix A, Ayme V, Moury B (2009) The durability of plant major resistance genes to pathogens depends on the genetic background, experimental evidence and consequences for breeding strategies. New Phytol 183:190–199

Reeves G, Monroy-Barbosa A, Bosland PW (2013) A novel Capsicum gene inhibits host-specific disease resistance to *Phytophthora capsici*. Phytopathology 103:472–478

Schornack S, van Damme M, Bozkurt TO, Cano LM, Smoker M, Thines M, Gaulin E, Kamoun S, Huitema E (2010) Ancient class of translocated oomycete effectors targets the host nucleus. Proc Natl Acad Sci USA 107:17421–17426

Stam R, Jupe J, Howden AJ, Morris JA, Boevink PC, Hedley PE, Huitema E (2013a) Identification and characterisation CRN effectors in phytophthora capsici shows modularity and functional diversity. PLoS ONE 8:e59517

Stam R, Howden AJM, Delgado-Cerezo M, Amaro T, Motion GB, Pham J, Huitema E (2013b) Characterisation of cell death inducing *Phytophthora capsici* CRN effectors suggests diverse activities in the host nucleus. Front Plant Sci 4:387

Torto T, Li S, Styer A, Huitema E, Testa A, Gow NAR, van West P, Kamoun S (2003) EST mining and functional expression assays identify extracellular effector proteins from *Phytophthora*. Genome Res 13:1675–1685

Tyler BM, Tripathy S, Zhang X, Dehal P, Jiang RHY, Aerts A, Arredondo FD, Baxter L, Bensasson D, Beynon JL, Chapman J, Damasceno CMB, Dorrance AE, Dou D, Dickerman AW, Dubchak IL, Garbelotto M, Gijzen M, Gordon SG, Govers F, Grunwald NJ, Huang W, Ivors KL, Jones RW, Kamoun S, Krampis

K, Lamour KH, Lee M, Maclean DJ, McDonald WH, Medina M, Meijer HJG, Morris PF, Nordberg EK, Ospina-Giraldo MD, Phuntumart V, Putnam NH, Rash S, Rose JKC, Sakihama Y, Salamov AA, Savidor A, Scheuring CF, Smith BM, Sobral BWS, Terry A, Torto-Alalibo TA, Win J, Xu Z, Zhang H, Grigoriev IV, Rokhsar DS, Boore JL (2006) *Phytophthora* genome sequences uncover evolutionary origins and mechanisms of pathogenesis. Science 1261–1266

Wang Y, Bouwmeester K, Van de Mortel JE, Shan W, Govers F (2013) A novel Arabidiopsis-oomycete pathosystem: differential interactions with *Phytophthora capsici* reveal a role for camalexin, indole glucosinolates and salicylic acid in defence. Plant Cell Environ 36:1192–1203

# 7 The *Phytophthora sojae* Genome Sequence: Foundation for a Revolution

Brett M. Tyler and Mark Gijzen

## 7.1 Introduction to *Phytophthora sojae*

### 7.1.1 Biology and Pathology

*Phytophthora sojae* is a widespread plant pathogen and soil-borne microorganism that causes stem and root rot of soybean (Kaufmann and Gerdemann 1958; Hildebrand 1959). This pathogen is highly host specific and is able to infect and kill soybean plants at any stage of growth, but it is most damaging to seedlings and younger plants. Poor drainage, soil compaction, no-till practices, and clay soils tend to provide favorable conditions for the disease (Schmitthenner 1985, 2000; Dorrance et al. 2009). *Phytophthora sojae* is endemic in North America where it leads to annual losses in the range of $10^9$ kg of grain (Wrather and Koenning 2006). The market value of this varies with the traded price of soybeans but *P. sojae* is responsible for hundreds of millions of dollars of losses each year in North America alone. The disease is not limited to North America, however as it has been reported in all major soybean growing areas on every continent (Schmitthenner 2000).

A current hypothesis is that *P. sojae* originated in North America and performed a successful "host jump" to soybeans when this crop was introduced and extensively cultivated. There are three main arguments that support this hypothesis. First, the greatest amount of genetic diversity of *P. sojae* appears to reside in North America, where all four "progenitor strains" of the pathogen are extensively distributed (Förster et al. 1994; Wang et al. 2007; Dorrance and Grünwald 2009). Second, at least 26 different species of lupines (*Lupinus* spp.) including many that are native to North America are susceptible to *P. sojae* (Jones 1969; Jones and Johnson 1969). Third, root rot caused by *P. sojae* appears to be a new disease in Asia, where soybeans have been grown since their domestication there more than 3,000 years ago (Sugimoto et al. 2012). However, there are uncertainties about each of these arguments and so the source of *P. sojae* is not definite.

Despite the uncertainty of the origin of *P. sojae*, remarkable progress has been made over the last 40 years in understanding the greater taxonomy and proper phylogenic placement of the organism in relation to other eukaryotes. *Phytophthora sojae* is one of the ~120 species of plant pathogenic organisms within the genus *Phytophthora* (Cooke et al. 2000; Kroon et al. 2012; Erwin and Ribeiro 1996). Examples of related genera include *Pythium, Peronospora*, and *Albugo*. All these organisms are oomycetes and were previously classified together with the

B. M. Tyler (✉)
Department of Botany and Plant Pathology and Center for Genome Research and Biocomputing, Oregon State University, Corvallis, OR 97331, USA
e-mail: brett.tyler@oregonstate.edu

M. Gijzen
Agriculture and Agri-Food Canada, 1391 Sandford Street, London, ON N5V 4T3, Canada

R. A. Dean et al. (eds.), *Genomics of Plant-Associated Fungi and Oomycetes: Dicot Pathogens*,
DOI: 10.1007/978-3-662-44056-8_7, © Springer-Verlag Berlin Heidelberg 2014

fungi because of their physical resemblances. Over the past 25 years, however, analysis of DNA sequences from oomycetes and fungi have laid to rest any lingering doubts as to whether these two groups of organisms should be considered separate. It is clear oomycetes and fungi diverged during the early stages of eukaryotic radiation and current similarities are the result of convergent evolution, except perhaps for rare occurrences of horizontal gene transfer. Similar to most fungi, oomycetes are filamentous osmotrophic microorganisms that propagate by means of sexual and asexual spores. However, oomycetes are bikonts (two flagella present on motile spores) whereas true fungi are unikonts (single flagellum on motile spores). In fact, the oomycetes belong to a special group of bikont organisms known as the heterokonts that possess unique "tinsel" and "whiplash" flagella on their zoospores. Oomycetes also differ from fungi in the composition of their cell walls. Polymers of beta-glucan comprise the largest fraction of oomycete cell wall material as opposed to chitin, which predominates in the fungi. Another fundamental difference is that oomycetes are diploid organisms throughout most of their life cycle whereas the fungi are often haploid or dikaryotic. The diploid nature of oomycetes has important consequences that affect their evolution and adaptability.

The life cycle of *P. sojae* includes three different spore types for propagation and dissemination. These are the oospore, zoospore, and chlamydospore. The oospore is a sexual spore that results from fusion of the female gametophyte, the oogonium, with the male gametophyte, the antheridium (Erwin and Ribeiro 1996; Judelson 2009). Meiosis occurs in the oogonium and the antheridium and these are the only haploid stages within the life cycle. *P. sojae* is a homothallic organism, meaning oospores can be formed efficiently from self-fertilization of a single strain, as well as by outcrosses between different strains. There are no known mating types. However, from a practical standpoint certain pairs of strains appear to be more sexually compatible in the laboratory than others. In the field, the thick-walled oospores are abundantly produced in infected host tissues and are released into the soil when the plant decomposes, where they may remain viable for years. Under favorable conditions, oospores can germinate and infect soybean roots directly or develop a sporangium that releases multiple zoospores. Sporangia also develop from hyphae in infection sites, or even from germinating zoospores. Zoospores are asexual propagules that are flagellated and water-motile. Zoospores are kidney-shaped and do not possess a cell wall. *P. sojae* is also capable of producing asexual chlamydospores that are spherical and that have thinner walls than the oospores.

The oospores are responsible for the persistence of *P. sojae* in soils, whereas the zoospores provide a means for rapid spread of the pathogen in wet conditions. Zoospores are attracted to soybean roots where the spore encysts and adheres to the root surface (Morris and Ward 1992; Tyler et al. 1996). Encysted zoospores develop a germ tube that penetrates into the root, often forming an appressorium-like structure at the point of entry (Enkerli et al. 1997). The infection hyphae develop haustoria for close association with host cells. The pathogen ramifies throughout the cortical cells and eventually breeches the root endodermis to enter the stele, whereupon the infection accelerates and spreads quickly (Enkerli et al. 1997). An infected soybean plant can wilt and die with a matter of days. Dead plants generally remain standing with their leaves attached.

### 7.1.2 Genetic Manipulation of *P. sojae*

Tools for genetically manipulating *P. sojae* have been instrumental for making use of the genomics resources that have been developed for this organism. In particular genetic crosses have been used to map avirulence genes, which facilitated cloning of those genes (Whisson et al. 1994, 1995; Förster et al. 1994; Tyler et al. 1995; Gijzen et al. 1996; May et al. 2002; MacGregor et al. 2002; reviewed in Jiang and

Tyler 2012; see also below). *Phytophthora sojae* is homothallic, and therefore, molecular markers are generally used to identify $F_1$ progeny that can self-fertilize to produce segregating populations (Whisson et al. 1994, 1995; Förster et al. 1994; Tyler et al. 1995; May et al. 2002).

Stable DNA transformation of *P. sojae* was first reported by Judelson et al. (1993). With further improvements in the procedure to generate protoplasts (McLeod et al. 2008) and introduction of the antibiotic G418 as a selection marker (Dou et al. 2008b), DNA transformation became useful as a routine tool for testing the functions of genes identified from the genome sequence. Following similar technical developments in *P. infestans*, stable silencing of genes in *P. sojae* using sense, anti-sense, or inverted repeat constructs has become routine (Dou et al. 2010; Li et al. 2010a, 2013; Liu et al. 2011; Wang et al. 2011; Yu et al. 2012; Chen and Wang 2012; Qiao et al. 2013; Song et al. 2013; Lu et al. 2013). Transient silencing of *P. sojae* by transformation of protoplasts with double-stranded RNA has also become an extremely useful tool for rapidly assessing gene function (Wang et al. 2011).

## 7.2 Genome Sequence of *P. sojae*

### 7.2.1 Genome Sequencing

The draft genome sequence of *P. sojae*, together with that of the forest pathogen *P. ramorum*, was completed in 2004 and published in 2006 (Tyler et al. 2006). The *P. sojae* assembly was generated from 9-fold coverage of Sanger reads comprising paired-end sequences from small insert plasmids (~2–4 kb), medium insert (~8 kb) plasmids, and large insert (~36 kb) fosmids. A bacterial artificial chromosome (BAC) library of 8,681 clones was assembled by fingerprinting and integrated with the sequence assembly by end-sequencing a collection of 1,440 BAC clones that defined a minimum tiling path (Zhang et al. 2006). In a subsequent project to advance the *P. sojae* sequence to near-finished quality, a total of 15,360 BAC clones from two new libraries were end-sequenced and integrated with previous data to produce a much improved assembly. Gaps and regions of poor sequence were filled by automated primer walking, and 100 BAC clones spanning poor quality regions were finished by transposon insertion sequencing. A total of 50 million RNA sequence tags from *P. sojae* mycelia and from *P. sojae* growing *in planta* were produced using ABI Solid$^{TM}$ technology and were used to improve the gene prediction. Finally, draft assemblies of three additional strains of *P. sojae* were generated from 13-fold coverage by 454 pyrosequencing (Wang et al. 2011).

The genome sizes of *P. sojae* and *P. ramorum* were estimated at 95 and 65 Mb, respectively, while the genome assemblies contained 78 and 54 Mb of sequences, respectively. After finishing, the *P. sojae* assembly spanned 82.6 Mb containing 79.3 Mb of sequence. The substantial discrepancy between the scaffold size and the size estimated from the assembly statistics has been a feature of all oomycete genomes sequenced to date, and likely arises from errors in assembling numerous tandem repeat regions that are present in the genome (Mao and Tyler 1996; Jiang et al. 2006a, b; Qutob et al. 2009). For example, coverage analysis shows that the gene *Avh426* is present in 54 copies, but the assembly contains just one copy (Qutob et al. 2009). The initial estimate of gene numbers were 19,027 predicted genes in *P. sojae* and 15,743 in *P. ramorum*. Prediction of the genes was aided by two sets of expressed sequence tags (ESTs) (Qutob et al. 2000; Torto-Alalibo et al. 2007). A subsequent revision that removed many previously unidentified transposons identified 16,988 and 14,451 predicted genes in each genome, respectively (Haas et al. 2009). More recently, with the completion of the *P. sojae* genome finishing project including deep RNA-seq data, 9,596 additional genes were added to the genome, bringing the latest estimate to 26,584 (Tripathy in prep; http://genome.jgi-psf.org/Physo3/Physo3.info.html).

Comparison of the initial *P. sojae* gene set with that of *P. ramorum* identified 9,768 orthologous genes (51 % of the initial *P. sojae* gene set), 1,755 genes (9 %) unique to *P. sojae*, and 7,503 (40 %) that belong to families in common with *P. ramorum* (Tyler et al. 2006). Inclusion of the *P. infestans* gene set identified 9,550 *P. sojae* genes with orthologs in both *P. ramorum* and *P. infestans* (Haas et al. 2009). Around 39 % of *P. sojae* genome was identified as repetitive, compared to 28 and 74 % for *P. ramorum* and *P. infestans*, respectively (Haas et al. 2009).

### 7.2.2 Bipartite Genome Organization of *P. sojae*

The genome sequences of *P. sojae* and of other oomycetes have a highly bipartite organization. Well-conserved, orthologous, housekeeping genes are closely spaced, even overlapping in places (Jiang et al. 2006a; Tyler et al. 2006; Haas et al. 2009; Baxter et al. 2010). Their order along the genome is highly conserved among *Phytophthora* species and downy mildews (Fig. 7.1). These syntenic blocks contain around 78 % of the orthologous *P. sojae* genes (Haas et al. 2009). These blocks are, however, interrupted by gene-poor, transposon-rich regions that vary greatly in length, sequence, and composition among different species (Jiang et al. 2006a; Tyler et al. 2006; Haas et al. 2009; Baxter et al. 2010). In *P. sojae*, the transposon content of the transposon-rich regions was 60 % whereas the transposon content of gene-rich regions was only 28 % (Haas et al. 2009). Genes involved in infection, especially rapidly evolving families encoding small secreted proteins, are, however, over-represented in these dynamic regions. Genes in these regions show high frequencies of inversions, transpositions, gene conversions, duplications, deletions, and gene silencing. The distinction between the highly conserved syntenic blocks of orthologs that are conserved even between *Phytophthora* and downy mildew species, and highly dynamic regions that are not conserved, even among *Phytophthora* species, is striking. The distinction suggests that specific mechanisms are in play to generate diversity within the transposon-rich regions, driven perhaps by the transposons themselves, while at the same time protecting the genes encoding the core proteome from change.

A major discovery from comparing the genome sequences of *P. sojae* and *P. ramorum* was the presence of many large and rapidly evolving families of proteins potentially involved in infection (Fig. 7.2). As mentioned above, as much as 49 % of the *P. sojae* gene set does not have identifiable orthologs in *P. ramorum* or *P. infestans*, indicating that those genes are rapidly evolving. Many of those genes belong to families of proteases, lipases, glycosyl hydrolases, other lyases and hydrolases, transporters, lipid transfer proteins, protein toxins, and effector proteins, all of which are potentially involved in infection. Some of the most striking examples of gene expansions include the necrosis and ethylene-inducing peptide 1 (Nep1)-like protein (NLP) toxin family, the crinkler effector family, and the RxLR effector family. The NLP and RxLR families are described in detail below. The crinkler family is described in detail elsewhere (Haas et al. 2009; Shen et al. 2013).

### 7.2.3 High-Frequency Gene Conversion in the *P. sojae* Genome

*Phytophthora sojae* can use parasexual mechanisms for generating somatic variation and genetic diversity (Fig. 7.3). When oomycete hybrids and crosses were first tracked with DNA markers in the 1990s, it soon became apparent that progeny from outcrosses underwent a loss of heterozygosity over time, as the cultures were clonally propagated (Francis et al. 1994; Forster and Coffey 1990). In *P. sojae*, the mapping of the *Avr1b* and *Avr1a* loci led to the observation of unusual segregation patterns for many of the DNA markers that were polymorphic in the parental strains. There was a general deficiency of heterozygous $F_2$ progeny and a corresponding overrepresentation of homozygous genotypes, for most codominant DNA markers scored in progeny (MacGregor et al. 2002; Chamnanpunt

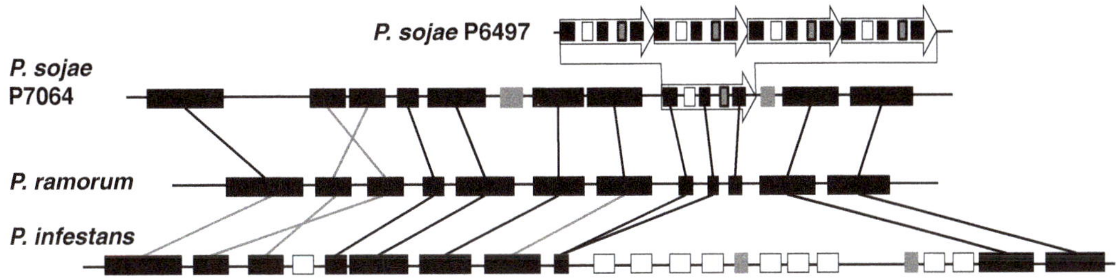


**Fig. 7.1** Conserved and dynamic regions of the *P. sojae* genome. A region of the *P. sojae* genome spanning the Avr3a RxLR effector gene, illustrating conserved gene order of orthologous housekeeping genes (*black*) and dynamic arrangements of RxLR effector genes (*gray*) including Avr3a (*gray* with *black border*). A block of 5 genes spanning Avr3a is tandemly repeated four times in one *P. sojae* strain (P6497) but is present in one copy in a second strain (P7064). Nonconserved genes of unknown function are shown in *white*. *Black lines* join orthologs. Gray lines join orthologs with reversed orientations. The regions shown correspond to *P. sojae* scaffold 80, nucleotides 351400 (*left*) to 288300 (*right*) of v1.1 of the genome sequence; *P. ramorum* scaffold 24 nucleotides 553100 (*left*) to 519900 (*right*) of v1.0 of the genome sequence; *P. infestans* supercontig 1.11 nucleotides 3189700 (*left*) to 3351310 (*right*) of v1.0 of the sequence. Adapted from Fig. 8b of Qutob et al. (2009)

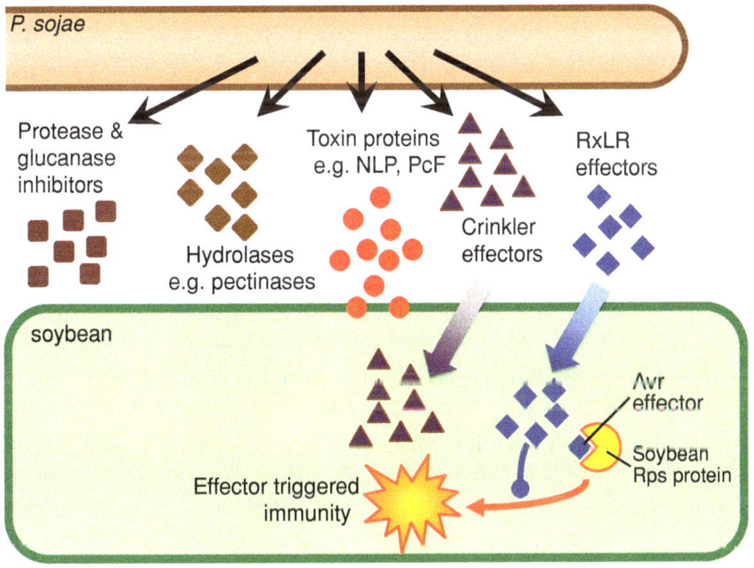


**Fig. 7.2** Virulence protein families discovered from the genome sequence. Many large families of proteins are secreted by *P. sojae* and other *Phytophthora* species to facilitate infection. Some of the most important ones are illustrated in the diagram. Many families of inhibitor proteins and hydrolases target external host macromolecules. Several families of toxins trigger cell death presumably by interacting with the membrane, though some subfamilies of toxin proteins have evolved to alter host physiology without causing cell death. Crinkler and RxLR effectors can enter host cells via their crinkler and RxLR domains, in order to manipulate host physiology in favor of infection. All cloned *P. sojae* avirulence genes encode RxLR effectors that are recognized directly or indirectly by Rps gene products, resulting in effector-triggered immunity (ETI). Many RxLR effectors inhibit ETI but it is unknown if crinkler effectors suppress ETI

et al. 2001). This loss of heterozygosity was studied in-depth using crosses among different strains of *P. sojae* (Chamnanpunt et al. 2001). The study led to the conclusion that high-frequency mitotic gene conversion is the cause of the skewed segregation results, and that this is a potential driver of genetic diversity in the species. Similar observations have recently been

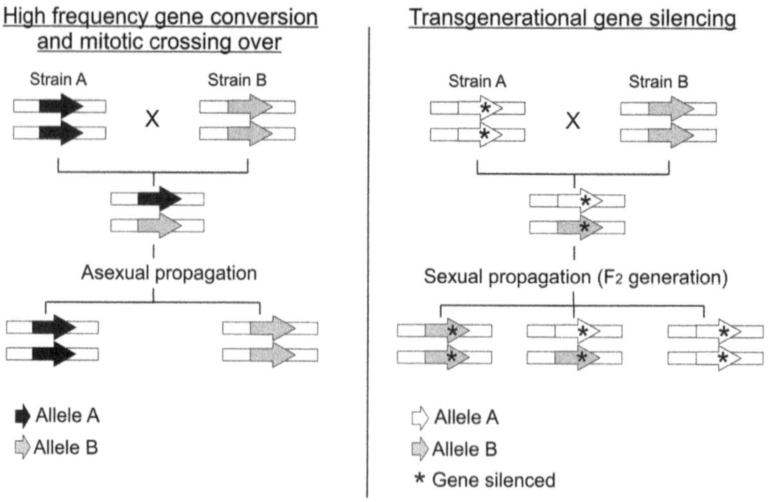


**Fig. 7.3** Schematic illustration showing unusual mechanisms for generating genetic and phenotypic variation in *P. sojae*. High-frequency gene conversion and mitotic crossing over can result in loss of heterozygosity in clonal lineages. Gene conversion affects short-localized regions (Chamnanpunt et al. 2001; MacGregor et al. 2002) while mitotic crossing over affects all regions distal to the crossover from the centromere (Lamour et al. 2012). Transgenerational gene silencing is an epigenetic phenomenon that results in non-Mendelian patterns of phenotypic inheritance. It can occur following sexual reproduction (as diagrammed) (Qutob et al. 2013) or following asexual fusions that result in heterokaryons (van West et al. 1999)

described in the related species *Phytophthora capsici* (Lamour et al. 2012) and *Saprolegnia parasitica* (Jiang et al. 2013). Thus, hyphal sectors or clonally produced propagules such as zoospores or chlamydospores may be genetically diverse based upon nuclear lineage-specific gene conversions that occur after a hybridization event. This parasexual method of generating diversity offers clear advantages to a pathogen that is under constant and heavy selection pressure for host adaptation.

### 7.2.4 Insights into the Evolution of Oomycetes and the Kingdom Stramenopila

Oomycetes belong to the kingdom Stramenopila, which includes several large phyla of photosynthetic algae, including golden-brown algae, brown algae, and diatoms (Harper et al. 2005). The kingdom also includes a diverse array of nonphotosynthetic organisms. The chloroplasts of the photosynthetic members of the kingdom derive from an ancient red algal endosymbiont (Li et al. 2006; Keeling 2010). The chloroplasts of two closely related kingdoms, the alveolates and the haptophytes, also derive from red algal endosymbionts. Therefore, a major evolutionary question has been whether the chloroplasts of the three kingdoms have a single common origin, the so-called "chromalveolate hypothesis" (Keeling 2010). The presence of genes in the *Phytophthora* genome that have close evolutionary affinity with orthologs from photosynthetic organisms, such as 2-isopropylmalate synthase and N-phosphoribosyl-carboxy-aminoimidazole (NCAIR) mutase, suggests that the ancestor of the stramenopiles was already photosynthetic, supporting the chromalveolate hypothesis (Tyler et al. 2006). More recently, as much larger amounts of genome sequence data have accumulated, a more complex picture of the evolution of these secondarily photosynthetic lineages has emerged. This picture involves repeated transient associations of evolving stramenopile organisms with both red and green algae, and the incorporation of genes from those algae into the stramenopile genome during each encounter (Keeling 2010; Deschamps and Moreira 2012).

In this context it is uncertain whether the algal-like genes in *Phytophthora* derive from the same organism(s) that contributed the chloroplasts of the photosynthetic stramenopiles or from other more transient endosymbionts.

### 7.2.5 Role of Horizontal Gene Transfer in Shaping the *P. sojae* Genome

Horizontal gene transfer has emerged as a powerful force in eukaryotic evolution, including oomycete plant pathogens (Morris et al. 2009; Richards and Talbot 2007; Richards et al. 2011). Around 7.6 % of the *Phytophthora ramorum* secretome was estimated to derive from horizontal gene transfer from fungi (Richards and Talbot 2007; Richards et al. 2011). The most striking example of probable horizontal gene transfer is provided by the NLP family. NLPs are toxins that appear to be associated with plant pathogenicity (described in more detail below). NLPs are present in the genomes of many plant pathogenic fungi, where there are generally one to four copies of the gene per genome (Gijzen and Nurnberger 2006). NLP genes are also present in a few bacteria. In the bacterial plant pathogen *Erwinia carotovora*, there is a single NLP gene that is required for pathogenicity (Mattinen et al. 2004). In the oomycetes, NLP genes are found only within plant pathogenic oomycetes in the class Peronosporomycetidae (*Phytophthora*, *Pythium*, and downy mildews). In the genomes of these organisms, however, NLP genes have rapidly expanded in number and have diversified. In *P. sojae* there are 33 NLP genes and in *P. ramorum* there are 40 genes (Tyler et al. 2006). Many of these genes form species-specific clades, indicating that the gene family is rapidly diversifying.

### 7.2.6 Database Resources for *P. sojae* and Other Oomycetes

Several database resources have been created to house genome sequence data from *P. sojae* and other oomycetes. The initial draft sequences of *P. sojae* and *P. ramorum* were made available via the DOE JGI's genome portal. The initial draft sequence is available at http://genome.jgi-psf.org/Physo1_1/Physo1_1.home.html, and the improved sequence is available at http://genome.jgi-psf.org/Physo3/Physo3.home.html. To better enable comparative genomics analyses, an integrated oomycete genomics resource (eumicrobedb.org) has been created that contains both genome and transcriptome sequences from *P. sojae*, *P. ramorum*, and *Hyaloperonospora arabidopsidis*. The site also provides for blast searches of a wide array of oomycete genome and transcriptome sequences (Baxter et al. 2010; Tripathy et al. 2006, 2012). This database is the primary repository for manual annotations of the *P. sojae* sequence from the oomycete community. More recently, an instance of the EuPathDB comparative genomics resource (EuPathDB.org; Aurrecoechea et al. 2010), named FungiDB (FungiDB.org) has been established (Stajich et al. 2012) which includes the genome sequences of *P. sojae* and five other oomycetes. Like EuPathDB, FungiDB provides a rich data-mining interface to the comparative and functional genomic data of multiple species of fungi and oomycetes. The *P. sojae* genome sequence has also been included in the Comparative Fungal Genomics Platform (http://cfgp.riceblast.snu.ac.kr/main.php; Park et al. 2008) and transcriptomics data are also available at the *Phytophthora* transcriptional database (PTD) (phy.njau.edu.cn/ptd; Ye et al. 2011).

## 7.3 Major Advances Emerging from the Sequencing of *P. sojae*

### 7.3.1 Functional Genomics of *P. sojae*

The sampling of ESTs from soybean tissues infected with *P. sojae* represented one of the earliest EST experiments to employ a cDNA library derived from an actual infection site of a host-pathogen interaction (Qutob et al. 2000). It was controversial at the time because some scientists questioned the utility of randomly sequencing transcripts, especially if more than one organism was present in the source material used for gene cloning. Since then, genome and transcriptome sequencing have pervaded biology and have become standard start-points for biological investigations. The concept of the interaction transcriptome has become well established and the related area of metagenomics has evolved into a discipline in itself. It is clear that changes in DNA sequencing technology have contributed to this revolution in thinking, but it also required a change in attitude among scientists in how to prioritize research activities.

The initial EST study on *P. sojae* generated some 3,000 sequence reads, whereas today hundreds of millions of reads can be obtained with far less work. Nonetheless, this early study established that for most sequences one could predict with high confidence whether it originated from the host or the pathogen, based upon nucleotide (GC) composition and best BLAST match (Qutob et al. 2000). The work also provided raw materials for generating a microarray platform containing gene targets from both organisms (Moy et al. 2004). The EST and microarray studies characterized the gene expression changes that accompanied the transition from biotrophic to necrotrophic growth of *P. sojae* during the course of infection. They showed that pathogen biomass rapidly accumulates in diseased tissues. By 24 h postinfection, *P. sojae* transcripts accounted for 40 % of the total RNA, reaching close to 70 % by 48 h (Moy et al. 2004). The proportion of host and pathogen biomass could be tracked by monitoring the large subunit ribosomal RNA, since these RNAs are polymorphic in size between *P. sojae* and soybean. Similar observations have been documented for the biotrophic downy mildew pathogen, *Hyaloperonospora arabidopsidis* (Cabral et al. 2011).

A more extensive EST study (Torto-Alalibo et al. 2007) generated 24,000 additional ESTs from a variety of libraries including the mycelial, zoospore, and interaction libraries created by Qutob et al. (2000). Additional libraries were generated from germinating cysts and from mycelia grown under nutrient-replete and nutrient-limiting conditions. These libraries identified a variety of genes specific for each growth stage with a statistical confidence higher than was possible with the original set of 3,000 ESTs. This EST set was also invaluable in the original annotation of the *P. sojae* genome sequence, and the design of an Affymetrix chip carrying *P. sojae* and soybean probes, both of which occurred in 2004 prior to the spread of next-generation sequencing technology.

The early EST and microarray work also provided insights into overall biological responses of the host and pathogen during the course of infection, in the context of a susceptible interaction. Massive changes in host gene expression occur that result in down-regulation of photosynthesis and hyperactivation of salicylic acid (SA)-based defense responses (Moy et al. 2004). On the pathogen side, gene expression analysis indicates that *P. sojae* relies on glycolysis and mixed alcohol and formic acid fermentation for energy generation during growth in the host (Qutob et al. 2000; Torto-Alalibo et al. 2007). The distribution of *P. sojae* ESTs in the libraries from infected soybean tissue most closely resembled those in the library from mycelia grown in defined medium rich in sugar and amino-acids (Torto-Alalibo et al. 2007). In contrast, mycelia and cysts growing on regular or dilute V8 broth appeared to derive their carbon from proteins and lipids, and transcripts characteristic of gluconeogenesis were strongly up-regulated (Torto-Alalibo et al. 2007).

The Affymetrix gene chip designed in 2004 contained probes for 15,800 *P. sojae* genes and more than 37,500 soybean genes, as well as 7,500 probe sets for soybean cyst nematode transcripts. This platform was used for extensive analyses of the interaction between *P. sojae* and soybean. One study showed that more than 98 % of soybean transcript levels were significantly perturbed during *P. sojae* infection, showing that the presence of the pathogen had a profound impact on the transcriptional program of the host (Zhou et al. 2009). Up to 37 % of the changes had a magnitude of over two-fold. Elevation in transcript levels was particularly extensive among host genes involved in metabolism, as well as those in disease and defense. In contrast, host genes involved in transcription were generally depressed.

More recently, Ye et al. (2011) used Digital gene expression to examine the expression of *P. sojae* genes across five tissue types and five infection time points. Nearly 98 million raw tags were generated, resulting in over 63 million clean 21 nucleotide (nt) tags mapped to the *P. sojae* genome. The results revealed that the *P. sojae* transcriptome during infection is distinctly different than any other life stage and is characterized by elevated expression of glucosidases, cellulases, and proteases, especially during later infection. Expression of RxLR effectors and NLP toxin proteins spiked in germinating cysts just prior to infection, though some classes of these transcripts may have been undercounted due to the presence of closely related paralogs in the genome. Expression of cutinases, polygalacturonases, and proteins protecting against oxidative stress were induced in both germinating cysts and during infection.

### 7.3.2 Role of Epigenetics in Shaping the Expression of the *P. sojae* Genome

The sequencing of the *P. sojae* genome has provided the basis for investigating the long-suspected involvement of epigenetics in the adaptability and variability of this and other oomycetes. What are commonly referred to as "epigenetic" phenomena involve two different but overlapping concepts, namely epigenetic inheritance and epigenetic marks. Epigenetic inheritance is defined as a type of non-Mendelian, transgenerational inheritance whereby traits and phenotypes are determined by some other self-propagating factor besides the information contained in the DNA sequence itself. In epigenetic inheritance, the gene as it is conventionally known is not the unit of inheritance, at least not its DNA sequence. Epigenetic inheritance is difficult to understand because it contradicts fundamental rules that are ingrained into biologists, such as the gene as the unit of inheritance, with its DNA sequence being both necessary and sufficient for information transfer across generations. Nonetheless, a growing number of examples of transgenerational epigenetic inheritance are emerging in animals, plants, fungi, and other organisms including oomycetes.

Epigenetic marks refer to covalent modifications of the DNA or its associated proteins that alter the transcriptional activity of genes. Epigenetic marks include methylation of DNA bases and modifications of chromatin proteins such as cysteine methylation, histone methylation, and histone acetylation. These changes can modify the expression state of a gene by interfering with transcriptional machinery or by altering the chromatin structure of the associated DNA. Epigenetic marks are known to be influenced by development and environment, but they typically are thought to be reset at meiosis. If one observes epigenetic inheritance, epigenetic marks are likely to be associated with the inherited differences, but there are exceptions since prions can cause epigenetic inheritance that is not associated with any epigenetic marks (Halfmann et al. 2012). Likewise, epigenetic marks do not necessarily lead to epigenetic inheritance, and in fact they rarely do so.

Apparent epigenetic effects on gene expression are often observed in the laboratory when organisms undergo transformation with recombinant DNA or RNA molecules. For example, inter-nuclear gene silencing was observed in *P. infestans* cultures transformed with the elicitin gene *infl* (van West et al. 1999). Transformed cultures were chimeric, with some nuclei carrying the transgene and others not; however, transcription was silenced in all nuclei. It was proposed that silencing was propagated by small RNA (sRNA) molecules that in turn directed histone modifications and chromatin changes, resulting in gene silencing (van West et al. 1999, 2008). It has been observed that transgene-based silencing can spread along the chromosome to genes in the vicinity of the target, through chromatin changes (Ah-Fong et al. 2008;

**Table 7.1** Identified *Avr* genes of *Phytophthora sojae*

| *Avr* gene | *Rps* specificity | Gain of virulence mutations | References |
|---|---|---|---|
| *Avr1a* | *Rps*1a and *Rps*1c | Gene deletion; loss of transcript | (Qutob et al. 2009; Na et al. 2014) |
| *Avr1b* | *Rps*1b, *Rps*1k | Amino acid changes; loss of transcript | (Shan et al. et al. 2004) |
| *Avr1c* | *Rps*1c | Gene deletion; loss of transcript | (Na et al. 2014) |
| *Avr1d* | *Rps*1d | Gene deletion | (Yin et al. 2013; Na et al. 2013) |
| *Avr1k* | *Rps*1k | Frame shift/truncation | (Song et al. 2013) |
| *Avr3a/5* | *Rps*3a, *Rps*5 | Amino acid changes; loss of transcript | (Qutob et al. 2009; Dong et al. 2011b) |
| *Avr3b* | *Rps*3b | Amino acid changes; frame shift/ truncation | (Dong et al. 2011a) |
| *Avr3c* | *Rps*3c | Amino acid changes | (Dong et al. 2009) |
| *Avr4/6* | *Rps*4, *Rps*6 | Changes to 5′ un-translated region | (Dou et al. 2010) |

Judelson and Tani 2007). Silencing of an *Avr* effector gene by transformation with a construct containing short interspersed elements (SINE) suggests that transposon-based silencing systems may be co-opted for generating variation in effector gene expression (Whisson et al. 2012).

There are fewer examples of naturally occurring epigenetic phenomenon in oomycetes, but those that do exist are remarkable because they affect virulence and host adaptation. As detailed in Table 7.1, a common gain of virulence mechanism is by loss of the gene transcript. In many instances, there are no sequence differences whatsoever between *Avr* alleles that differ in expression. Alleles that are sequence-identical but that differ in expression are known as epialleles. For *P. sojae Avr* genes, epiallelles have been observed for *Avr1a*, *Avr1b*, *Avr1c*, and *Avr3a/5* (Shan et al. 2004; Qutob et al. 2009, 2013; Dong et al. 2011a). In *P. infestans* the *Avr2* locus displays epiallelic variation (Gilroy et al. 2011), and likely others do as well (Vetukuri et al. 2012). It has been proposed that epigenetic control of effector gene expression is pervasive and is mediated through sRNA molecules, and that the sRNA characteristics may vary depending on the effector family (Vetukuri et al. 2012; Fahlgren et al. 2013).

Proof for epigenetic control of effector gene expression was demonstrated by genetic analysis of the *P. sojae Avr3a/5* locus (Qutob et al. 2013). In crosses between *P. sojae* strains that differ in expression of *Avr3a/5*, it was shown that certain *Avr3a/5* alleles display non-Mendelian interactions that result in epigenetic, transgenerational gene silencing (Qutob et al. 2013) (Fig. 7.3). Results indicated that the *Avr3a/5* gene itself segregates according to Mendelian rules, but none of the $F_1$, $F_2$, or $F_3$ progeny showed any expression of either of the parental alleles, regardless of genotype. Thus, the genes that affect virulence are actually passed on in a Mendelian way, but the trait (virulence) that they determine is not. This is because the expression of the *Avr3a/5* gene is changed, and this change is heritable and independent from the gene itself, or from any other genes for that matter. This result implies that some other self-propagating factor besides the *Avr3a/5* alleles determines the trait. Experiments suggest that the relevant self-propagating factor is sRNA molecules (Qutob et al. 2013). The transgenerational gene silencing of *P. sojae Avr3a/5* represents a novel gain of virulence mechanism and a striking example of naturally occurring epigenetic inheritance (Fig. 7.3).

This example of naturally occurring epigenetic inheritance is exceptional because it affects a trait (virulence) that is of paramount importance to a pathogenic organism such as *P. sojae*. It has implications for how we understand the evolution of virulence in this pathogen and others like it. In fact, results from the related species *P. infestans* and *P. ramorum* substantiate the view that epigenetic mechanisms play a major role in the evolution and adaptation of these pathogens toward their hosts (Raffaele et al. 2010; Kasuga et al. 2012). It is known from whole genome sequencing of *Phytophthora* species that a bipartite organization, consisting of highly conserved regions interspersed with repetitive and polymorphic regions, is a recognizable structural feature of their genomes (Haas et al. 2009; Tyler et al. 2006; Gijzen 2009) (Fig. 7.1). By re-sequencing closely related sister species to *P. infestans* and looking for sequence and structural signatures that are associated with rapidly evolving gene families that preferentially occur in polymorphic regions, it was determined that genes controlling epigenetic marks such as histone methylation fell into this category (Raffaele et al. 2010). For *P. ramorum*, experiments suggest that host adaptation is mediated by transposon derepression and large-scale gene expression reprogramming in clonal lineages (Kasuga et al. 2012). Thus, clonally generated epigenetic variation in *P. ramorum* may be crucial to enabling its wide host range and invasiveness. Whether these epigenetic changes are stable or are reset during sexual outcrossing of *P. ramorum* is not known, but would be interesting to test. If the changes are heritable, it would represent a rare example of environmentally induced epigenetic change that acts transgenerationally.

Why do species of *Phytophthora* rely on epigenetic mechanisms for generating variation in key traits such as virulence and host adaptation? The answer may lie in the responsiveness and reversibility of epigenetic phenomena. Epigenetics is often considered a biological way for an organism to hedge its bet on future environments or selective pressures, because it does not require irreversible, code-based changes to the DNA sequence. This is advantageous as the DNA sequences were originally selected for in environments or hosts that could easily be encountered in the future. Likewise, transgenerational inheritance offers a way for beneficial expression patterns to spread quickly in populations. Moreover, because the resting structure and most persistent spore type in many oomycete pathogens including *P. sojae* is the sexually produced oospore, a mechanism of epigenetic inheritance that can persist through meiosis is strategically beneficial in an organism that is subject to seasonal and episodic availability of its hosts.

### 7.3.3 Necrosis-Inducing Proteins as Major Players in *P. sojae* Infection

The discovery of the *P. sojae* necrosis-inducing protein (PsojNIP) resulted from a functional screen of candidates selected from early EST data (Qutob et al. 2002). The basis of the screen was ectopic expression in *Nicotiana benthamiana* leaves using the potato virus X (PVX) vector (Torto et al. 2003). The outcome from *PVX::PsojNIP* expression was rapidly spreading and severe necrosis, beginning at 7 d after inoculation, whereas controls were asymptomatic. Transient expression of PsojNIP in soybean tissues also resulted in cell death. Measurement of *PsojNIP* transcripts during the course of infection showed that the gene is most highly expressed during the necrotrophic stages of growth, 48 h after inoculation (Qutob et al. 2002). We now know that PsojNIP belongs to a large and diverse group of proteins that occur in fungi, bacteria, and oomycetes, especially in plant pathogenic species (Gijzen and Nurnberger 2006). The original example for this group, from *Fusarium oxysporum* f. sp. *erythroxyli*, was named as the necrosis and ethylene-inducing peptide 1, or simply Nep1 (Bailey et al. 1997). In recognition of this, proteins in this family are now called Nep1-like proteins (NLP), and PsojNIP has been renamed PsNLP1.

The *P. sojae* NLP family comprises an estimated 70 genes, but 37 of these are thought to be pseudogenes. The remaining 33 functional genes are diverse but can be placed into six different groups based upon sequence comparisons (Dong et al. 2012). Only two of the six groups include NLP sequences with an intact heptapeptide motif Gly-His-Arg-His-Asp-Trp-Glu (GHRHDWE). This motif, in addition to selected residues at other sites, is necessary for the necrosis-inducing activity when the protein is expressed or infiltrated in plant tissues (Fellbrich et al. 2002; Ottmann et al. 2009). Thus, the majority of *P. sojae* NLPs do not cause visible necrotic symptoms in plant tissues. An additional requirement for necrosis-inducing activity when the NLP is ectopically expressed in plant cells is that it possesses a functional leader peptide for secretion (Qutob et al. 2006). This result implies that a receptor or site of action for NLPs occurs on the surface of the plant cell plasma membrane. Indeed, a role for NLPs as cytolytic toxins has been proposed based upon their activity and three-dimensional protein structural analysis (Ottmann et al. 2009). However, this hypothesis raises questions as to the function of NLPs that lack any detectable necrosis inducing or cytolytic activity, or why NLPs occur in exclusively biotrophic pathogens such as *Hyaloperonospora arabidopsidis* (Cabral et al. 2012). One explanation is that the large size of the NLP family causes redundancy and provides the genetic space for relaxed selection, enabling the sub- or neofunctionalization of duplicated NLP genes to emerge over time. The evidence suggests that NLPs lacking necrosis-inducing activity have evolved new roles because two of the relevant NLP groups are under strong diversifying (positive) selection pressure, based upon amino acid substitution (Ka/Ks) analysis (Dong et al. 2012).

Varying patterns of expression of the NLP family in *P. sojae* also suggest varied functional roles; however, expression patterns do not appear to correlate with sequence phylogeny or even necrosis-inducing activity (Dong et al. 2012). Two basic patterns of expression are observed. Individual NLPs peak in expression either during cyst germination and early infection or during late infection stages and necrotrophic growth, as was originally observed for PsNLP1. The NLPs active in necrosis-inducing activity fall into both categories as they are expressed at early and late infection stages. This is surprising because host cell death and necrosis are counterproductive to *P. sojae* during the early biotrophic stages of growth. It has been proposed that additional secreted effectors, such as particular RxLR proteins, can suppress host cell death and counteract the necrosis-inducing activity of NLPs that are expressed during early infection (Wang et al. 2011; Kelley et al. 2010). Why would a pathogen secrete a potentially cytolytic toxin and at the same time suppress its effects with additional secreted effectors? This makes sense if one assumes that host cells actually detect some NLP activities as danger signals of infection, resulting in defense-related cell death. If NLPs have a role in providing nutrient release from host cells and fostering pathogen growth, then this would be beneficial for both biotrophic and necrotrophic growth modes (Fig. 7.2). On the other hand, perturbations to host cell homeostasis caused by NLPs could be subjected to surveillance and response by host defense systems.

### 7.3.4 Cell-Entering RxLR Effectors that Suppress Host Immunity

One of the most important discoveries to emerge from the genome sequence of *P. sojae* and from subsequent oomycete genome sequencing projects has been a large rapidly evolving class of effector proteins named RxLR effectors (Jiang et al. 2008; Tyler et al. 2006) (Fig. 7.2). These proteins are small secreted hydrophilic proteins that have the ability to enter the cells of soybeans and other plants in order to promote infection by the pathogen (Dou et al. 2008b; Kale et al. 2010; Whisson et al. 2007; Tyler et al. 2013). The amino acid sequences of the proteins are highly variable, but several sequence motifs are conserved among the proteins. One motif is the RxLR motif, which most commonly takes the form Arg-any residue-Leu-Arg, and which gives this effector class its name (Jiang et al. 2008; Tyler et al. 2006). A second motif is the dEER motif, which typically consists of a string of acidic residues followed by an arginine residue, for example, Asp-Glu-Glu- Arg (Jiang et al. 2008; Tyler et al. 2006). Together the RxLR and dEER motifs define an N-terminal domain common to these effectors that mediates the entry of the effectors into plant cells (Dou et al.

2008b; Kale et al. 2010; Whisson et al. 2007). Current evidence supports the hypothesis that RxLR effectors enter plant cells by endocytosis following binding of the RxLR-dEER domain, and possibly other residues, to the cell surface lipid phosphatidyinositol-3-phosphate (PI3P) (Kale et al. 2010; Sun et al. 2013). Entry does not appear to require any machinery from the pathogen (Dou et al. 2008b; Kale et al. 2010; Tyler et al. 2013); though it is possible that *P. sojae* may secrete PI3P in order to facilitate effector entry (Lu et al. 2013).

The RxLR-dEER cell entry domain of the effectors contains other nonrandom amino acid residues in addition to the two motifs (Jiang et al. 2008; Tyler et al. 2006). These residues enabled the development of hidden markov models that can accurately identify RxLR effector genes in oomycete genomes (Jiang et al. 2008). Using these models, together with some manual curation, large numbers of RxLR effectors have been identified in the genome of *P. sojae* (395) (Jiang et al. 2008; Tyler et al. 2006) as well as *P. infestans* (550) (Haas et al. 2009), *P. ramorum* (350) (Jiang et al. 2008; Tyler et al. 2006), and *Hyaloperonospora arabidopsidis* (130) (Baxter et al. 2010). Three C-terminal motifs, namely the W, Y, and L motifs, have also been identified and occur in roughly half of the RxLR effectors (Jiang et al. 2008; Dou et al. 2008a). They have recently been shown to correspond to a structural fold common to those effectors (Win et al. 2012).

Despite the common motifs, the RxLR effector superfamily is very diverse. Whereas proteins of core metabolism typically show 80–99 % amino acid identity between species (e.g., *P. sojae* and *P. ramorum*), RxLR effectors typically show 20–70 % identity or less (Jiang et al. 2008). The rapid divergence of RxLR effectors is likely driven by their involvement in infection, which selects for rapid adaptation. The great majority of RxLR effector genes are located in highly dynamic repeat-rich regions of the genomes, and it seems likely that these locations facilitate the rapid generation of diversity in the RxLR genes (Jiang et al. 2006a, 2008; Tyler et al. 2006; Haas et al. 2009; Win et al. 2007). Clusters of RxLR genes are characterized by frequent duplications, deletions, gene conversions, and numerous pseudogenes (Jiang et al. 2006a, 2008; Tyler et al. 2006; Haas et al. 2009).

The importance of RxLR effectors for the *P. sojae*-soybean interaction is underlined by the finding that all eleven avirulence genes identified thus far from *P. sojae* (Table 7.1) encode RxLR effectors (discussed below). Numerous avirulence genes that have been cloned from *P. infestans* and *Hyaloperonospora arabidopsidis* also encode RxLR effectors.

Several publications have demonstrated that RxLR effectors act to suppress plant immunity (reviewed in Jiang and Tyler 2012). For example, overexpression of *P. sojae* RxLR effector Avr1b in *P. sojae* transformants increased the virulence of the transformants toward soybean (Dou et al. 2008a), while silencing of effectors Avh238, Avh172 (Wang et al. 2011), Avh241 (Yu et al. 2012), and Avr3b (Dong et al. 2011a) reduced the virulence of *P. sojae* transformants. Overlapping sets of plant immune responses are triggered by pathogen-associated molecular patterns (PAMPs)—PAMP-triggered immunity (PTI) and by the presence of pathogen effectors—effector-triggered immunity (ETI) (Jones and Dangl 2006). Programmed cell death is often associated with ETI and sometimes with PTI. In a wide survey of 169 *P. sojae* RxLR effectors, 127 including Avr1b could suppress programmed cell death (Wang et al. 2011). In a more extensive characterization of 49 *P. sojae* effectors, 92 % could suppress cell death triggered by other effectors and 47 % could suppress cell death triggered by the PAMP INF1. Thus, a large percentage of the *P. sojae* RxLR repertoire is devoted to suppression of cell death. Similar observations have been made for RxLR effectors from *P. infestans* (Bos et al. 2006, 2010; Oh et al. 2009) and *Hyaloperonospora arabidopsidis* (Sohn et al. 2007; Fabro et al. 2011). *P. sojae* RxLR effector Avr3b has a unique mechanism for suppressing immunity (Dong et al. 2011a). This effector is a nudix pyrophosphorylase that targets ADP-ribose and NADPH, and strongly suppresses the accumulation of reactive oxygen species (Dong et al. 2011a).

Despite their apparent functional similarities in suppressing host cell death, *P. sojae* RxLR effectors display varying transcriptional programs (Wang et al. 2011). To begin with, 80 % of RxLR effectors show very low transcript levels, even during infection, suggesting that they generally do not make a large contribution to virulence. Of the remaining 20 %, some RxLR effectors are transcribed at moderately high levels prior to contact with the plant, and are moderately elevated (2–10 fold) during the first 12 h of infection (Wang et al. 2011). These effectors, that include Avr1b, Avr1d, and Avh172, have been dubbed "immediate-early" effectors and are enriched in effectors than can suppress ETI (Wang et al. 2011). Other effectors are expressed weakly prior to contact with the plant, but are induced 20–200 fold during the first 12 h of infection. These "early effectors" include Avh238, Avh94, Avh240, and Avh181, and are enriched in effectors than can suppress PTI (Wang et al. 2011). Effectors expressed primarily during the necrotrophic phase are dubbed "late effectors."

### 7.3.5 Characterization of *P. sojae* Avirulence Genes that Mediate Soybean Resistance to the Pathogen

The molecular factors that determine compatibility between soybean plants and *P. sojae* have long been central to studies of this host-pathogen interaction (Gijzen and Qutob 2009; Tyler 2002; Ward 1990). In fact, soon after the disease and the organism itself were discovered, the first soybean resistance genes to *P. sojae* (*Rps* genes) emerged from breeding programs and were deployed in the field. Breeding for resistance remains the primary strategy for managing the disease.

In addition to *Rps* genes, which are more fully discussed below, soybean plants may also exhibit partial resistance to *P. sojae* (Dorrance et al. 2003; Buzzell and Anderson 1982). Partial resistance differs from *Rps* gene-mediated resistance because it is dependent on quantitative trait loci (QTL) that are additive and that do not appear to be strain-specific. Partial resistance is believed to be more durable than *Rps* gene resistance but there is no real understanding of the underlying mechanism. Various QTL that condition partial resistance have been mapped; some are reproducible across genotypes whereas others appear to be cultivar or genotype specific (Mideros et al. 2007; Wang et al. 2012; Weng et al. 2007; Han et al. 2008; Li et al. 2010b; Nguyen et al. 2012; Lee et al. 2013). Root suberin and, in particular the omega-hydroxy fatty acid components of suberin have been proposed to influence partial resistance but these results have not been connected to any actual QTL, so this hypothesis remains unproven (Ranathunge et al. 2008; Thomas et al. 2007). It is necessary to map QTL that determine metabolite composition, such as suberin or components thereof, and demonstrate cosegregation with QTL controlling partial resistance.

In contrast to partial resistance, *Rps*-mediated resistance is genetically simpler because these genes act as single dominant factors that condition strain-specific resistance to *P. sojae.* Thus far, at least 16 different *Rps* specificities at 10 loci have been proposed. For example, the *Rps*1 locus includes *Rps*1a, *Rps*1b, *Rps*1c, *Rps*1d, and *Rps*1k; other proposed *Rps* genes or alleles are *Rps*2, *Rps*3a, *Rps*3b, *Rps*3c, *Rps*4, *Rps*5, *Rps*6, *Rps*7, *Rps*8, *Rps*9, and *Rps*10 (Sugimoto et al. 2012; Wu et al. 2011; Zhang et al. 2013c). The *Rps* genes that share the same locus number behave genetically as alleles. Potentially novel alleles or loci that remain unnumbered include *Rps*YB30, *Rps*YD25, *Rps*ZS18, *Rps*SN10, *Rps*Waseshiroge, RpsYD29, RpsUn1, and RpsUn2 (Sun et al. 2011; Zhu et al. 2010; Yao et al. 2010; Sugimoto et al. 2011; Fan et al. 2009; Zhang et al. 2013a; Lin et al. 2013). There is a good evidence that the *Rps*1k, *Rps*2, *Rps*4, and *Rps*6 genes encode canonical nucleotide-binding, leucine-rich repeat (NLR) type of cytoplasmic immune receptors (Sandhu et al. 2004; Graham et al. 2002; Gao and Bhattacharyya 2008). Indeed, the characteristics of most known

*Rps* genes are consistent with their being cytoplasmic NLR receptors, except for *Rps*10 which might encode a serine/threonine protein kinase.

The specificity of *Rps* genes means that they are subjected to defeat in the field when new *P. sojae* strains arise and spread in soybean growing regions. There is an ongoing battle between deploying effective *Rps* genes and pathogen adaptation to the cultivars, as is true for numerous crops that depend on NLR-type *R*-genes for resistance to pathogens. The *Rps* gene specificity of particular *P. sojae* strains is determined by *Avr* genes; plant immunity is triggered when a soybean Rps protein interacts, directly or indirectly, with an Avr factor from the pathogen. Thus, the *Rps* genes are effective only against *P. sojae* strains with corresponding *Avr* genes, and gain of virulence mutations of *Avr* genes can defeat *Rps* genes. For a long time, this model of soybean *Rps* and *P. sojae Avr* gene interaction was simply a theory with little hard evidence to support it. Claims that race-specific molecules could be purified from *P. sojae* cultures were controversial and led to arguments rather than the identification of any actual Avr factor (Wade and Albersheim 1979; Keen 1975). It was not until the *P. sojae* genome sequence became available that progress in the area of *Avr* gene characterization accelerated.

Identification of *P. sojae Avr* genes has been enabled by a high-quality reference genome assembly and by powerful new genetic and bioinformatic approaches to gene identification (Jiang and Tyler 2012; Tyler 2007). The ability to generate segregating $F_2$ populations has aided map-based approaches of *Avr* gene identification in *P. sojae* (Gijzen et al. 1996; Tyler et al. 1995; MacGregor et al. 2002; Whisson et al. 1994, 1995; May et al. 2002). Furthermore, the qualitative nature of *Rps* gene-specific virulence and the reliability of scoring phenotypes is a definite advantage in tracking down *Avr* genes. Phenotypes that are as robust and dependable as those conferred by *Avr* genes are exceptional in biological systems.

The simplest interpretation of the *Rps-Avr* gene interaction model predicts that for each known *Rps* gene, there is a corresponding *Avr* gene in *P. sojae*. The 15 different *Rps* gene specificities (for those *Rps* genes that are numbered) should then be matched by 15 different *Avr* genes in *P. sojae*. Presently, at least 11 of the 15 predicted *Avr* genes have been identified, though the relationships between the *Rps* genes and *Avr* genes eventually proved more complicated than the initial simple gene-for-gene model (Table 7.1). The first *P. sojae Avr* gene identified was *Avr1b* (Shan et al. 2004; Tyler 2002). The identification of *Avr1b* and those that followed led to the understanding that *P. sojae Avr* genes encode relatively small, secreted RxLR effector proteins (described above).

Early *Avr* gene mapping work suggested that different *Avr* genes could in fact be alleles (Gijzen et al. 1996; Tyler et al. 1995; Whisson et al. 1994). We now know this to be true in several cases; different soybean *Rps* genes can trigger immunity through their interaction with identical RxLR effector proteins. For example, the highly polymorphic *Avr3a/5* locus (Fig. 7.1) includes alleles such as $Avr3a^{P6497}$ that trigger immunity on *Rps*3a and *Rps*5, whereas the $Avr3a^{P7076}$ allele triggers immunity on *Rps*3a but escapes detection by *Rps*5 (Dong et al. 2011b). Additional alleles such as $Avr3a^{P7064}$ do not produce mRNA transcripts and are able to defeat *Rps*3a and *Rps*5. The *Rps*4 and *Rps*6 genes also interact with a single *P. sojae* effector locus, *Avr4/6* (Dou et al. 2010).

Different Rps proteins can recognize the same effector but the converse is also true; the same Rps protein can apparently recognize different effectors. In fact, most of the Rps proteins appear to recognize sequence variants of their effector targets. However, *Rps*1k is unusual because it appears to recognize distinctly different effectors; *Rps*1k interacts with Avr1b and *Avr1k*, which is somewhat surprising because the predicted Avr1b and Avr1k proteins are quite unrelated in their amino acid sequences though they are located only 5 kb apart (Song et al. 2013). This result suggests that *Rps*1k senses changes are caused by Avr1b and *Avr1k*, rather than directly interacting with the Avr factors. Alternatively, distinct Rps proteins encoded within the *Rps*1k locus might be responsible for the two specificities.

The great variety of *Rps* and *Avr* specificities, sometimes overlapping, is partly a consequence of selection pressures that cause expansion of the *R*- and effector-gene repertoires in the host and pathogen. In *P. sojae*, variable tandem arrays in the genome were recognized early on (Mao and Tyler 1996) and at least three different *Avr* genes occur in tandem arrays of near identical copies, that display sequence polymorphisms and copy number variation (CNV) among strains. The *Avr* loci that are known to reside in tandem arrays and display CNV include *Avr1a/1c* (Qutob et al. 2009) and *Avr3a/5* (Qutob et al. 2009; Dong et al. 2011b) (Fig. 7.1); whereas *Avr3c* occurs in a tandem array but does not show CNV (Dong et al. 2009). Multiple copies of effector genes may arise through positive selection because they enable infection in the absence of a corresponding *R*-gene in the host. However, once a tandem array of duplicated effector genes is "born," the individual copies within the array may be subject to relaxed selection because of their redundancy. Moreover, individual or multiple copies within the tandem array may easily be lost through homologous recombination, as evidently happened at the *Avr1a/1c* and *Avr3a/5* loci (Fig. 7.1). Tandem arrays also provide for sequence exchange opportunities among the different copies, as is evident at the *Avr3c* locus. Thus, the plasticity of tandem arrays provides opportunities for different types of gain of virulence mutations in the event of effector recognition by an R-protein.

In fact, the variety of gain of virulence mutations of *P. sojae Avr* genes go beyond those enabled by tandem arrays, as itemized in Table 7.1. There could be more gain of virulence mutations yet to discover since there are at least four (and likely more) possible *P. sojae Avr* genes that remain unidentified, including *Avr2*, *Avr7*, *Avr8*, and *Avr9*. The multiplicity of means to escape detection by *Rps* genes illustrates the powerful but somewhat random forces of natural selection. Whatever works will be selected for and transmitted to the following generations, however unusual the mechanism may appear. Exactly how it is transmitted to the following generations is also variable as epigenetic inheritance of *Avr* gene expression has been described, as discussed earlier in this review.

## 7.4 Future Perspectives

### 7.4.1 Comparative Genomics of *P. sojae* and Its Close Relatives

Comparative genomics provides a powerful tool for elucidating gene function. The *P. sojae* and *P. ramorum* sequences were generated at the same time (Tyler et al. 2006). As well as helping to correct assemblies and gene models, the comparison of the two sequences provided a rich set of information on conservation of gene structure and location, and on the rates of evolution of the encoded proteins. For example, the comparison identified RxLR and crinkler effectors and NLP and PcF toxins as evolving far more rapidly than housekeeping genes, underlining their potential roles in pathogenicity. The sequencing of three additional strains of *P. sojae* also provided valuable insights into the most polymorphic genes, especially RxLR effectors (Wang et al. 2011). In the future, even richer resources for inferring *P. sojae* gene function could be obtained by sequencing the genomes of the closely related species in clade 7 (Kroon et al. 2012), and eventually the entire genus. Clade 7 contains narrow host range legume pathogens such as *P. pisi*, *P. vignae*, and *P. cajani*, but also broad host range pathogens such as *P. cinnamomi* and *P. niederhauserii* (Kroon et al. 2012). Comparisons of the genome sequences of these species with *P. sojae* will aid the identification of genes associated with host range diversification. *P. niederhauserii* is likely to be of great interest as it is closely related to *P. sojae* (Blair et al. 2008), but has a broad host range (Faedda et al. 2013). Comparison of a range of sequences varying in distance from *P. sojae* will also enable identification of functional nonprotein-coding sequences, including promoter elements and regulatory RNAs. It may also enable proteins to be identified that are under diversifying selection across the clade.

### 7.4.2 Manipulating the *P. sojae* Genome with Improved Gene Silencing and Knockout Technology

The ability to overexpress, silence, knockout, or replace targeted genes is an essential tool for functionally testing the roles of genes found in the genome. Transformation of *P. sojae* (Judelson et al. 1993) and *P. infestans* (Judelson et al. 1991) opened the possibility of genetic manipulation of these species. Silencing is especially useful when complete knockout of a gene may be lethal. Silencing of genes in *P. infestans* (Ah-Fong et al. 2008) and *P. sojae* (Wang et al. 2011) has been proven useful, however, there are many genes that are difficult to silence. The ability to transiently trigger gene silencing in *P. infestans* (Whisson et al. 2005) and *P. sojae* (Wang et al. 2011) using double-stranded RNA has expanded the usefulness of the technique. The use of artificial miRNA sequences (Schwab et al. 2006) or trans-acting silencing (ta-si) RNA technologies (de la Luz Gutierrez-Nava et al. 2008) may further extend the usefulness of silencing approaches.

Since insertion of transgenes in *Phytophthora* occurs exclusively by nonhomologous end-joining, it has never been possible to carry out gene knockouts or gene replacements. The recent emergence of tailored nucleases that can cleave endogenous genomic DNA at specific sites (Gaj et al. 2013) raises the possibility that in the near future gene knockouts and replacements may become possible in *P. sojae* and other *Phytophthora* species. These nucleases, that include zinc finger nucleases (Dreyer and Cathomen 2012; Kim and Kim 2011), TALE nucleases (Zu et al. 2013; Zhang et al. 2013b), and CRSPR nucleases (Mali et al. 2013; Cong et al. 2013; Dicarlo et al. 2013) have been used for gene replacements in a wide diversity of organisms where they had long proved difficult or impossible such as plants and vertebrates (Gaj et al. 2013). Silencing or knockout of genes involved in nonhomologous end-joining, such as Ku70 and Ku80, has greatly improved the frequency of homologous recombination in fungi (Arentshorst et al. 2012), and may prove a useful approach in *P. sojae* and other oomycetes.

### 7.4.3 Applications of the Genome for Disease Control

The availability of genome sequences from *P. sojae* and other oomycetes, and the characterization of the genes discovered therein have made possible a wide range of applications useful for controlling disease caused by these organisms.

The most basic application of genome sequences has been the development of DNA-based diagnostic tests. Typically these have been targeted toward species identification using barcode sequences such as rRNA internal transcribed spacer (ITS) or cytochrome oxidase (COX) regions (Robideau et al. 2011). However, the cloning of many avirulence genes (Table 7.1) has opened the possibility of DNA- or RNA-based tests for the ability of isolates to overcome *Rps* resistance genes; seedling inoculation for this test is tedious and variable despite many decades of use. Furthermore, with the identification of a gene for mefenoxam resistance from *P. infestans* (Whisson et al. 2013), DNA-based tests for fungicide resistance may also become possible.

Breeding of soybean varieties incorporating *Rps* resistance genes is also hampered by the unreliability of the seedling inoculation assay. The cloning of many avirulence genes from *P. sojae* based on the genome sequence (Table 7.1) opens the possibility of using effector transient expression in planta (Kale and Tyler 2011), delivery from bacteria (Armstrong et al. 2005; Rentel et al. 2008) or direct treatment with protein (Kale et al. 2010) as a rapid and reliable assay for the presence of specific *Rps* genes.

The same assays that could be used to screen for the presence of known resistance genes could also be used to screen for the presence of novel resistance genes that detect particular effectors. Effectors that are essential to the pathogen are particularly attractive as novel targets. This approach has proven useful in identifying R genes against *P. infestans* (Oh et al. 2009). With the identification of essential effectors in *P. sojae* (e.g., Yu et al. 2012; Wang et al. 2011; Dong et al. 2011a), this approach could be used to identify new, potentially durable resistance genes against *P. sojae* in soybean.

The genome sequence of *P. sojae* and other oomycetes provides a rich resource for identifying specific proteins that could be targeted by new fungicides. A particularly interesting variation on this theme is host-induced gene silencing (HIGS) in which small silencing RNAs are generated in plant tissues and somehow transfer into the fungal cytoplasm (Nunes and Dean 2012). Disease control may be possible if the silencing RNAs target a gene that is essential for the pathogen. HIGS has yet to be demonstrated against an oomycete but has interesting potential.

Another intriguing strategy to emerge from the *P. sojae* genome sequence and subsequent characterization is targeted against RxLR effectors. Kale et al. (2010) showed that PI3P-binding proteins could block the entry of RxLR effectors into plant cells. This observation raises the possibility that plants secreting PI3P-binding proteins might be able to protect themselves against *Phytophthora* infection.

## 7.5 Conclusions

The sequencing of the *P. sojae* and *P. ramorum* genomes was performed some 10 years ago and the outcome for research into the molecular basis of oomycete-plant interactions has been revolutionary. Genome sequence information has driven research progress to an extent that far exceeded many of the original project goals. New technologies have certainly contributed, but the quality of the original sequence data and the timing of its release were major factors in propelling oomycete research to the forefront of molecular plant pathology. These facts make a strong case for arguments that biological investigations should begin with the acquisition of genome sequence information. For pathogenic and invasive organisms with limited genetic tools and that cause economic or ecosystem damage, such as species of *Phytophthora*, the case is even more compelling. Priorities need to begin with obtaining whole genome sequence data of target organisms. The cost of this has come down dramatically over the past decade though bottlenecks remain in bioinformatics and data archiving.

**Acknowledgments** We thank Jeannine Cropley for manuscript assistance. BMT was supported by grant 2011-68004-30104 from the Agriculture and Food Research Institute of the National Institute of Food and Agriculture of the USDA; MG was supported by Agriculture and Agri-Food Canada GRDI program.

## References

Ah-Fong AM, Bormann-Chung CA, Judelson HS (2008) Optimization of transgene-mediated silencing in *Phytophthora infestans* and its association with small-interfering RNAs. Fungal Genet Biol 45(8):1197–1205. doi:10.1016/j.fgb.2008.05.009

Arentshorst M, Ram AF, Meyer V (2012) Using non-homologous end-joining-deficient strains for functional gene analyses in filamentous fungi. Methods Mol Biol 835:133–150. doi:10.1007/978-1-61779-501-5_9

Armstrong MR, Whisson SC, Pritchard L, Bos JI, Venter E, Avrova AO, Rehmany AP, Bohme U, Brooks K, Cherevach I, Hamlin N, White B, Fraser A, Lord A, Quail MA, Churcher C, Hall N, Berriman M, Huang S, Kamoun S, Beynon JL, Birch PR (2005) An ancestral oomycete locus contains late blight avirulence gene *Avr3a*, encoding a protein that is recognized in the host cytoplasm. Proc Natl Acad Sci USA 102(21):7766–7771

Aurrecoechea C, Brestelli J, Brunk BP, Fischer S, Gajria B, Gao X, Gingle A, Grant G, Harb OS, Heiges M, Innamorato F, Iodice J, Kissinger JC, Kraemer ET, Li W, Miller JA, Nayak V, Pennington C, Pinney DF, Roos DS, Ross C, Srinivasamoorthy G, Stoeckert CJ, Jr., Thibodeau R, Treatman C, Wang H (2010) EuPathDB: a portal to eukaryotic pathogen databases. Nucleic Acids Res 38 (Database issue):D415–419. doi:10.1093/nar/gkp941

Bailey BA, Jennings JC, Anderson JD (1997) The 24-kDa protein from *Fusarium oxysporum* f.sp. *erythroxyli*:

occurrence in related fungi and the effect of growth medium on its production. Can J Microbiol 43(1):45–55

Baxter L, Tripathy S, Ishaque N, Boot N, Cabral A, Kemen E, Thines M, Ah-Fong A, Anderson R, Badejoko W, Bittner-Eddy P, Boore JL, Chibucos MC, Coates M, Dehal P, Delehaunty K, Dong S, Downton P, Dumas B, Fabro G, Fronick C, Fuerstenberg SI, Fulton L, Gaulin E, Govers F, Hughes L, Humphray S, Jiang RHY, Judelson HS, Kamoun S, Kyung K, Meijer H, Minx P, Morris P, Nelson J, Phuntumart V, Qutob D, Rehmany A, Rougon A, Ryden P, Torto-Alalibo T, Studholme D, Wang Y, Win J, Wood J, Clifton SW, Rogers J, Ackerveken GVd, Jones JDG, McDowell JM, Beynon J, Tyler BM (2010) Signatures of adaptation to obligate biotrophy in the *Hyaloperonospora arabidopsidis* genome. Science 330(6010):1549–1551

Blair JE, Coffey MD, Park SY, Geiser DM, Kang S (2008) A multi-locus phylogeny for *Phytophthora* utilizing markers derived from complete genome sequences. Fungal Genet Biol 45(3):266–277. doi:10.1016/j.fgb.2007.10.010

Bos JIB, Armstrong MR, Gilroy EM, Boevink PC, Hein I, Taylor RM, Zhendong T, Engelhardt S, Vetukuri RR, Harrower B, Dixelius C, Bryan G, Sadanandom A, Whisson SC, Kamoun S, Birch PR (2010) *Phytophthora infestans* effector AVR3a is essential for virulence and manipulates plant immunity by stabilizing host E3 ligase CMPG1. Proc Natl Acad Sci USA 107 (21):9909–9914. doi:10.1073/pnas.0914408107

Bos JIB, Kanneganti T-D, Young C, Cakir C, Huitema E, Win J, Armstrong M, Birch PRJ, Kamoun S (2006) The C-terminal half of *Phytophthora infestans* RXLR effector AVR3a is sufficient to trigger R3a-mediated hypersensitivity and suppress INF1-induced cell death in *Nicotiana benthamiana*. Plant J 48:165–176

Burzell RI, Anderson TR (1982) Plant loss response of soybean cultivars to *Phytophthora megasperma* f. sp. *glycinea* under field conditions. Plant Dis 66(12): 1146–1148

Cabral A, Oome S, Sander N, Kufner I, Nurnberger T, Van den Ackerveken G (2012) Nontoxic Nep1-like proteins of the downy mildew pathogen *Hyaloperonospora arabidopsidis*: repression of necrosis-inducing activity by a surface-exposed region. Mol Plant-Microbe Interact 25(5):697–708

Cabral A, Stassen JH, Seidl MF, Bautor J, Parker JE, Van den Ackerveken G (2011) Identification of *Hyaloperonospora arabidopsidis* transcript sequences expressed during infection reveals isolate-specific effectors. PLoS ONE 6 (5):e19328. doi:10.1371/journal.pone.0019328, PONE-D-10-06310

Chamnanpunt J, Shan WX, Tyler BM (2001) High frequency mitotic gene conversion in genetic hybrids of the oomycete *Phytophthora sojae*. Proc Natl Acad Sci USA 98(25):14530–14535

Chen XR, Wang YC (2012) Advances on the methods used for the functional analysis of oomycete genes. J Agric Biotechnol 20(5):568–575

Cong L, Ran FA, Cox D, Lin S, Barretto R, Habib N, Hsu PD, Wu X, Jiang W, Marraffini LA, Zhang F (2013) Multiplex genome engineering using CRISPR/Cas systems. Science 339(6121):819–823. doi:10.1126/science.1231143

Cooke DEL, Drenth A, Duncan JM, Wagels G, Brasier CM (2000) A molecular phylogeny of *Phytophthora* and related oomycetes. Fungal Genet Biol 30(1):17–32

de la Luz Gutierrez-Nava M, Aukerman MJ, Sakai H, Tingey SV, Williams RW (2008) Artificial trans-acting siRNAs confer consistent and effective gene silencing. Plant Physiol 147(2):543–551. doi:10.1104/pp.108.118307

Deschamps P, Moreira D (2012) Reevaluating the green contribution to diatom genomes. Genome Biol Evol 4(7):683–688. doi:10.1093/gbe/evs053

Dicarlo JE, Norville JE, Mali P, Rios X, Aach J, Church GM (2013) Genome engineering in *Saccharomyces cerevisiae* using CRISPR-Cas systems. Nucleic Acids Res 41(7):4336–4343. doi:10.1093/nar/gkt135

Dong S, Kong G, Qutob D, Yu X, Tang J, Kang J, Dai T, Wang H, Gijzen M, Wang Y (2012) The NLP toxin family in *Phytophthora sojae* includes rapidly evolving groups that lack necrosis-inducing activity. Mol Plant-Microbe Interact 25(7):896–909

Dong S, Yin W, Kong G, Yang X, Qutob D, Chen Q, Kale SD, Sui Y, Zhang Z, Dou D, Zheng X, Gijzen M, Tyler BM, Wang Y (2011a) *Phytophthora sojae* avirulence effector Avr3b is a secreted NADH and ADP-ribose pyrophosphorylase that modulates plant immunity. PLoS Pathog 7(11):e1002353

Dong S, Yu D, Cui L, Qutob D, Tedman-Jones J, Kale SD, Tyler BM, Wang Y, Gijzen M (2011b) Sequence variants of the *Phytophthora sojae* RXLR effector Avr3a/5 are differentially recognized by *Rps*3a and *Rps*5 in soybean. PLoS ONE 6(7):e20172. doi:10.1371/journal.pone.0020172

Dong SM, Qutob D, Tedman-Jones J, Kuflu K, Wang YC, Tyler BM, Gijzen M (2009) The *Phytophthora sojae* avirulence locus *Avr3c* encodes a multi-copy RXLR effector with sequence polymorphisms among pathogen strains. PLoS ONE 4(5). doi:10.1371/journal.pone.0005556

Dorrance A, Grünwald NJ (2009) *Phytophthora sojae*: diversity among and within populations. In: Lamour K, Kamoun S (eds) Oomycete genetics and genomics: diversity, interactions, and research tools. Wiley, pp 197–212. doi:10.1002/9780470475898.ch10

Dorrance AE, McClure SA, St. Martin SK (2003) Effect of partial resistance on *phytophthora* stem rot incidence and yield of soybean in Ohio. Plant Dis 87(3):308–312

Dorrance AE, Robertson AE, Cianzo S, Giesler LJ, Grau CR, Draper MA, Tenuta AU, Anderson TR (2009) Integrated management strategies for *Phytophthora sojae* combining host resistance and seed treatments. Plant Dis 93(9):875–882

Dou D, Kale SD, Liu T, Tang Q, Wang X, Arredondo FD, Basnayake S, Whisson S, Drenth A, Maclean D, Tyler BM (2010) Different domains of *Phytophthora sojae* effector Avr4/6 are recognized by soybean resistance genes *Rps*4 and *Rps*6. Mol Plant Microbe Interact 23(4):425–435

Dou D, Kale SD, Wang X, Chen Y, Wang Q, Wang X, Jiang RHY, Arredondo FD, Anderson R, Thakur P, McDowell J, Wang Y, Tyler BM (2008a) Carboxy-terminal motifs common to many oomycete RXLR effectors are required for avirulence and suppression of BAX-mediated programmed cell death by *Phytophthora sojae* effector Avr1b. Plant Cell 20(4): 1118–1133

Dou D, Kale SD, Wang X, Jiang RHY, Bruce NA, Arredondo FD, Zhang X, Tyler BM (2008b) RXLR-mediated entry of *Phytophthora sojae* effector Avr1b into soybean cells does not require pathogen-encoded machinery. Plant Cell 20(7):1930–1947

Dreyer AK, Cathomen T (2012) Zinc-finger nucleases-based genome engineering to generate isogenic human cell lines. Methods Mol Biol 813:145–156. doi:10.1007/978-1-61779-412-4_8

Enkerli K, Hahn MG, Mims CW (1997) Ultrastructure of compatible and incompatible interactions of soybean roots infected with the plant pathogenic oomycete *Phytophthora sojae*. Can J Bot 75(9):1494–1508

Erwin DC, Ribeiro OK (1996) *Phytophthora* diseases worldwide. The American Phytopathological Society, St. Paul

Fabro G, Steinbrenner J, Coates M, Ishaque N, Baxter L, Studholme DJ, Körner E, Allen RL, Piquerez SJM, Rougon-Cardoso A, Greenshields D, Lei R, Badel JL, Caillaud M-C, van den Ackerveken G, Parker JE, Beynon J, Jones JDG (2011) Multiple candidate effectors from the oomycete pathogen *Hyaloperonospora arabidopsidis* suppress host plant immunity. PLoS Pathog 7(11):e1002348

Faedda R, Cacciola SO, Pane A, Martini P, Odasso M, Magnano di San Lio G (2013) First report of *Phytophthora* taxon *niederhauserii* causing root and stem rot of mimosa in Italy. Plant Dis 97(5):688

Fahlgren N, Bollmann SR, Kasschau KD, Cuperus JT, Press CM, Sullivan CM, Chapman EJ, Hoyer JS, Gilbert KB, Grunwald NJ, Carrington JC (2013) *Phytophthora* have distinct endogenous small RNA populations that include short interfering and microRNAs. PLoS ONE 8(10):e77181. doi:10.1371/journal.pone.0077181

Fan A-Y, Wang X-M, Fang X-P, Wu X-F, Zhu Z-D (2009) Molecular identification of *Phytophthora* resistance gene in soybean cultivar Yudou 25. Acta Agron Sinica 35(10):1844

Fellbrich G, Romanski A, Varet A, Blume B, Brunner F, Engelhardt S, Felix G, Kemmerling B, Krzymowska M, Nurnberger T (2002) NPP1, a *Phytophthora*-associated trigger of plant defense in parsley and *Arabidopsis*. Plant J 32(3):375–390

Förster H, Coffey MD (1990) Mating behavior of *Phytophthora parasitica*: Evidence for sexual recombination in oospores using DNA restriction fragment length polymorphisms as genetic markers. Exp Mycol 14(4):351–359. doi:http://dx.doi.org/10.1016/0147-5975(90)90058-2

Förster H, Tyler BM, Coffey MD (1994) *Phytophthora sojae* races have arisen by clonal evolution and by rare outcrosses. Mol Plant-Microbe Interact 7(6):780–791

Francis DM, Gehlen MF, St Clair DA (1994) Genetic variation in homothallic and hyphal swelling isolates of *Pythium ultimum* var. *ultimum* and *P. ultimum* var. *sporangiferum*. Mol Plant-Microbe Interact 7(6): 766–775

Gaj T, Gersbach CA, Barbas CF 3rd (2013) ZFN, TALEN, and CRISPR/Cas-based methods for genome engineering. Trends Biotechnol. doi:10.1016/j.tibtech.2013.04.004

Gao H, Bhattacharyya MK (2008) The soybean-*Phytophthora* resistance locus *Rps*1-k encompasses coiled coil-nucleotide binding-leucine rich repeat-like genes and repetitive sequences. BMC Plant Biol 8(1):29. doi:10.1186/1471-2229-8-29

Gijzen M (2009) Runaway repeats force expansion of the *Phytophthora infestans* genome. Genome Biol 10(10):241. doi:10.1186/gb-2009-10-10-241

Gijzen M, Forster H, Coffey MD, Tyler BM (1996) Cosegregation of *Avr4* and *Avr6* in *Phytophthora sojae*. Can J Bot 74(5):800–802

Gijzen M, Nurnberger T (2006) Nep1-like proteins from plant pathogens: recruitment and diversification of the NPP1 domain across taxa. Phytochemistry 67(16):1800–1807. doi:10.1016/j.phytochem.2005.12.008, S0031-9422(05)00675-8

Gijzen M, Qutob D (2009) *Phytophthora sojae* and soybean. In: Lamour K, Kamoun S (eds) Oomycete genetics and genomics: diversity, interactions, and research tools. Wiley, pp 303–329. doi:10.1002/9780470475898.ch15

Gilroy EM, Breen S, Whisson SC, Squires J, Hein I, Kaczmarek M, Turnbull D, Boevink PC, Lokossou A, Cano LM, Morales J, Avrova AO, Pritchard L, Randall E, Lees A, Govers F, van West P, Kamoun S, Vleeshouwers VG, Cooke DE, Birch PR (2011) Presence/absence, differential expression and sequence polymorphisms between *PiAVR2 and PiAVR2-like* in *Phytophthora infestans* determine virulence on *R2* plants. New Phytol 191(3):763–776

Graham MA, Marek LF, Shoemaker RC (2002) Organization, expression and evolution of a disease resistance gene cluster in soybean. Genetics 162(4):1961–1977

Haas BJ, Kamoun S, Zody MC, Jiang RH, Handsaker RE, Cano LM, Grabherr M, Kodira CD, Raffaele S, Torto-Alalibo T, Bozkurt TO, Ah-Fong AM, Alvarado L, Anderson VL, Armstrong MR, Avrova A, Baxter L, Beynon J, Boevink PC, Bollmann SR, Bos JI, Bulone V, Cai G, Cakir C, Carrington JC, Chawner M, Conti L, Costanzo S, Ewan R, Fahlgren N, Fischbach MA, Fugelstad J, Gilroy EM, Gnerre S, Green PJ, Grenville-Briggs LJ, Griffith J, Grunwald NJ, Horn K, Horner NR, Hu CH, Huitema E, Jeong DH, Jones AM, Jones JD, Jones RW, Karlsson EK, Kunjeti SG, Lamour K, Liu Z, Ma L, Maclean D, Chibucos MC, McDonald H, McWalters J, Meijer HJ, Morgan W, Morris PF, Munro CA, O'Neill K, Ospina-Giraldo M, Pinzon A, Pritchard L,

Ramsahoye B, Ren Q, Restrepo S, Roy S, Sadanandom A, Savidor A, Schornack S, Schwartz DC, Schumann UD, Schwessinger B, Seyer L, Sharpe T, Silvar C, Song J, Studholme DJ, Sykes S, Thines M, van de Vondervoort PJ, Phuntumart V, Wawra S, Weide R, Win J, Young C, Zhou S, Fry W, Meyers BC, van West P, Ristaino J, Govers F, Birch PR, Whisson SC, Judelson HS, Nusbaum C (2009) Genome sequence and analysis of the Irish potato famine pathogen *Phytophthora infestans*. Nature 461(7262):393–398. doi:10.1038/nature08358

Halfmann R, Jarosz DF, Jones SK, Chang A, Lancaster AK, Lindquist S (2012) Prions are a common mechanism for phenotypic inheritance in wild yeasts. Nature 482(7385):363–368

Han Y, Teng W, Yu K, Poysa V, Anderson T, Qiu L, Lightfoot DA, Li W (2008) Mapping QTL tolerance to *Phytophthora* root rot in soybean using microsatellite and RAPD/SCAR derived markers. Euphytica 162(2):231–239. doi:10.1007/s10681-007-9558-4

Harper JT, Waanders E, Keeling PJ (2005) On the monophyly of chromalveolates using a six-protein phylogeny of eukaryotes. Int J Syst Evol Microbiol 55(Pt 1):487–496

Hildebrand AA (1959) A root and stalk rot of soybeans caused by *Phytophthora megasperma* Drechsler var. *sojae* var. nov. Can J Bot 37:927–957

Jiang RH, Tyler BM, Govers F (2006a) Comparative analysis of *Phytophthora* genes encoding secreted proteins reveals conserved synteny and lineage-specific gene duplications and deletions. Mol Plant-Microbe Interact 19(12):1311–1321

Jiang RH, Tyler BM, Whisson SC, Hardham AR, Govers F (2006b) Ancient origin of elicitin gene clusters in *Phytophthora* genomes. Mol Biol Evol 23(2):338–351

Jiang RHY, de Bruijn I, Haas BJ, Belmonte R, Löbach L, Christie J, van den Ackerveken G, Bottin A, Dumas B, Fan L, Gaulin E, Govers F, Grenville-Briggs LJ, Horner NR, Levin JZ, Mammella M, Meijer HJG, Morris P, Nusbaum C, Oome S, Rooyen Dv, Saraiva M, Secombes CJ, Seidl MF, Snel B, Stassen J, Sykes S, Tripathy S, van den Berg H, Vega-Arreguin JC, Wawra S, Young S, Zeng Q, Dieguez-Uribeondo J, Russ C, Tyler BM, van West P (2013) Distinctive expansion of potential virulence genes in the genome of the oomycete fish pathogen *Saprolegnia parasitica*. PLoS Genet 9(6):e1003272

Jiang RHY, Tripathy S, Govers F, Tyler BM (2008) RXLR effector reservoir in two *Phytophthora* species is dominated by a single rapidly evolving superfamily with more than 700 members. Proc Natl Acad Sci USA 105(12):4874–4879

Jiang RHY, Tyler BM (2012) Mechanisms and evolution of virulence in oomycetes. Ann Rev Phytopath 50:295–318

Jones JD, Dangl JL (2006) The plant immune system. Nature 444(7117):323–329

Jones JP (1969) Reaction of *Lupinus* species to *Phytophthora megasperma* var. *sojae*. Plant Dis Report 53:907–909

Jones JP, Johnson HW (1969) Lupine, a new host for *Phytophthora megasperma* var. *sojae*. Phytopathology 59:504–507

Judelson H, Tyler BM, Michelmore RW (1991) Transformation of the oömycete pathogen, *Phytophthora infestans*. Mol Plant-Microbe Interact 4(6):602–607

Judelson HS (2009) Sexual reproduction in oomycetes: biology, diversity, and contributions to fitness. In: Lamour K, Kamoun S (eds) Oomycete genetics and genomics: diversity, interactions, and research tools. Wiley, pp 121–138. doi:10.1002/9780470475898.ch6

Judelson HS, Coffey MD, Arredondo F, Tyler BM (1993) Transformation of the oömycete pathogen *Phytophthora megasperma* f.sp. *glycinea* occurs by DNA integration into single or multiple chromosomes. Curr Genet 23:211–218

Judelson HS, Tani S (2007) Transgene-induced silencing of the zoosporogenesis-specific NIFC gene cluster of *Phytophthora infestans* involves chromatin alterations. Eukaryot Cell 6(7):1200–1209

Kale SD, Gu B, Capelluto DGS, Dou D-L, Feldman E, Rumore A, Arredondo FD, Hanlon R, Fudal I, Rouxel T, Lawrence CB, Shan W-X, Tyler BM (2010) External lipid PI-3-P mediates entry of eukaryotic pathogen effectors into plant and animal host cells. Cell 142(2):284–295

Kale SD, Tyler BM (2011) Assaying effector function in planta using double-barreled particle bombardment. In: McDowell JM (ed) Methods in molecular biology. The plant immune response. Humana, Totowa, pp 153–172

Kasuga T, Kozanitas M, Bui M, Huberli D, Rizzo DM, Garbelotto M (2012) Phenotypic diversification is associated with host-induced transposon derepression in the sudden oak death pathogen *Phytophthora ramorum*. PLoS ONE 7(4):e34728

Kaufmann MJ, Gerdemann JW (1958) Root and stem rot of soybeans caused by *Phytophthora sojae* n. sp. Phytopathology 48:201–208

Keeling PJ (2010) The endosymbiotic origin, diversification and fate of plastids. Philos Trans R Soc Lond B Biol Sci 365(1541):729–748. doi:10.1098/rstb.2009.0103

Keen NT (1975) Specific elicitors of plant phytoalexin production: detenninants of race specificity in pathogens? Science 187(4171):74–75

Kelley BS, Lee SJ, Damasceno CM, Chakravarthy S, Kim BD, Martin GB, Rose JK (2010) A secreted effector protein (SNE1) from *Phytophthora infestans* is a broadly acting suppressor of programmed cell death. Plant J 62(3):357–366

Kim S, Kim JS (2011) Targeted genome engineering via zinc finger nucleases. Plant Biotechnol Rep 5(1):9–17. doi:10.1007/s11816-010-0161-0

Kroon L, Henk B, Decock A, Govers F (2012) The *Phytophthora* genus anno 2012. Phytopathology 102(4):348–364. doi:10.1094/PHYTO-01-11-0025

Lamour KH, Mudge J, Gobena D, Hurtado-Gonzales OP, Schmutz J, Kuo A, Miller NA, Rice BJ, Raffaele S, Cano LM, Bharti AK, Donahoo RS, Finley S,

Huitema E, Hulvey J, Platt D, Salamov A, Savidor A, Sharma R, Stam R, Storey D, Thines M, Win J, Haas BJ, Dinwiddie DL, Jenkins J, Knight JR, Affourtit JP, Han CS, Chertkov O, Lindquist EA, Detter C, Grigoriev IV, Kamoun S, Kingsmore SF (2012) Genome sequencing and mapping reveal loss of heterozygosity as a mechanism for rapid adaptation in the vegetable pathogen *Phytophthora capsici*. Mol Plant-Microbe Interact 25(10):1350–1360

Lee S, Mian MA, McHale LK, Wang H, Wijeratne AJ, Sneller CH, Dorrance AE (2013) Novel quantitative trait loci for partial resistance to *Phytophthora sojae* in soybean PI 398841. Theor Appl Genet. doi:10.1007/s00122-013-2040-x

Li A, Wang Y, Tao K, Dong S, Huang Q, Dai T, Zheng X (2010a) PsSAK1, a stress-activated MAP kinase of *Phytophthora sojae*, is required for zoospore viability and infection of soybean. Mol Plant Microbe Interact 23(8):1022–1031. doi:10.1094/MPMI-23-8-1022

Li X, Han Y, Teng W, Zhang S, Yu K, Poysa V, Anderson T, Ding J, Li W (2010b) Pyramided QTL underlying tolerance to *Phytophthora* root rot in mega-environments from soybean cultivars 'Conrad' and 'Hefeng 25'. Theor Appl Genet 121(4):651–658. doi:10.1007/s00122-010-1337-2

Li D, Zhao Z, Huang Y, Lu Z, Yao M, Hao Y, Zhai C, Wang Y (2013) PsVPS1, a dynamin-related protein, is involved in cyst germination and soybean infection of *Phytophthora sojae*. PLoS ONE 8(3):e58623. doi:10.1371/journal.pone.0058623

Li S, Nosenko T, Hackett JD, Bhattacharya D (2006) Phylogenomic analysis identifies red algal genes of endosymbiotic origin in the chromalveolates. Mol Biol Evol 23(3):663–674

Lin F, Zhao M, Ping J, Johnson A, Zhang B, Abney TS, Hughes TJ, Ma J (2013) Molecular mapping of two genes conferring resistance to *Phytophthora sojae* in a soybean landrace PI 567139B. Theor Appl Genet 126(8):2177–2185. doi:10.1007/s00122-013-2127-4

Liu TL, Ye WW, Ru YY, Yang XY, Gu B, Tao K, Lu S, Dong SM, Zheng XB, Shan WX, Wang YC, Dou DL (2011) Two host cytoplasmic effectors are required for pathogenesis of *Phytophthora sojae* by suppression of host defenses. Plant Phys 155(1):490–501

Lu S, Chen L, Tao K, Sun N, Wu Y, Lu X, Wang Y, Dou D (2013) Intracellular and extracellular phosphatidylinositol 3-phosphate produced by *Phytophthora* species are important for infection. Mol Plant. doi:10.1093/mp/sst047

MacGregor T, Bhattacharyya M, Tyler BM, Bhat R, Schmitthenner AF, Gijzen M (2002) Genetic and physical mapping of *Avr1a* in *Phytophthora sojae*. Genetics 160:949–959

Mali P, Yang L, Esvelt KM, Aach J, Guell M, DiCarlo JE, Norville JE, Church GM (2013) RNA-guided human genome engineering via Cas9. Science 339(6121): 823–826. doi:10.1126/science.1232033

Mao Y, Tyler BM (1996) The *Phytophthora sojae* genome contains tandem repeat sequences which vary from strain to strain. Fungal Genet Biol 20(1):43–51

Mattinen L, Tshuikina M, Mae A, Pirhonen M (2004) Identification and characterization of Nip, necrosis-inducing virulence protein of *Erwinia carotovora* subsp. *carotovora*. Mol Plant Microbe Interact 17(12):1366–1375

May KJ, Whisson SC, Zwart RS, Searle IR, Irwin JAG, Maclean DJ, Carroll BJ, Drenth A (2002) Inheritance and mapping of eleven avirulence genes in *Phytophthora sojae*. Fungal Genet Biol 37:1–12

McLeod A, Fry BA, Zuluaga AP, Myers KL, Fry WE (2008) Toward improvements of oomycete transformation protocols. J Eukaryot Microbiol 55(2):103–109. doi:10.1111/j.1550-7408.2008.00304.x

Mideros S, Nita M, Dorrance AE (2007) Characterization of components of partial resistance, *Rps2*, and root resistance to *Phytophthora sojae* in soybean. Phytopathology 97(5):655–662

Morris PF, Ward EWB (1992) Chemoattraction of zoospores of the soybean pathogen, *P. sojae*, by isoflavones. Phys Mol Plant Pathol 40:17–22

Morris PF, Schlosser LR, Onasch KD, Wittenschlaeger T, Austin R, Provart N (2009) Multiple horizontal gene transfer events and domain fusions have created novel regulatory and metabolic networks in the oomycete genome. PLoS ONE 4(7):e6133. doi:10.1371/journal.pone.0006133

Moy P, Qutob D, Chapman BP, Atkinson I, Gijzen M (2004) Patterns of gene expression upon infection of soybean plants by *Phytophthora sojae*. Mol Plant-Microbe Interact 17(10):1051–1062

Na R, Yu D, Qutob D, Zhao J, Gijzen M (2013) Deletion of the *Phytophthora sojae* avirulence gene *Avr1d* causes gain of virulence on *Rps1d*. Mol Plant Microbe Interact 26(8):969–976. doi:10.1094/MPMI-02-13-0036-R

Na R, Yu D, Chapman BP, Zhang Y, Kuflu K, Austin R, Qutob D, Zhao J, Wang Y, Gijzen M (2014) Genome re-sequencing and functional analysis places the *Phytophthora sojae* avirulence genes *Avr1c* and *Avr1a* in a tandem repeat at a single locus. PLoS ONE 9(2):e89738

Nguyen VT, Vuong TD, VanToai T, Lee JD, Wu X, Rouf Mian MA, Dorrance AE, Shannon JG, Nguyen HT (2012) Mapping of quantitative trait loci associated with resistance to *Phytophthora sojae* and flooding tolerance in soybean. Crop Sci 52(6):2481–2493

Nunes CC, Dean RA (2012) Host-induced gene silencing: a tool for understanding fungal host interaction and for developing novel disease control strategies. Mol Plant Pathol 13(5):519–529. doi:10.1111/j.1364-3703.2011.00766.x

Oh SK, Young C, Lee M, Oliva R, Bozkurt TO, Cano LM, Win J, Bos JI, Liu HY, van Damme M, Morgan W, Choi D, Van der Vossen EA, Vleeshouwers VG, Kamoun S (2009) In planta expression screens of *Phytophthora infestans* RXLR effectors reveal diverse phenotypes, including activation of the *Solanum bulbocastanum* disease resistance protein Rpi-blb2. Plant Cell 21(9):2928–2947. doi:10.1105/tpc.109.068247

Ottmann C, Luberacki B, Kufner I, Koch W, Brunner F, Weyand M, Mattinen L, Pirhonen M, Anderluh G, Seitz HU, Nurnberger T, Oecking C (2009) A common toxin fold mediates microbial attack and plant defense. Proc Natl Acad Sci USA 106(25):10359–10364

Park J, Park B, Jung K, Jang S, Yu K, Choi J, Kong S, Kim S, Kim H, Kim JF, Blair JE, Lee K, Kang S, Lee YH (2008) CFGP: a web-based, comparative fungal genomics platform. Nucleic Acids Res 36 (Database issue):D562–571. doi:10.1093/nar/gkm758

Qiao Y, Liu L, Xiong Q, Flores C, Wong J, Shi J, Wang X, Liu X, Xiang Q, Jiang S, Zhang F, Wang Y, Judelson HS, Chen X, Ma W (2013) Oomycete pathogens encode RNA silencing suppressors. Nat Genet 45(3):330–333. doi:10.1038/ng.2525

Qutob D, Chapman BP, Gijzen M (2013) Transgenerational gene silencing causes gain of virulence in a plant pathogen. Nat Commun 4:1349. doi:10.1038/ncomms2354

Qutob D, Hraber P, Sobral B, Gijzen M (2000) Comparative analysis of expressed sequences in *Phytophthora sojae*. Plant Phys 123(1):243–253

Qutob D, Kamoun S, Gijzen M (2002) Expression of a *Phytophthora sojae* necrosis-inducing protein occurs during transition from biotrophy to necrotrophy. Plant J 32(3):361–373

Qutob D, Kemmerling B, Brunner F, Kufner I, Engelhardt S, Gust AA, Luberacki B, Seitz HU, Stahl D, Rauhut T, Glawischnig E, Schween G, Lacombe B, Watanabe N, Lam E, Schlichting R, Scheel D, Nau K, Dodt G, Hubert D, Gijzen M, Nurnberger T (2006) Phytotoxicity and innate immune responses induced by Nep1-like proteins. Plant Cell 18(12):3721–3744

Qutob D, Tedman-Jones J, Dong S, Kuflu K, Pham H, Wang Y, Dou D, Kale SD, Arredondo FD, Tyler BM, Gijzen M (2009) Copy number variation and transcriptional polymorphisms of *Phytophthora sojae* RXLR effector genes *Avr1a* and *Avr3a*. PLoS ONE 4(4):e5066

Raffaele S, Farrer RA, Cano LM, Studholme DJ, MacLean D, Thines M, Jiang RH, Zody MC, Kunjeti SG, Donofrio NM, Meyers BC, Nusbaum C, Kamoun S (2010) Genome evolution following host jumps in the Irish potato famine pathogen lineage. Science 330(6010):1540–1543

Ranathunge K, Thomas RH, Fang XX, Peterson CA, Gijzen M, Bernards MA (2008) Soybean root suberin and partial resistance to root rot caused by *Phytophthora sojae*. Phytopathology 98(11):1179–1189. doi:10.1094/phyto-98-11-1179

Rentel MC, Leonelli L, Dahlbeck D, Zhao B, Staskawicz BJ (2008) Recognition of the *Hyaloperonospora parasitica* effector ATR13 triggers resistance against oomycete, bacterial, and viral pathogens. Proc Natl Acad Sci USA 105(3):1091–1096. doi:10.1073/pnas.0711215105

Richards TA, Soanes DM, Jones MD, Vasieva O, Leonard G, Paszkiewicz K, Foster PG, Hall N, Talbot NJ (2011) Horizontal gene transfer facilitated the evolution of plant parasitic mechanisms in the oomycetes. Proc Natl Acad Sci USA 108(37):15258–15263. doi:10.1073/pnas.1105100108

Richards TA, Talbot NJ (2007) Plant parasitic oomycetes such as *Phytophthora* species contain genes derived from three eukaryotic lineages. Plant Signal Behav 2(2):112–114

Robideau GP, De Cock AW, Coffey MD, Voglmayr H, Brouwer H, Bala K, Chitty DW, Desaulniers N, Eggertson QA, Gachon CM, Hu CH, Kupper FC, Rintoul TL, Sarhan E, Verstappen EC, Zhang Y, Bonants PJ, Ristaino JB, Levesque CA (2011) DNA barcoding of oomycetes with cytochrome c oxidase subunit I and internal transcribed spacer. Mol Ecol Resour 11(6):1002–1011. doi:10.1111/j.1755-0998.2011.03041.x

Sandhu D, Gao H, Cianzio S, Bhattacharyya MK (2004) Deletion of a disease resistance nucleotide-binding-site leucine-rich- repeat-like sequence is associated with the loss of the *Phytophthora* resistance gene *Rps4* in soybean. Genetics 168(4):2157–2167. doi:10.1534/genetics.104.032037, 168/4/2157

Schmitthenner AF (1985) Problems and progress in control of *Phytophthora* root rot of soybean. Plant Dis 69(4):362–368

Schmitthenner AF (2000) Phytophthora Rot of Soybean. In: Hartman GL, Sinclair JB, and Rupe JC (eds) Compendium of soybean diseases, 4th Edn, 1999. The American Phytopathological Society, St. Paul, pp 39–42

Schwab R, Ossowski S, Riester M, Warthmann N, Weigel D (2006) Highly specific gene silencing by artificial microRNAs in *Arabidopsis*. Plant Cell 18(5):1121–1133. doi:10.1105/tpc.105.039834

Shan W, Cao M, Leung D, Tyler BM (2004) The *Avr1b* locus of *Phytophthora sojae* encodes an elicitor and a regulator required for avirulence on soybean plants carrying resistance gene *Rps1b*. Mol Plant-Microbe Interact 17(4):394–403

Shen D, Liu T, Ye W, Liu L, Liu P, Wu Y, Wang Y, Dou D (2013) Gene duplication and fragment recombination drive functional diversification of a superfamily of cytoplasmic effectors in *Phytophthora sojae*. PLoS ONE 8(7):e70036. doi:10.1371/journal.pone.0070036

Sohn KH, Lei R, Nemri A, Jones JD (2007) The downy mildew effector proteins ATR1 and ATR13 promote disease susceptibility in *Arabidopsis thaliana*. Plant Cell 19(12):4077–4090

Song T-Q, Kale SD, Arredondo FD, Shen D-Y, Su L-M, Liu L, Wu Y-R, Wang Y-C, Dou D-L, Tyler BM (2013) Two RxLR avirulence genes in *Phytophthora sojae* determine soybean *Rps1k*-mediated disease resistance. Mol Plant-Microbe Interact 26(7):711–720

Stajich JE, Harris T, Brunk BP, Brestelli J, Fischer S, Harb OS, Kissinger JC, Li W, Nayak V, Pinney DF, Stoeckert CJ, Jr., Roos DS (2012) FungiDB: an integrated functional genomics database for fungi. Nucleic Acids Res 40 (Database issue):D675–681. doi:10.1093/nar/gkr918

Sugimoto T, Kato M, Yoshida S, Matsumoto I, Kobayashi T, Kaga A, Hajika M, Yamamoto R, Watanabe

K, Aino M, Matoh T, Walker DR, Biggs AR, Ishimoto M (2012) Pathogenic diversity of *Phytophthora sojae* and breeding strategies to develop *Phytophthora*-resistant soybeans. Breed Sci 61(5):511–522

Sugimoto T, Yoshida S, Kaga A, Hajika M, Watanabe K, Aino M, Tatsuda K, Yamamoto R, Matoh T, Walker DR, Biggs AR, Ishimoto M (2011) Genetic analysis and identification of DNA markers linked to a novel *Phytophthora sojae* resistance gene in the Japanese soybean cultivar Waseshiroge. Euphytica 182(1):133–145. doi:10.1007/s10681-011-0525-8

Sun F, Kale SD, Azurmendi HF, Li D, Tyler BM, Capelluto DGS (2013) Structural basis for interactions of the *Phytophthora sojae* RXLR effector Avh5 with phosphatidylinositol 3-phosphate and for host cell entry. Mol Plant-Microbe Interact 26(3):330–344. http://dx.doi.org/10.1094/MPMI-07-12-0184-R

Sun S, Wu XL, Zhao JM, Wang YC, Tang QH, Yu DY, Gai JY, Xing H (2011) Characterization and mapping of RpsYu25, a novel resistance gene to *Phytophthora sojae*. Plant Breed 130(2):139–143. doi:10.1111/j.1439-0523.2010.01794.x

Thomas R, Fang X, Ranathunge K, Anderson TR, Peterson CA, Bernards MA (2007) Soybean root suberin: anatomical distribution, chemical composition, and relationship to partial resistance to *Phytophthora sojae*. Plant Phys 144(1):299–311

Torto TA, Li S, Styer A, Huitema E, Testa A, Gow NA, van West P, Kamoun S (2003) EST mining and functional expression assays identify extracellular effector proteins from the plant pathogen *Phytophthora*. Genome Res 13(7):1675–1685

Torto-Alalibo T, Tripathy S, Smith BM, Arredondo F, Zhou L, Li H, Qutob D, Gijzen M, Mao C, Sobral BWS, Waugh ME, Mitchell TK, Dean RA, Tyler BM (2007) Expressed sequence tags from *Phytophthora sojae* reveal genes specific to development and infection. Mol Plant Microbe Interact 20(7):781–793

Tripathy S, Deo T, Tyler BM (2012) Oomycete transcriptomics database: a resource for oomycete transcriptomes. BMC Genom 13:303

Tripathy S, Pandey VN, Fang B, Salas F, Tyler BM (2006) VMD: a community annotation database for microbial genomes. Nucleic Acids Res 34:D379–D381. doi:10.1093/nar/gkj042

Tyler BM, Wu M-H, Wang J-M, Cheung WWS, Morris PF (1996) Chemotactic preferences and strain variation in the response of *Phytophthora sojae* zoospores to host isoflavones. Appl Environ Microbiol 62(8):2811–2817

Tyler BM (2002) Molecular basis of recognition between *Phytophthora* pathogens and their hosts. Annu Rev Phytopathol 40:137–167

Tyler BM (2007) *Phytophthora sojae*: root rot pathogen of soybean and model oomycete. Mol Plant Pathol 8(1):1–8

Tyler BM, Forster H, Coffey MD (1995) Inheritance of avirulence factors and restriction fragment length polymorphism markers in outcrosses of the oomycete *Phytophthora sojae*. Mol Plant-Microbe Interact 8:515–523

Tyler BM, Tripathy S, Zhang X, Dehal P, Jiang RH, Aerts A, Arredondo FD, Baxter L, Bensasson D, Beynon JL, Chapman J, Damasceno CM, Dorrance AE, Dou D, Dickerman AW, Dubchak IL, Garbelotto M, Gijzen M, Gordon SG, Govers F, Grunwald NJ, Huang W, Ivors KL, Jones RW, Kamoun S, Krampis K, Lamour KH, Lee MK, McDonald WH, Medina M, Meijer HJ, Nordberg EK, Maclean DJ, Ospina-Giraldo MD, Morris PF, Phuntumart V, Putnam NH, Rash S, Rose JK, Sakihama Y, Salamov AA, Savidor A, Scheuring CF, Smith BM, Sobral BW, Terry A, Torto-Alalibo TA, Win J, Xu Z, Zhang H, Grigoriev IV, Rokhsar DS, Boore JL (2006) *Phytophthora* genome sequences uncover evolutionary origins and mechanisms of pathogenesis. Science 313(5791):1261–1266

Tyler BM, Kale SD, Wang Q, Tao K, Clark HR, Drews K, Antignani V, Rumore A, Hayes T, Plett JM, Fudal I, Gu B, Chen Q, Affeldt KJ, Berthier E, Fischer GJ, Dou D, Shan W, Keller N, Martin F, Rouxel T, Lawrence CB (2013) Microbe-independent entry of oomycete RxLR effectors and fungal RxLR-like effectors into plant and animal cells is specific and reproducible. Mol Plant-Microbe Interact 26(7):611–616

van West P, Kamoun S, van 't Klooster JW, Govers F (1999) Internuclear gene silencing in *Phytophthora infestans*. Mol Cell 3(3):339–348

van West P, Shepherd SJ, Walker CA, Li S, Appiah AA, Grenville-Briggs LJ, Govers F, Gow NA (2008) Internuclear gene silencing in *Phytophthora infestans* is established through chromatin remodelling. Microbiology 154(Pt 5):1482–1490

Vetukuri RR, Asman AK, Tellgren-Roth C, Jahan SN, Reimegard J, Fogelqvist J, Savenkov E, Soderbom F, Avrova AO, Whisson SC, Dixelius C (2012) Evidence for small RNAs homologous to effector-encoding genes and transposable elements in the oomycete *Phytophthora infestans*. PLoS ONE 7(12):e51399

Wade M, Albersheim P (1979) Race-specific molecules that protect soybeans from *Phytophthora megasperma var. sojae*. Proc Natl Acad Sci USA 76(9):4433–4437

Wang H, Wijeratne A, Wijeratne S, Lee S, Taylor CG, St Martin SK, McHale L, Dorrance AE (2012) Dissection of two soybean QTL conferring partial resistance to *Phytophthora sojae* through sequence and gene expression analysis. BMC Genomics 13:428. doi:10.1186/1471-2164-13-428

Wang Q, Han C, Ferreira AO, Yu X, Ye W, Tripathy S, Kale SD, Gu B, Sheng Y, Sui Y, Wang X, Zhang Z, Cheng B, Dong S, Shan W, Zheng X, Dou D, Tyler BM, Wang Y (2011) Transcriptional programming and functional interactions within the *Phytophthora sojae* RXLR effector repertoire. Plant Cell 23(6):2064–2086

Wang Z, Wang Y, Zhang Z, Zheng X (2007) Genetic relationships among Chinese and American isolates of *Phytophthora sojae* by ISSR markers. Biodivers Sci 15(3):215–223

Ward EWB (1990) The interaction of soya beans with *Phytophthora megasperma* f.sp. *glycinea*: pathogenicity. In: Hornby D (ed) Biological control of soil-borne plant pathogens. CAB International, Wallingford, pp 311–327

Weng C, Yu K, Andersen TR, Poysa V (2007) A quantitative trait locus influencing tolerance to *Phytophthora* root rot in the soybean cultivar 'Conrad'. Euphytica 158:81–86

Whisson S, Vetukuri R, Avrova A, Dixelius C (2012) Can silencing of transposons contribute to variation in effector gene expression in *Phytophthora infestans*? Mob Genet Elem 2(2):110–114

Whisson SC, Avrova AO, VANW P, Jones JT (2005) A method for double-stranded RNA-mediated transient gene silencing in *Phytophthora infestans*. Mol Plant Pathol 6(2):153–163. doi:10.1111/j.1364-3703.2005.00272.x

Whisson SC, Boevink PC, Moleleki L, Avrova AO, Morales JG, Gilroy EM, Armstrong MR, Grouffaud S, West Pv, Chapman S, Hein I, Toth IK, Pritchard L, Birch PRJ (2007) A translocation signal for delivery of oomycete effector proteins into host plant cells. Nature 450:115–119

Whisson SC, Drenth A, Maclean DJ, Irwin JA (1994) Evidence for outcrossing in *Phytophthora sojae* and linkage of a DNA marker to two avirulence genes. Curr Genet 27(1):77–82

Whisson SC, Drenth A, Maclean DJ, Irwin JAG (1995) *Phytophthora sojae* avirulence genes, RAPD and RFLP markers used to construct a detailed genetic linkage map. Mol Plant-Microbe Interact 8(6):988–995

Whisson SC, Randall E, Young V, Birch PRJ, Cooke DEL, Csukai M (2013) Involvement of RNA polymerase I in mefenoxam insensitivity in *Phytophthora infestans*. In: Oomycete molecular genetics network meeting, Asilomar, California, 2013

Win J, Krasileva KV, Kamoun S, Shirasu K, Staskawicz BJ, Banfield MJ (2012) Sequence divergent RXLR effectors share a structural fold conserved across plant pathogenic oomycete species. PLoS Pathog 8(1):e1002400

Win J, Morgan W, Bos J, Krasileva KV, Cano LM, Chaparro-Garcia A, Ammar R, Staskawicz BJ, Kamoun S (2007) Adaptive evolution has targeted the C-terminal domain of the RXLR effectors of plant pathogenic oomycetes. Plant Cell 19(8):2349–2369

Wrather JA, Koenning SR (2006) Estimates of disease effects on soybean yields in the United States 2003 to 2005. J Nematol 38(2):173–180

Wu XL, Zhang BQ, Sun S, Zhao JM, Yang F, Guo N, Gai JY, Xing H (2011) Identification, genetic analysis and mapping of resistance to *Phytophthora sojae* of Pm28 in soybean. Agric Sci China 10(10):1506–1511. doi:10.1016/s1671-2927(11)60145-4

Yao HY, Wang XL, Wu X, Xiao Y, Zhu Z (2010) Molecular mapping of *Phytophthora* resistance gene in soybean cultivar Zaoshu 18. J Plant Genet Resour 2:213–217

Ye W, Wang X, Tao K, Lu Y, Dai T, Dong S, Dou D, Gijzen M, Wang Y (2011) Digital gene expression profiling of the *Phytophthora sojae* transcriptome. Mol Plant-Microbe Interact 24(12):1530–1539. doi:10.1094/MPMI-05-11-0106

Yin W, Dong S, Zhai L, Lin Y, Zheng X, Wang Y (2013) The *Phytophthora sojae Avr1d* gene encodes an RxLR-dEER effector with presence and absence polymorphisms among pathogen strains. Mol Plant Microbe Interact 26(8):958–968. doi:10.1094/MPMI-02-13-0035-R

Yu X, Tang J, Wang Q, Ye W, Tao K, Duan S, Lu C, Yang X, Dong S, Zheng X, Wang Y (2012) The RxLR effector Avh241 from *Phytophthora sojae* requires plasma membrane localization to induce plant cell death. New Phytol 196(1):247–260. doi:10.1111/j.1469-8137.2012.04241.x

Zhang J, Xia C, Wang X, Duan C, Sun S, Wu X, Zhu Z (2013a) Genetic characterization and fine mapping of the novel *Phytophthora* resistance gene in a Chinese soybean cultivar. Theor Appl Genet 126(6):1555–1561. doi:10.1007/s00122-013-2073-1

Zhang X, Xu Z, Tripathy S, Lee M-K, Scheuring C, Ko A, Tian K, Arredondo F, Zhang H-B, Tyler BM (2006) An integrated BAC and genome sequence physical map of *Phytophthora sojae*. Mol Plant-Microbe Interact 19(12):1302–1310

Zhang Y, Zhang F, Li X, Baller JA, Qi Y, Starker CG, Bogdanove AJ, Voytas DF (2013b) Transcription activator-like effector nucleases enable efficient plant genome engineering. Plant Phys 161(1):20–27. doi:10.1104/pp.112.205179

Zhang J, Xia C, Duan C, Sun S, Wang X, Wu X, Zhu Z (2013c) Identification and candidate gene analysis of a novel *Phytophthora* resistance gene *Rps*10 in a Chinese soybean cultivar. PLoS ONE 8(7):e69799. doi:10.1371/journal.pone.0069799

Zhou L, Mideros SX, Bao L, Hanlon R, Arredondo FD, Tripathy S, Krampis K, Jerauld A, Evans C, St. Martin SK, Maroof SMA, Hoeschele I, Dorrance AE, Tyler BM (2009) Infection and genotype remodel the entire soybean transcriptome. BMC Genomics 10:49

Zhu ZD, Huo YL, Wang XM, Huang JB, Wu XF (2010) Molecular identification of a novel *Phytophthora* resistance gene in soybean. Acta Agron Sinica 33(2):154–157. http://en.cnki.com.cn/Article_en/CJFDTotal-XBZW200701027.htm

Zu Y, Tong X, Wang Z, Liu D, Pan R, Li Z, Hu Y, Luo Z, Huang P, Wu Q, Zhu Z, Zhang B, Lin S (2013) TALEN-mediated precise genome modification by homologous recombination in zebrafish. Nat Methods 10(4):329–331. doi:10.1038/nmeth.2374

# Phytophthora ramorum

**8**

Sydney E. Everhart, Javier F. Tabima, and Niklaus J. Grünwald

## 8.1 Introduction

### 8.1.1 Sudden Oak Death and Ramorum Blight

*Phytophthora ramorum* is a recently emerged plant pathogen and causal agent of one of the most destructive and devastating diseases currently affecting US horticulture and forests (Rizzo et al. 2002, 2005). This oomycete pathogen was discovered in Marin County, California, in the mid-1990s, causing sudden oak death on coast live oak (*Quercus agrifolia*) and tanoak (*Notholithocarpus densiflorus*) and simultaneously discovered in Europe causing foliar blight on *Rhododendron* and *Viburnum* (Rizzo et al. 2002; Werres et al. 2001). It is now known to affect more than 100 plant species, including economically important nursery and forest host species (Frankel 2008; Rizzo et al. 2005; Tooley et al. 2004; Tooley and Kyde 2007).

This pathogen has two distinct disease symptom classes (Grünwald et al. 2008). On some plant species, mostly woody ornamentals such as *Viburnum* and *Rhododendron*, symptoms are nonlethal and show foliar or twig blight, which allow prolific production of aerial sporangia. In contrast, on coast live oak, tanoak, and Japanese larch (*Larix kaempferi*), infections can be lethal and include bleeding bole cankers (Fig. 8.1). Sudden oak death has resulted in about 80 % mortality of tanoaks in portions of the Los Padres National Forest in California, killing 119,000 tanoaks across approximately 3,200 ha (Rizzo et al. 2005). The high mortality of dominant oak species and foliar blight of understory shrubs has permanently altered natural forest ecosystems in the western US. Until recently, tree infections in Europe have been comparatively rare and primarily affected native beech and non-native oak species, however an outbreak in 2009 caused widespread mortality of Japanese larch in plantations in SW England. By 2010, an outbreak in Wales led to approximately half a million trees felled over 1,300 ha, and by 2013 the most recent surveys indicate the outbreak has expanded to an additional 1,800 ha of newly infected trees in south Wales (COMTF 2013).

Infested nursery plants offer a very effective means of dispersing the pathogen. This has happened twice in the United States with shipments of infected camellias from California that resulted in 1.6 million potentially infected plants detected in 175 infested sites in over 20 states (see California Oak Mortality Task Force web site http://www.suddenoakdeath.org for current status). State, national, and international quarantines have been imposed on all host plant species grown in affected areas, including eradication of affected

S. E. Everhart · J. F. Tabima · N. J. Grünwald (✉)
Department of Botany and Plant Pathology, Oregon State University, Corvallis, OR 97331, USA
e-mail: grunwaln@science.oregonstate.edu

N. J. Grünwald
Horticultural Crops Research Laboratory, USDA ARS, Corvallis, OR 97331, USA

R. A. Dean et al. (eds.), *Genomics of Plant-Associated Fungi and Oomycetes: Dicot Pathogens*,
DOI: 10.1007/978-3-662-44056-8_8, © Springer-Verlag Berlin Heidelberg (outside the USA) 2014

**Fig. 8.1** **a** Sudden oak death epidemic on tanoak in Marin County, California (*courtesy* S. Frankel). **b** Sudden oak death symptoms showing necrosis found beneath bleeding cankers produced on tanoak in native forests (*courtesy* S. Everhart)

nursery stock. Similarly, *P. ramorum* has been reported in 21 European countries, where emergency phytosanitary measures have been implemented since 2002 for member countries of the European Union (Walters et al. 2009). These trade regulations and phytosanitary measures can directly impact commercial nurseries and retailers. For example, horticultural nurseries across the US have lost millions of dollars from destruction of infected stock and suffer further losses from disrupted and lost markets. Furthermore, combining direct loss of nursery and ornamental crops, decrease in property values with dead/dying trees, and the costs of disease tracking and management, total economic losses are in the tens of millions of dollars (Cave et al. 2005; Frankel 2008). Risk analysis for the US has shown that if the pathogen spreads to forests on the East Coast as well as into new production systems on the West Coast, the economic impact of quarantine regulations on trade in conifer and hardwood products, logs, Christmas trees, and tree seedlings will be even greater (Cave et al. 2005; Frankel 2008). Rapid spread of disease since the mid-1990s and simultaneous discovery in Europe led to intensive research efforts to characterize the pathogen, which resulted in whole-genome sequencing only 3 years after formal description of the pathogen in 2001.

### 8.1.2 Taxonomy

Formally described in 2001, *P. ramorum* is a filamentous, diploid protozoan that is one of 117 currently recognized *Phytophthora* species commonly known as water molds. *Phytophthora* species are in the phylum Oomycota, which includes several notable plant pathogens, such as *Phytophthora infestans, P. sojae, P. cinnamomi, P. capsici,* and *Pythium* spp. This group is a member of the Stramenopiles and is most closely related to the golden-brown algae and diatoms. Current phylogenetic analysis of *Phytophthora* species based on seven nuclear loci places *P. ramorum* in clade 8c (Blair et al. 2008). The other close relatives of *P. ramorum in* clade 8c include *P. lateralis* that causes root rot of Port Orford-cedar (*Chamaecyparis lawsoniana*), *P. hibernalis* that primarily causes citrus brown rot and leaf/twig blight, and *P. foliorum* that causes leaf spot of *Rhododendron* spp. (Fig. 8.2). Although the origin of *P. ramorum* is unknown, the discovery of *P. lateralis* in Taiwan suggests that the origin of this clade may be in Eastern Asia (Brasier et al. 2010), but more direct evidence in support of this hypothesis is needed.

### 8.1.3 Life Cycle

In nature, the life cycle of *P. ramorum* currently includes only an asexual phase although a sexual phase is theoretically possible, where asexual spores include sporangia and chlamydospores and the sexual phase yields oospores. Asexual sporangia are produced from infected leaf tissue and germinate directly, producing a germ tube and appressorium to infect the host tissue, or indirectly by release of zoospores in the presence

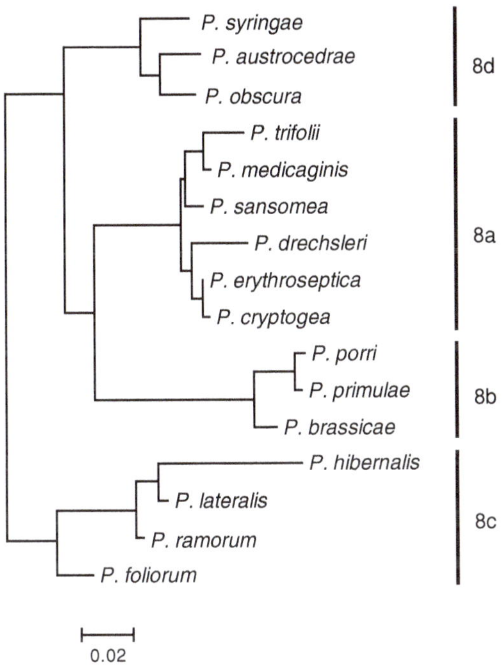


**Fig. 8.2** Maximum likelihood phylogeny for *Phytophthora* clade 8 based on ITS sequence (adapted from Grünwald et al. 2012b). *Phytophthora ramorum* is a clade 8c taxon with *P. lateralis*, *P. hibernalis*, and *P. foliorum* as sister-taxa

of free water. *P. ramorum* readily produces thick-walled chlamydospores that can serve both as primary inoculum and provide a means of surviving adverse conditions such as drought. When favorable conditions return, the chlamydospore will germinate directly to infect the host or produce sporangia. *P. ramorum* is readily distinguished from other *Phytophthora* species through its formation of abundant, large chlamydospores (Werres et al. 2001). *P. ramorum* is heterothallic (out-crossing) and sexual mating in the laboratory, thus requires contact between individuals of compatible mating types, A1 and A2. Based on knowledge about other species, we expect that pheromones released by the opposite mating type induce formation of mating structures, oogonia and antheridia. Fertilization of oogonia by antheridia, followed by karyogamy, would lead to the development of oospores; however, this final process is not observed in nature. This species is currently considered solely clonal in its known range. Molecular evidence of sexual recombination has not been detected and laboratory pairings among different mating types only leads to occasional formation of gametangia and, where meiosis has occurred, leads to aneuploid progeny of reduced fitness (Boutet et al. 2010, Vercauteren et al. 2011, Van Poucke et al. 2012). Oospores produced under controlled laboratory conditions appear aberrant and alleles do not segregate in a Mendelian fashion (Boutet et al. 2010; Brasier and Kirk 2004; Vercauteren et al. 2011).

### 8.1.4 Genetics

Like many *Phytophthora* species, *P. ramorum* is a diploid, clonal organism. There are four genetically and phenotypically distinct clonal lineages of *P. ramorum* (NA1, NA2, EU1, and EU2) that were named after the continent (North America or Europe) where each was first detected. Genetic variation that discriminates the four clonal lineages include AFLP, SSR microsatellites, and SNPs, whereas ITS and isozyme profiles are indistinct, consistent with the conspecific nature of the lineages. Phenotypic traits associated with fitness, such as growth rate, aggressiveness, and colony stability have been shown to vary with respect to lineage, where EU1 and NA2 are considered more aggressive than NA1, which also showed uniform and irregular growth types that are also called non-wild type (*nwt*) or senescent phenotypes (Braiser et al. 2006; Elliott et al. 2011).

In North America, three lineages are present, where NA1 genotypes are by far the most frequent and diverse in the US (Goss et al. 2009a, b; Ivors et al. 2006; Ivors et al. 2004; Mascheretti et al. 2008; Prospero et al. 2007, 2009). In Europe, the pathogen exists as the EU1 clonal lineage (Ivors et al. 2006; Vercauteren et al. 2010) and more recently, a fourth clonal lineage, EU2, was identified in Northern Ireland and subsequently in Western Scotland, causing epidemic disease on Japanese larch trees (*Larix kaempferi*; Van Poucke et al. 2012). Population genetic analyses using microsatellite markers have been used to characterize population

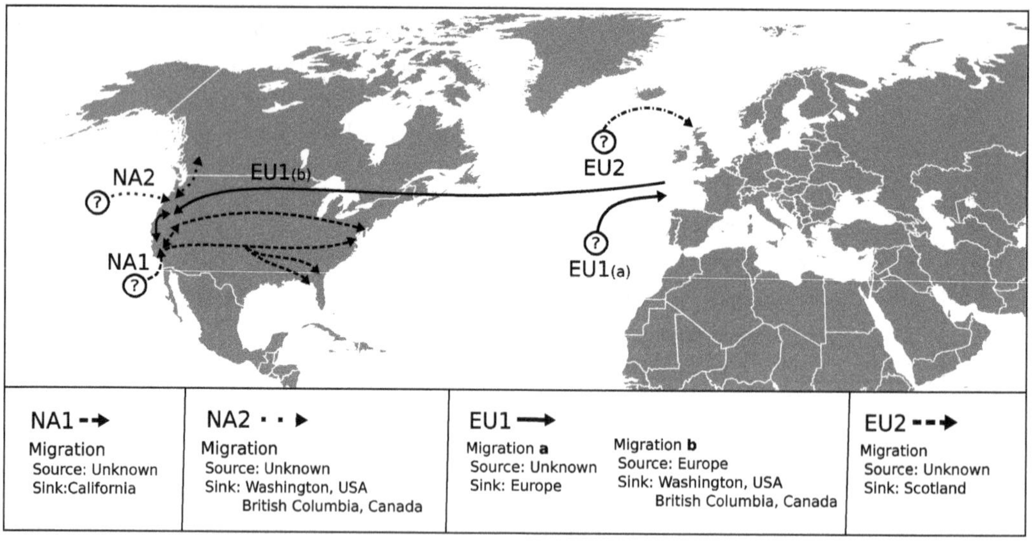


**Fig. 8.3** Scenarios depicting repeated emergence and migration of the sudden oak death pathogen *P. ramorum* (after Grünwald et al. 2012). Five intercontinental migrations of *P. ramorum* are supported by population genetic and evolutionary studies. Shown are the most likely scenarios for repeated introduction of the three known clonal lineages *NA1* and *NA2* into North America, *EU1* into North America and Europe, and *EU2* into Europe

structure and migration of the pathogen in North America and Europe (Fig. 8.3). Microsatellite genetic markers have been used to examine the population structure in the United States after a multi-state outbreak in 2004, with supporting evidence provided by trace-forward shipping records (Goss et al. 2009b), and also used to determine the direction and rate of migration within North America and from Europe by inferring migration using coalescent and Bayesian approaches (Goss et al. 2011). Finally, although the population structure of the EU2 lineage has not been explored, the relatively lower amount of genetic variation as compared to the EU1 lineage suggests that introduction has been more recent (Van Poucke et al. 2012).

## 8.2 Genome

### 8.2.1 Genome Structure

Whole-genome shotgun sequencing was performed on strain Pr-102 (ATCC MYA-2949). This strain was isolated in 2004 from a coast live oak in Marin County, CA, and was selected because it had the same multilocus microsatellite genotype as the majority of strains genotyped previously (Ivors et al. 2004). Sequence data was generated from a combination of paired-end reads of small and medium insert plasmids (2–4 and 8 kb) and large insert fosmids (36 kb), with an estimated sequence depth of 7.7 fold coverage (Tyler et al. 2006). Over 1 million reads were assembled with the JGI assembler, Jazz, which formed 2,576 scaffolds (N50 = 308 kb) with total scaffold length of 66.7 Mb, and is available for download via the JGI website for *P. ramorum* (http://genome.jgi-psf.org/Phyra1_1/Phyra1_1.info.html).

The number of chromosomes in *P. ramorum* is unknown, but the genome structure of *P. ramorum* is similar to that of other *Phytophthora* spp., with a large portion of conserved, syntenic, gene-dense regions (Haas et al. 2009). Compared to *P. sojae* and *P. infestans,* there are 8,492 orthologous clusters in the *P. ramorum* genome that contains 9,664 genes that are orthologues or close paralogues, of which, 7,113 are strict orthologues (1:1:1), thus comprising a core genome. This region of the genome encodes genes involved in cellular processes (i.e., DNA replication, transcription, and protein translation), with relatively fewer related to defense mechanisms (Haas et al. 2009). Unique regions of the genome, outside of the core genome, are thought to be related to host specificity and the lifestyle of the pathogen.

### 8.2.2 Repetitive DNA

The genomes of *Phytophthora* spp. are known to have highly repetitive regions, where the amount of repetitive DNA is correlated with the size of the genome. For example, *P. ramorum* has approximately 28 % repetitive DNA and a smaller genome as compared to *P. sojae* and *P. infestans* that have 39 and 74 % repetitive DNA, respectively (Haas et al. 2009). The amount of repetitive DNA is also unequally distributed throughout the genome, with a greater percentage of repeats found outside collinear blocks, thus suggesting that the genome size difference among species is largely due to repeat-driven genome expansion in these gene-sparse regions. Indeed, further investigations into the most likely mechanism of dramatic genome expansion in *P. infestans* showed these regions were enriched for transposons, where a striking 29 % of the genome was identified as corresponding to Gypsy-element type retrotransposons (Haas et al. 2009). The majority of these sequences show high similarity, suggesting a high rate of recent activity that has also been demonstrated experimentally (Haas et al. 2009). In addition, the most widespread of these retrotransposons shows similar GC content and codon usage with genes in *Phytophthora* species. For example, the codon usage of three dominant retrotransposons (*CopiaPr-1*, *GypsyPr-0*, and *GypsyPr-2*) showed a positive correlation with codon usage estimated from 10,000 open reading frames (ORFs) randomly selected from the *P. ramorum* genome. Codon usage is influenced by mutational bias and selection pressure, thus suggesting that genome invasion by these retrotransposons was an ancient event, prior to the divergence of *P. ramorum*, *P. sojae*, and *P. infestans* (Jiang and Govers 2006).

Interestingly, gene families encoding host-defense related effector proteins (specifically the RxLR and CRN gene families) show correlation in the number of genes and size of the flanking intergenic regions, with retrotransposons flanking RxLR-type effectors significantly more frequently than average genes (Jiang et al. 2008; Haas et al. 2009). Further research examining the evolutionary dynamics of genes within these highly repetitive regions has shown that effector genes in the *P. ramorum* genome have the expected rapid birth–death rate expected from other well-studied genomes (Goss et al. 2009a; Grünwald and Goss 2009; Tyler et al. 2006). Evidence suggests that repetitive regions in both fungal and oomycete genomes provide evolutionary advantages including, for example, more rapid adaptation to new environments or conditions as well as rapid recombination and reshuffling of existing domains into genes with novel or revised functions (Raffaele and Kamoun 2012; Haas et al. 2009).

### 8.2.3 Comparison with Other Oomycete Genomes

To date, seven oomycete plant pathogen genomes have been sequenced and published, including *P. ramorum, P. infestans, P. sojae, P. capsici, Pythium ultimum, Albugo candida,* and *Hyaloperonospora arabidopsidis* (Table 8.1). Genome sizes range from 42.8 to 240 Mb, where *P. infestans* has the largest genome, roughly four times the size of *P. ramorum*. The number of predicted genes has substantially less variation across species, ranging from approximately 14.5–19 k genes, while most of the genome plasticity is found in the repetitive, gene-sparse regions of the genomes.

Notable features of the *P. ramorum* genome are evidenced in comparisons with other sequenced *Phytophthora* species. Pairwise comparison with *P. sojae* and *P. infestans* showed

**Table 8.1** Comparison of features for Stramenopile genomes sequenced to date

| Oomycete plant pathogens | Lifestyle | Mating system | Genome size (Mb) | Predicted genes | Repetitive sequence (%) | Effector proteins | | Citation |
|---|---|---|---|---|---|---|---|---|
| | | | | | | RxLR | CRN | |
| *Phytophthora ramorum*<br>Sudden oak death and foliar blight | Hemibiotroph | Out-crossing | 65 | 15,743 | 28 | 350 | 19 | Tyler et al. (2006) |
| *Phytophthora sojae*<br>Root/stem rot of soybean | Hemibiotroph | Selfing | 95 | 19,027 | 39 | 350 | 100 | Tyler et al. (2006) |
| *Phytophthora infestans*<br>Late blight of potato | Hemibiotroph | Out-crossing | 240 | 17,797 | 74 | 563 | 196 | Haas et al. (2009) |
| *Phytophthora capsici*<br>Phytophthora blight | Hemibiotroph | Out-crossing | 64 | 17,123 | 19 | 357 | 29 | Lamour et al. (2012) |
| *Pythium ultimum*<br>Damping off | Necrotroph | Selfing (?) | 42.8 | 15,323 | 7 | 0 | 26 | Lévesque et al. (2010) |
| *Hyaloperonospora arabidopsidis*<br>Downy mildew of *Arabidopsis* | Biotroph | Selfing | 100 | 14,543 | 42 | 134 | 20 | Baxter et al. (2010) |
| *Albugo candida*<br>White rust | Biotroph | Selfing | 45.3 | 15,824 | 17 | 26 | 6 | Links et al. (2011) |
| Sister-taxa in the Stramenopiles | | | | | | | | |
| *Thalassiosira pseudonana*<br>Diatom | Marine phytoplankton | Unknown | 34 | 11,242 | 2 | 0 | 0 | Armbrust et al. (2004) |
| *Phaeodactylum tricornutum*<br>Diatom | Marine phytoplankton | Unknown | 27.4 | 10,402 | 5 | 0 | 0 | Bowler et al. (2008) |

approximately 37 Mb of the *P. ramorum* genome falls into collinear blocks (Haas et al. 2009). These collinear regions have higher gene density (median intergenic spacing 633 bp inside vs. 1.5 kb outside collinear blocks) and fewer repetitive elements (13 vs. 56 %). Despite several local rearrangements and non-orthologous genes, *P. ramorum* was found to have an expanded/diverse NPP1 gene family with 40 genes, whereas several fungal plant pathogens contain only 2–4 genes (Tyler et al. 2006). In contrast to RxLR and CRN proteins that function during the biotrophic phase of invasion, NPP1 proteins induce cell death and are part of the necrotrophic phase. *Phytophthora ramorum* is a hemibiotroph, which means that the initial infection process involves host invasion, followed by a switch to necrotrophy after establishment. This dual-form lifestyle was evidenced in the genome in the form of genes encoding cell-wall degrading enzymes (hydrolases and proteases) and may also be responsible for the diversity of the NPP1 gene family.

*Phytophthora ramorum* has more than 27 times the number of heterozygous single nucleotide polymorphic sites as *P. sojae*, which has only 499. This is likely attributed to their differing modes of reproduction, where *P. ramorum* is at least ancestrally out-crossing (heterothallic) and *P. sojae* is inbreeding (homothallic; Tyler et al. 2006; Goss et al. 2009a).

### 8.2.4 Evolutionary History and Clonality

*Phytophthora ramorum* emerged only recently as a plant pathogen in the Western hemisphere, but evolutionary analysis of the known clonal lineages indicates an ancient divergence from ancestrally sexual populations occurred approximately 150,000–500,000 years ago based on very crude assumptions given the lack of a molecular clock for oomycetes (Goss et al. 2009a). Both the European and US clonal lineages are gradually diverging from the ancestral, invasive NA1, NA2, or EU1 clones in a process of clonal divergence that is driven by the accumulation of random mutations and genetic drift, although selection acting on adaptive mutations while not documented to date might also be active (Goss et al. 2009a, 2011; Vercauteren et al. 2010; Fig. 8.3). This pattern is apparent in all lineages as a result of using hypervariable loci for genotyping but is most prominent for NA1, possibly due to the larger population size (Goss et al. 2009a, 2011; Mascheretti et al. 2008, 2009; Prospero et al. 2007, 2009; Vercauteren et al. 2010). Despite the emergence of new genotypes in forest ecosystems, there is no clear evidence that selection is favoring any genotypes in established infestations, where dominant genotypes have remained unchanged over the last 10 years and minor genotypes emerge and disappear stochastically due to drift (Goss et al. 2009b). Changes in dominance of genotypes in new forest and nursery infestations have been documented and have been ascribed to founder effects during establishment, followed by genetic drift (Goss et al. 2009a; Mascheretti et al. 2009; Prospero et al. 2007).

### 8.2.5 Effectors

*Phytophthora* genomes have several classes of genes known as effectors that are involved in host-pathogen coevolution and provide signatures of selection (Kamoun 2006; Tyler 2009; Tyler et al. 2006). Effectors are small, secreted proteins produced by the pathogen that target host plant molecules and/or alter host plant processes. General types of effectors include cytoplasmic effectors, such as the RxLR and CRN ("crinkler") gene families and apoplastic effectors that include several hydrolytic enzymes (proteases, lipases, glycosylases, enzyme inhibitors for host defenses, and necrotizing toxins). These families of proteins were discovered after availability of whole-genome sequences and since then have generated a large amount of scientific interest and research in effector biology.

The RxLR proteins, a superfamily of approximately 350 genes in the *P. ramorum* genome, is the largest class of host-translocated proteins and are typically found in areas of the genome that are highly repetitive. The amino acids motif RxLR (Arg-X-Leu-Arg) and in some but not all cases dEER (Asp-Glu-Glu-Arg, with variability in the Asp) located in the N-terminal region of the proteins are highly conserved and expected signatures of this class of effectors (Tyler et al. 2006). Members of this gene class have been cloned and validated to be avirulence genes corresponding to host R genes, including *Avr1b-1* from *P. sojae* (Shan et al. 2004), *Avr3a* from *P. infestans* (Armstrong et al. 2005), and *ATR1* (Rehmany et al. 2005) and *ATR13* (Allen et al. 2004) from *H. arabidopsidis*. Figure 8.4 shows the species-specific birth and death process typically observed for effectors that are rapidly coevolving in an evolutionary arms race. While 35 effectors are conserved across *P. infestans, P. sojae,* and *P. ramorum,* individual clades show species-specific expansions that are thought to be the result of recombination and gene duplication.

Another group of important host-translocated effector proteins belong to the crinkler gene superfamily (CRN), which received this name because these proteins were first shown to cause crinkling and necrosis of leaves when overexpressed in the plant *Nicotiana benthamiana* (Torto et al. 2003). This group contains more than 60 different gene families found in all species of oomycetes and are characterized by the presence of a LXFLAK amino acid domain (Leu-X-Phe-Leu-Ala-Lys; Kamoun 2007; Schornack et al. 2010). These genes encode modular proteins that are translocated into the cytoplasm and localize to the nucleus of the host cell by the way of an N-terminal signaling peptide and subsequently induce various responses in the host cell through a functional interaction with the C-terminal peptide (Schornack et al. 2010; Stam et al. 2013). The function of the CRN genes is not well defined, where some are known to induce cell death and function to translocate proteins into the cell that perturb host nuclear processes. CRN genes are abundant and diverse within the genus

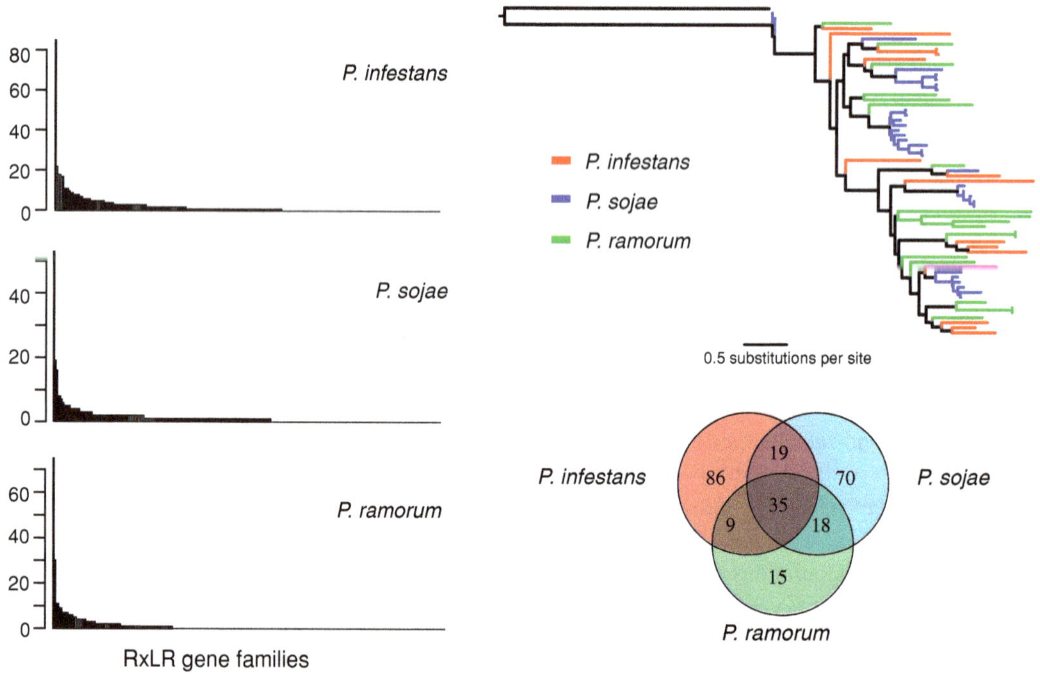


**Fig. 8.4** RxLR family genes in *P. ramorum* (*green*) compared to *P. infestans* (*red*) and *P. sojae* (*blue*). The distribution of genes within RxLR gene families shows *P. ramorum* has a lower number of genes and RxLR families as compared to *P. infestans* and *P. sojae*. Phylogenetic analysis of the second most abundant RxLR family (RxLR Family 2) shows a diverse relationship among genes belonging to each *Phytophthora* species. The number of unique and common families among the three species shows a core set of 35 genes that is shared and *P. ramorum* has the lowest proportion of unique RxLR gene families

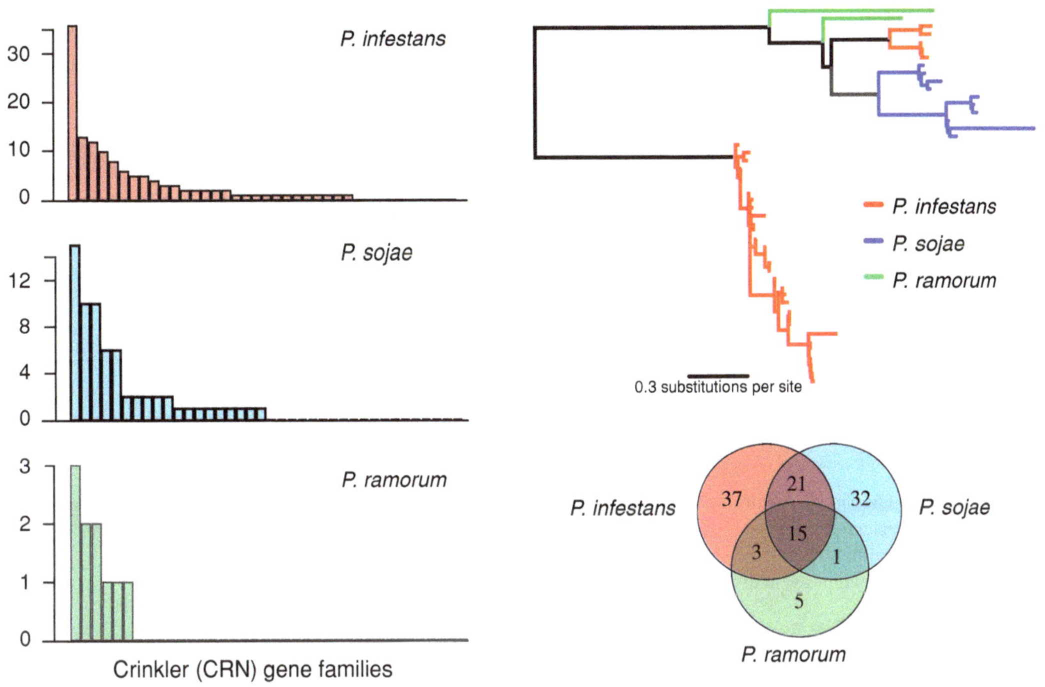


**Fig. 8.5** Crinkler family genes in *P. ramorum* (*green*) compared to *P. infestans* (*red*) and *P. sojae* (*blue*). The distribution of genes within CRN gene families shows *P. ramorum* has fewer genes than *P. infestans* and *P. sojae*. Phylogenetic analysis of the second most abundant CRN family (LFLAKdom DWL DXZ CRN family) shows two divergent sets of CRN family genes, where one branch of diverse CRN genes belong only to *P. infestans* and the other branch contains related genes found within all three species. There is a core set of 15 genes shared among these three species, where *P. ramorum* has only five genes not found in the other species

*Phytophthora*, with 76, 69, and 24 genes reported within *P. infestans*, *P. sojae*, and *P. ramorum*, respectively (Fig. 8.5). The origin of these genes, as shown in Fig. 8.5, likely first arose within a common ancestor and later independently diversified within each species. Interestingly, *P. infestans* shows a unique clade only observed in that species.

The necrosis inducing *Phytophthora* protein (NPP1) gene family is another group of effectors that have been described in *P. ramorum*. The NPP1 gene family was discovered in 2002 as a gene associated with triggering of plant defenses by *P. parasitica* (Fellbrich et al. 2002). These genes have been found in a variety of distantly related taxonomic groups, such as oomycetes, fungi, and bacteria, suggesting an ancient origin to this gene family. In *P. ramorum*, there are more NPP1 genes than in *P. infestans* and *P. sojae* (Fig. 8.6). In contrast to RxLR and CRN proteins that function during the biotrophic phase of invasion, NPP1 proteins induce cell death and are part of the necrotrophic phase. Like the other effector gene classes NPP1 genes show an evolutionary signature of rapid birth–death differentiating species (Fig. 8.6).

### 8.2.6 Gene Expression

A striking amount of variation in colony morphology and virulence (lesion size) has been observed for *P. ramorum* in the NA1 lineage, even among isolates sharing the same multilocus genotype and lacking any known host adaptation (Huberli and Garbelotto 2011). One hypothesis to explain this is that *P. ramorum* is a generalist pathogen that responds to the host via epigenetic

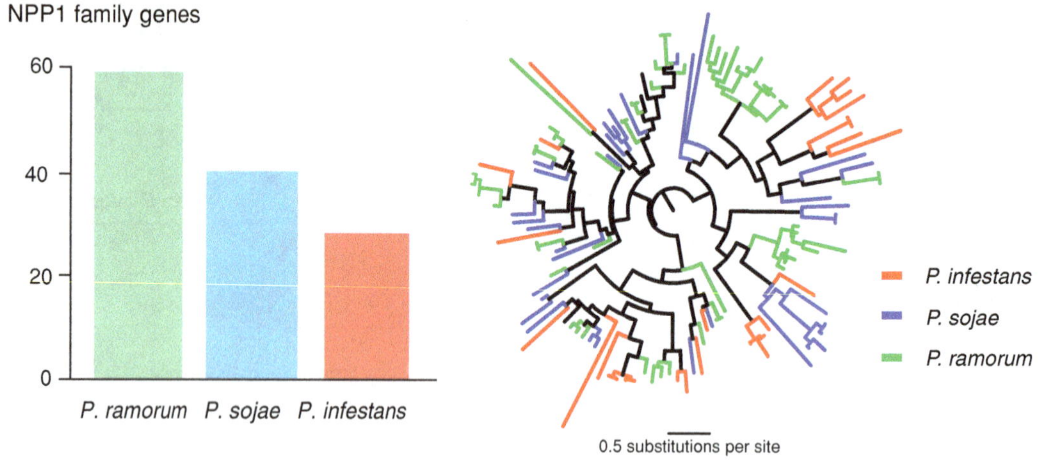


**Fig. 8.6** NPP1 genes in *P. ramorum* compared to other *Phytophthora* species. *P. ramorum* (*green*) has more NPP1 genes than either *P. sojae* (*blue*) or *P. infestans* (*red*), with nearly twice as many NPP1 genes in *P. ramorum* as *P. infestans*. Phylogenetic analysis of these genes shows a diverse relationship of NPP1 genes among the three *Phytophthora* species

regulation of gene expression, including virulence factors. To evaluate this, whole-genome expression was measured using a NimbleGen microarray designed with 61,963 probes for 15,488 gene models (approx. four probes per ORF), which was based on the published transcript of *P. ramorum* at the JGI website. Expression analysis of *P. ramorum* (NA1) following recovery from coast live oak, a dead-end host where infections are typically lethal and lead to little sporulation, was compared to the pathogen when recovered from bay laurel (*Umbellularia californica*), which is a transmissible host that yields prolific sporulation and is typically nonlethal to the host, showed significant difference in gene expression between isolates (Kasuga et al. 2012).

A total of 13 isolates were analyzed for whole-genome differential expression, where seven were from coast live oak and 6 were from bay laurel, with gene expression profiling performed on cultures grown in the lab, under dark conditions to reduce effects of circadian rhythm on gene expression. Results of gene expression profiling were first background corrected and normalized with a robust multi-array average (RMA) algorithm and then genes with low intensity were removed, thus resulting in a total of 12,516 genes used for further analysis. Isolates recovered from coast live oak (dead-end host) showed reduced virulence when subsequently inoculated onto bay laurel (transmissible host) and many also showed a senescence phenotype (non-wildtype, *nwt*). This phenotypic profile is more often derived from isolates derived from dead-end hosts and among the seven isolates involved in transcription profiling, isolates with the senescent phenotype were hierarchically clustered in a different group from all other isolates. Furthermore, isolates derived from bay laurel showed up-regulation of five CRN genes, whereas isolates recovered from coast live oak showed over 454 up-regulated genes, with more than half (297) encoding transposable elements (transposon derepression). Epigenetic regulation of transposable elements is known in eukaryotes to function via RNA interference (RNAi), heterochromatin formation, and DNA/histone methylation (Whisson et al. 2012; Zeh et al. 2009). In *P. infestans*, these mechanisms have been explored, where RNA interference via small interfering RNAs (siRNA) and heterochromatin formation are currently seen as the predominant mechanisms (Ah-Fong et al. 2008; Judelson and Tani 2007; Van West et al. 2008).

## 8.3 Applications Resulting from the Genome

### 8.3.1 Molecular Markers

Various types of molecular markers have been developed as a result of the genome sequence that have been applied for identification and detection needs and also used to examine the population variation and disease epidemiology. Markers developed for identification and detection of *P. ramorum* were developed by several groups relying on different approaches and genic regions. For example, one assay relies on PCR amplification of the *cox*I and II genes for detection and discrimination of *P. ramorum* from two common, native *Phytophthora* spp., *P. nemorosa* and *P. syringae* (Martin et al. 2004). A more sensitive method of detection was developed using quantitative real-time PCR to amplify the ITS region (Tooley et al. 2004). Yet another group developed a nested approach for detection and discrimination using quantitative real-time PCR on the ITS region (Hayden et al. 2004, 2006). To determine the best methods for routine application by state and federal agencies, a large-scale comparison of protocols was performed and the best protocols emerging from this study are now used routinely for testing samples from nursery environments in federal and state laboratories (Martin et al. 2009).

### 8.3.2 Population Biology

Following genome sequencing of *P. ramorum* in 2004 (Tyler et al. 2006), multilocus analyses confirmed the presence of distinct clonal lineages, which are readily distinguished by all molecular marker systems employed to date including amplified fragment length polymorphisms (AFLP), simple sequence repeats (SSRs), and DNA sequences for mitochondrial or nuclear loci (Grünwald et al. 2009). Microsatellite markers (SSRs) were used most prominently to document sources of contaminated material and spread of the pathogen after its emergence.

Currently, the NA1 lineage is the most widespread in North America and microsatellite markers developed from the sequenced genome (also NA1) show more clonal variation (six of nine are polymorphic) than other North American lineages, thus enabling detailed population genetic analysis (Goss et al. 2009b). The NA1 clone was most likely introduced by importation of infected nursery host plants to a local nursery in Scotts Valley, CA in the late twentieth century from an unknown source population (Mascheretti et al. 2008, 2009; Fig. 8.3). However, it became clear that the emergence of the NA1 clonal lineage has distinct dynamics in forest and nursery environments. Forest populations are genetically indistinguishable at distances of over 50 km (Mascheretti et al. 2009), which is not plausible using only natural dispersal mechanisms of large airborne and splash-dispersed sporangia (Davidson et al. 2005; Mascheretti et al. 2008). Thus, the observed population structure was likely the result of human-assisted movement of the pathogen on plants or soil (including soil on shoes or vehicle tires; Mascheretti et al. 2009). From the initial introduction foci, *P. ramorum* spread rapidly to native tanoaks (*Notholithocarpus densiflorus*) and oaks (*Quercus* spp.) presumably through foliar infection of California bay laurel, a host on which the pathogen sporulates profusely, and led to the sudden oak death epidemic in forest ecosystems. In the decade since discovery, this clonal lineage has subsequently spread to the current limits of the sudden oak death epidemic up and down the California coast (Meentemeyer et al. 2004; Fig. 8.3). Interestingly, the infestation in southern Oregon was likely the result of an introduction from infected nursery stock because this population cannot be genetically linked to any one California forest population (Mascheretti et al. 2008; Prospero et al. 2007).

The NA2 clonal lineage was first observed in 2004 in California and Washington (Ivors et al. 2006). However, it appears that this lineage was introduced to either British Columbia or Washington from an unknown source (Goss et al. 2011; Fig. 8.3). Currently, it is only rarely found in CA, more commonly found in WA and BC, and was only recently found in OR in 2012.

The EU1 clonal lineage was first detected in nurseries and established gardens in Germany and the Netherlands on ornamentals including *Rhododendron* and *Viburnum* (Werres et al. 2001). A few years later the pathogen was discovered in other areas of Europe. Currently, *P. ramorum* is present in many European countries, where it is mainly found in ornamental nurseries or gardens (Vercauteren et al. 2010). However, in 2009 there was an epidemic of 'sudden larch death' causing heavy dieback and death in plantations of Japanese larch (*Larix kaempferi*) in western Britain and Northern Ireland, leading to millions of trees being cut down. Neither the site nor the origin of the first introduction of *P. ramorum* in Europe has been determined.

The EU1 clonal lineage is also found in North America and appears to have been introduced into the Pacific Northwest (Grünwald and Goss 2011). A coalescent analysis with migration provided support for a unidirectional migration of the EU1 clonal lineages from Europe to North America (Goss et al. 2011; Grünwald and Goss 2011; Fig. 8.3). However, this effort could not establish whether the pathogen was introduced to British Columbia, Canada, or Washington, USA due to lack of power in the marker system and sample size used to infer migration routes. It has to be kept in mind that this pathogen cannot be sampled systematically like other pathogens as it is subject to eradication imposed by quarantine regulations thus making population genetic analysis more difficult. After its first introduction into the Pacific Northwest, EU1 has since migrated to California and Oregon (Grünwald et al. 2012a).

Another lineage, EU2, was recently described from Europe, first found in Northern Ireland in 2007 (Van Poucke et al. 2012). Thus far, it has only been recovered from a limited geographic region including Northern Ireland and Scotland, but has already been collected from several hosts (*Quercus robur, Larix kaempferi, Vaccinium myrtilus,* and *Rhododendron ponticum*). The EU2 mating type is the same as the EU1 lineage (A1 mating type) and genetic analysis (SSR and nuclear sequence) do not support the emergence of this lineage as the result of recent sexual or somatic recombination between lineages (Van Poucke et al. 2012). The origin of this lineage and spread from Northern Ireland to or from Scotland are currently unknown (Fig. 8.3).

In summary, it appears that the four known clonal lineages are the result of five distinct intercontinental migrations including NA1 and NA2 into North America, EU1 and EU2 into Europe, and EU1 from Europe to North America (Grünwald et al. 2012a; Fig. 8.3).

## 8.4 Future Perspectives

Rapid advances in technologies such as genome (re)sequencing and genotyping by sequencing, combined with rapidly dropping costs of using these technologies provide a bright future for studying pathogens such as *P. ramorum*. In contrast to *P. infestans* and *P. sojae*, *P. ramorum* has a very wide host range. We currently do not understand what genomic signatures and features are responsible for host adaptation. The novel sequencing technologies will provide tools for addressing host adaptation in the genus *Phytophthora*. In fact, efforts are under way for sequencing the majority of species in the genus *Phytophthora*. It is hoped that these efforts will provide novel insights into host adaptation, effector biology, and evolutionary processes in this important genus. Also underway is the use of genotyping by sequencing for large-scale, partial genome sequencing using restriction enzyme digestion prior to sequencing. This approach enables discovery of tens of thousands of SNPs in many individuals at a fraction of the cost of whole-genome resequencing. Preliminary results have found approximately 40,000 SNPs that will be interrogated for use in molecular genetic tractability (Everhart and Grünwald unpublished data).

Traditional approaches of disease management based on host resistance breeding or chemical control will not work for management of sudden oak death. Resistance will not work because the host range is too large and because breeding is not economically feasible for timber species of low value such as tanoak or for horticultural crops that include hundreds of cultivars in a single genus, such as *Rhododendron.* Given the fact that there is zero tolerance for infection of ornamentals in the nursery industry, use of fungicides is similarly not recommended as it can conceal latent infections. Use of fungicides is also cost prohibitive in natural forests. Novel means of control based on transgenic approaches should thus be explored as management alternatives.

*Phytophthora ramorum* has emerged globally at least four times over about two decades given the four distinct clonal lineages recognized to date. Despite recognition of this fact, we still lack knowledge of the geographic origin of these migrants. A concerted effort is needed to explore potential centers of origin of this pathogen. Knowledge of origins will provide tools for avoiding further migrations. The migrations of *P. ramorum* are driven by the movement of ornamental plants. Federal and state governments need to rethink the trade of ornamentals if further movements of exotic pathogens such as *P. ramorum* are to be avoided. Clearly, the repeated emergence of *P. ramorum* is a wake-up call that we should heed.

## References


Ah-Fong AMV, Bormann-Chung CA, Judelson HS (2008) Optimization of transgene-mediated silencing in *Phytophthora infestans* and its association with small-interfering RNAs. Fungal Genet Biol 45:1197–1205

Allen RL, Bittner-Eddy PD, Grenville-Briggs LJ, Meitz JC, Rehmany AP, Rose LE, Beynon JL (2004) Host-parasite coevolutionary conflict between *Arabidopsis* and downy mildew. Science 306:1957–1960

Armbrust EV, Berges JA, Bowler C, Green BR, Martinez D, Putnam NH, Zhou S, Allen AE, Apt KE, Bechner M, Brzezinski MA, Chaal BK, Chiovitti A, Davis AK, Demarest MS, Detter JC, Glavina T, Goodstein D, Hadi MZ, Hellsten U, Hildebrand M, Jenkins BD, Jurka J, Kapitonov VV, Kröger N, Lau WWY, Lane TW, Larimer FW, Lippmeier JC, Lucas S, Medina M, Montsant A, Obornik M, Parker MS, Palenik B, Pazour GJ, Richardson PM, Rynearson TA, Saito MA, Schwartz DC, Thamatrakoln K, Valentin K, Vardi A, Wilkerson FP, Rokhsar DS (2004) The Genome of the Diatom Thalassiosira Pseudonana. Ecology, Evolution, and Metabolism. Science 306:79–86. doi:10.1126/science.1101156

Armstrong MR, Whisson SC, Pritchard L, Bos JIB, Venter E, Avrova AO, Rehmany AP, Bohme U, Brooks K, Cherevach I, Hamlin N, White B, Fraser A, Lord A, Quail MA, Churcher C, Hall N, Berriman M, Huang S, Kamoun S, Beynon JL, Birch PRJ (2005) An ancestral oomycete locus contains late blight avirulence gene Avr3a, encoding a protein that is recognized in the host cytoplasm. Pro Natl Acad Sci U S A 102:7766–7771

Baxter L, Tripathy S, Ishaque N, Boot N, Cabral A, Kemen E, Thines M, Ah-Fong A, Anderson R, Badejoko W, Bittner-Eddy P, Boore JL, Chibucos MC, Coates M, Dehal P, Delehaunty K, Dong S, Downton P, Dumas B, Fabro G, Fronick C, Fuerstenberg SI, Fulton L, Gaulin E, Govers F, Hughes L, Humphray S, Jiang RHY, Judelson H, Kamoun S, Kyung K, Meijer H, Minx P, Morris P, Nelson J, Phuntumart V, Qutob D, Rehmany A, Rougon-Cardoso A, Ryden P, Torto-Alalibo T, Studholme D, Wang Y, Win J, Wood J, Clifton SW, Rogers J, Van den Ackerveken G, Jones JDG, McDowell JM, Beynon J, Tyler BM (2010) Signatures of Adaptation to Obligate Biotrophy in the Hyaloperonospora arabidopsidis Genome. Science 330:1549–1551. doi:10.1126/science.1195203

Blair JE, Coffey MD, Park S-Y, Geiser DM, Kang S (2008) A multi-locus phylogeny for *Phytophthora* utilizing markers derived from complete genome sequences. Fungal Genet Biol 45:266–277

Boutet X, Vercauteren A, Heungens K, Laurent F, Chandelier A (2010) Oospores progenies from *Phytophthora ramorum.* Fungal Biol 114:369–378

Bowler C, Allen AE, Badger JH, Grimwood J, Jabbari K, Kuo A, Maheswari U, Martens C, Maumus F, Otillar RP, Rayko E, Salamov A, Vandepoele K, Beszteri B, Gruber A, Heijde M, Katinka M, Mock T, Valentin K, Verret F, Berges JA, Brownlee C, Cadoret J P, Chiovitti A, Choi CJ, Coesel S, De Martino A, Detter JC, Durkin C, Falciatore A, Fournet J, Haruta M, Huysman MJJ, Jenkins BD, Jiroutova K, Jorgensen RE, Joubert Y, Kaplan A, Kroger N, Kroth PG, La Roche J, Lindquist E, Lommer M, Martin-Jezequel V, Lopez PJ, Lucas S, Mangogna M, McGinnis K, Medlin LK, Montsant A, Secq M-PO-L, Napoli C, Obornik M, Parker MS, Petit J-L, Porcel BM, Poulsen N, Robison M, Rychlewski L, Rynearson TA, Schmutz J, Shapiro H, Siaut M, Stanley M, Sussman MR, Taylor AR, Vardi A, von Dassow P, Vyverman W, Willis A, Wyrwicz LS, Rokhsar DS, Weissenbach J, Armbrust EV, Green BR, Van de Peer Y, Grigoriev IV (2008) The Phaeodactylum genome reveals the evolutionary history of diatom genomes. Nature 456:239–244. doi:10.1038/nature07410

Brasier C, Kirk S (2004) Production of gametangia by *Phytophthora ramorum* in vitro. Mycol Res 108:823–827

Brasier CM, Vettraino AM, Chang TT, Vannini A (2010) *Phytophthora lateralis* discovered in an old growth *Chamaecyparis* forest in Taiwan. Plant Pathol 59:595–603

Brasier C, Kirk S, Rose J (2006) Differences in phenotypic stability and adaptive variation between the main European and American lineages of Phytophthora ramorum. In: Proceedings of the Third International IUFRO Working Party (S07. 02.09) Meeting: Progress in Research on Phytophthora Diseases of Forest Trees. pp. 166–173

California Oak Mortality Task Force (2013) In: Palmieri KM, Frankel SJ (eds) California oak mortality task force report. Online, September 2013

Cave GL, Randall-Schadel B, Redlin SC (2005) Risk analysis for *Phytophthora ramorum* werres, de cock & in't veld, causal agent of phytophthora canker (sudden oak death), ramorum leaf blight, and ramorum dieback. USDA-ARS, Raleigh, p 77

Davidson JM, Wickland AC, Patterson HA, Falk KR, Rizzo DM (2005) Transmission of *Phytophthora ramorum* in mixed-evergreen forest in California. Phytopathology 95:587–596

Elliott M, Sumampong G, Varga A, Shamoun SF, James D, Masri S, Grunwald NJ (2011) Phenotypic differences among three clonal lineages of Phytophthora ramorum. For. Pathol. 41:7–14. doi:10.1111/j.1439-0329.2009.00627.x

Fellbrich G, Romanski A, Varet A, Blume B, Brunner F, Engelhardt S, Felix G, Kemmerling B, Krzymowska M, Nrnberger T (2002) NPP1, a Phytophthora-associated trigger of plant defense in parsley and Arabidopsis. Plant J. 32:375–390. doi:10.1046/j.1365-313X.2002.01454.x

Frankel SJ (2008) Sudden oak death and *Phytophthora ramorum* in the USA: a management challenge. Australas Plant Pathol 37:19–25

Goss EM, Carbone I, Grünwald NJ (2009a) Ancient isolation and independent evolution of the three clonal lineages of the exotic sudden oak death pathogen *Phytophthora ramorum*. Mol Ecol 18:1161–1174

Goss EM, Larsen M, Chastagner GA, Givens DR, Grünwald NJ (2009b) Population genetic analysis infers migration pathways of *Phytophthora ramorum* in US nurseries. PLoS Pathog 5:e1000583

Goss EM, Larsen M, Vercauteren A, Werres S, Heungens K, Grünwald NJ (2011) *Phytophthora ramorum* in Canada: evidence for migration within North America and from Europe. Phytopathology 101:166–171

Grünwald NJ, Garbelotto M, Goss EM, Heungens K, Prospero S (2012a) Emergence of the sudden oak death pathogen *Phytophthora ramorum*. Trends Microbiol 20:131–138

Grünwald NJ, Goss EM (2011) Evolution and population genetics of exotic and re-emerging pathogens: novel tools and approaches. Annu Rev Phytopathol 49:249–267

Grünwald NJ, Goss EM, Ivors K, Garbelotto M, Martin FN, Prospero S, Hansen E, Bonants PJM, Hamelin RC, Chastagner G, Werres S, Rizzo DM, Abad G, Beales P, Bilodeau GJ, Blomquist CL, Brasier C, Brière SC, Chandelier A, Davidson JM, Denman S, Elliott M, Frankel SJ, Goheen EM, de Gruyter H, Heungens K, James D, Kanaskie A, McWilliams MG, Man in 't Veld W, Moralejo E, Osterbauer NK, Palm ME, Parke JL, Perez Sierra AM, Shamoun SF, Shishkoff N, Tooley PW, Vettraino AM, Webber J, Widmer TL (2009) Standardizing the nomenclature for clonal lineages of the sudden oak death pathogen, *Phytophthora ramorum*. Phytopathology 99:792–795

Grünwald NJ, Goss EM (2009) Genetics and evolution of the sudden oak death pathogen *Phytophthora ramorum*. In: Lamour KH, Kamoun S (eds) Oomycete genetics and genomics: biology, interactions, and research tools. Wiley, Hoboken, New Jersey, pp 179–196

Grünwald NJ, Goss EM, Press CM (2008) *Phytophthora ramorum*: a pathogen with a remarkably wide host-range causing sudden oak death on oaks and ramorum blight on woody ornamentals. Mol Plant Pathol 9:729–740

Grünwald NJ, Werres S, Goss EM, Taylor CR, Fieland VJ (2012b) *Phytophthora obscura* sp. nov., a new species of the novel *Phytophthora* subclade 8d. Plant Pathol 61:610–622

Haas BJ, Kamoun S, Zody MC, Jiang RHY, Handsaker RE, Cano LM, Grabherr M, Kodira CD, Raffaele S, Torto-Alalibo T, Bozkurt TO, Ah-Fong AMV, Alvarado L, Anderson VL, Armstrong MR, Avrova A, Baxter L, Beynon J, Boevink PC, Bollmann SR, Bos JIB, Bulone V, Cai G, Cakir C, Carrington JC, Chawner M, Conti L, Costanzo S, Ewan R, Fahlgren N, Fischbach MA, Fugelstad J, Gilroy EM, Gnerre S, Green PJ, Grenville-Briggs LJ, Griffith J, Grünwald NJ, Horn K, Horner NR, Hu CH, Huitema E, Jeong DH, Jones AME, Jones JDG, Jones RW, Karlsson EK, Kunjeti SG, Lamour K, Liu Z, Ma L, Maclean D, Chibucos MC, McDonald H, McWalters J, Meijer HJG, Morgan W, Morris PF, Munro CA, O'Neill K, Ospina-Giraldo M, Pinzón A, Pritchard L, Ramsahoye B, Ren Q, Restrepo S, Roy S, Sadanandom A, Savidor A, Schornack S, Schwartz DC, Schumann UD, Schwessinger B, Seyer L, Sharpe T, Silvar C, Song J, Studholme DJ, Sykes S, Thines M, van de Vondervoort PJI, Phuntumart V, Wawra S, Weide R, Win J, Young C, Zhou S, Fry W, Meyers BC, van West P, Ristaino J, Govers F, Birch PRJ, Whisson SC, Judelson HS, Nusbaum C (2009) Genome sequence and analysis of the Irish potato famine pathogen *Phytophthora infestans*. Nature 461:393–398

Hayden K, Ivors K, Wilkinson C, Garbelotto M (2006) TaqMan Chemistry for *Phytophthora ramorum* detection and quantification, with a comparison of diagnostic methods. Phytopathology 96:846–854

Hayden KJ, Rizzo D, Tse J, Garbelotto M (2004) Detection and quantification of *Phytophthora ramorum* from California forests using a real-time polymerase chain reaction assay. Phytopathology 94:1075–1083

Huberli D, Garbelotto M (2011) *Phytophthora ramorum* is a generalist plant pathogen with differences in virulence between isolates from infectious and dead-end hosts. Forest Pathol 42:8–13

Ivors K, Garbelotto M, Vries IDE, Ruyter-Spira C, Hekkert BT, Rosenzweig N, Bonants P (2006) Microsatellite markers identify three lineages of *Phytophthora ramorum* in US nurseries, yet single lineages in US forest and European nursery populations. Mol Ecol 15:1493–1505

Ivors KL, Hayden KJ, Bonants PJM, Rizzo DM, Garbelotto M (2004) AFLP and phylogenetic analyses of North American and European populations of *Phytophthora ramorum*. Mycol Res 108:378–392

Jiang RHY, Govers F (2006) Nonneutral GC3 and retroelement codon mimicry in *Phytophthora*. J Mol Evol 63:458–472

Jiang RHY, Tripathy S, Govers F, Tyler BM (2008) RXLR effector reservoir in two *Phytophthora* species is dominated by a single rapidly evolving superfamily with more than 700 members. Proc Natl Acad Sci 12:4874–4879

Judelson HS, Tani S (2007) Transgene-induced silencing of the zoosporogenesis-specific NIFC gene cluster of *Phytophthora infestans* involves chromatin alterations. Eukaryot Cell 6:1200–1209

Kamoun S (2006) A catalogue of the effector secretome of plant pathogenic oomycetes. Annu Rev Phytopathol 44:41–60

Kamoun S (2007) Groovy times: filamentous pathogen effectors revealed. Curr Opin Plant Biol 10:358–365

Kasuga T, Kozanitas M, Bui M, Huberli D, Rizzo DM, Garbelotto M (2012) Phenotypic diversification is associated with host-induced transposon derepression in the sudden oak death pathogen *Phytophthora ramorum*. PLoS ONE 7:e34728. doi:10.1371/journal.pone.0034728

Lamour KH, Mudge J, Gobena D, Hurtado-Gonzales OP, Schmutz J, Kuo A, Miller NA, Rice BJ, Raffaele S, Cano LM, Bharti AK, Donahoo RS, Finley S, Huitema E, Hulvey J, Platt D, Salamov A, Savidor A, Sharma R, Stam R, Storey D, Thines M, Win J, Haas BJ, Dinwiddie DL, Jenkins J, Knight JR, Affourtit JP, Han CS, Chertkov O, Lindquist EA, Detter C, Grigoriev IV, Kamoun S, Kingsmore SF (2012) Genome Sequencing and Mapping Reveal Loss of Heterozygosity as a Mechanism for Rapid Adaptation in the Vegetable Pathogen Phytophthora capsici. Mol. Plant. Microbe Interact. 25:1350–1360. doi:10.1094/MPMI-02-12-0028-R

Links MG, Holub E, Jiang RH, Sharpe AG, Hegedus D, Beynon E, Sillito D, Clarke WE, Uzuhashi S, Borhan MH (2011) De novo sequence assembly of Albugo candida reveals a small genome relative to other biotrophic oomycetes. BMC Genomics 12:503. doi:10.1186/1471-2164-12-503

Lévesque CA, Brouwer H, Cano L, Hamilton JP, Holt C, Huitema E, Raffaele S, Robideau GP, Thines M, Win J, Zerillo MM, Beakes GW, Boore JL, Busam D, Dumas B, Ferriera S, Fuerstenberg SI, Gachon CM, Gaulin E, Govers F, Grenville-Briggs L, Horner N, Hostetler J, Jiang RH, Johnson J, Krajaejun T, Lin H, Meijer HJ, Moore B, Morris P, Phuntmart V, Puiu D, Shetty J, Stajich JE, Tripathy S, Wawra S, van West P, Whitty BR, Coutinho PM, Henrissat B, Martin F, Thomas PD, Tyler BM, Vries RPD, Kamoun S, Yandell M, Tisserat N, Buell CR (2010) Genome sequence of the necrotrophic plant pathogen Pythium ultimum reveals original pathogenicity mechanisms and effector repertoire. Genome Biol. 11:R73. doi:10.1186/gb-2010-11-7-r73

Martin FN, Coffey MD, Zeller K, Hamelin RC, Tooley P, Garbelotto M, Hughes KJD, Kubisiak T, Bilodeau GJ, Levy L, Blomquist C, Berger PH (2009) Evaluation of molecular markers for *Phytophthora ramorum* detection and identification: testing for specificity using a standardized library of isolates. Phytopathology 99:390–403

Martin FN, Tooley PW, Blomquist C (2004) Molecular Detection of *Phytophthora ramorum*, the causal agent of sudden oak death in California, and two additional species commonly recovered from diseased plant material. Phytopathology 94:621–631

Mascheretti S, Croucher PJ, Kozanitas M, Baker L, Garbelotto M (2009) Genetic epidemiology of the sudden oak death pathogen *Phytophthora ramorum* in California. Mol Ecol 18:4577–4590

Mascheretti S, Croucher PJP, Vettraino A, Prospero S, Garbelotto M (2008) Reconstruction of the sudden oak death epidemic in California through microsatellite analysis of the pathogen *Phytophthora ramorum*. Mol Ecol 17:2755–2768

Meentemeyer R, Rizzo D, Mark W, Lotz E (2004) Mapping the risk of establishment and spread of sudden oak death in California. For Ecol Manage 200:195–214

Van Poucke K, Franceschini S, Webber JF, Vercauteren A, Turner JA, McCracken AR, Heungens K, Brasier CM (2012) Discovery of a fourth evolutionary lineage of *Phytophthora ramorum*: EU2. Fungal Biol 116:1178–1191

Prospero S, Grünwald NJ, Winton LM, Hansen EM (2009) Migration patterns of the emerging plant pathogen *Phytophthora ramorum* on the west coast of the United States of America. Phytopathology 99:739–749

Prospero S, Hansen EM, Grünwald NJ, Winton LM (2007) Population dynamics of the sudden oak death pathogen *Phytophthora ramorum* in Oregon from 2001 to 2004. Mol Ecol 16:2958–2973

Raffaele S, Kamoun S (2012) Genome evolution in filamentous plant pathogens: why bigger can be better. Nat Rev Microbiol 10:417–430

Rehmany AP, Gordon A, Rose LE, Allen RL, Armstrong MR, Whisson SC, Kamoun S, Tyler BM, Birch PRJ, Beynon JL (2005) Differential recognition of highly divergent downy mildew avirulence gene alleles by RPP1 resistance genes from two *Arabidopsis* lines. Plant Cell 17:1839–1850

Rizzo DM, Garbelotto M, Davidson JM, Slaughter GW, Koike ST (2002) *Phytophthora ramorum* as the cause of extensive mortality of *Quercus* spp. and *Lithocarpus densiflorus* in California. Plant Dis 86:205–214

Rizzo DM, Garbelotto M, Hansen EM (2005) *Phytophthora ramorum*: integrative research and management of an emerging pathogen in California and Oregon forests. Annu Rev Phytopathol 43:309–335

Schornack S, Van Damme M, Bozkurt TO, Cano LM, Smoker M, Thines M, Gaulin E, Kamoun S, Huitema E (2010) Ancient class of translocated oomycete effectors targets the host nucleus. Proc Natl Acad Sci U S A 107:17421–17426

Shan W, Cao M, Leung D, Tyler BM (2004) The Avr1b locus of *Phytophthora sojae* encodes an elicitor and a regulator required for avirulence on soybean plants carrying resistance gene Rps1b. Mol Plant Microbe Interact 17:394–403

Stam R, Jupe J, Howden AJM, Morris JA, Boevink PC, Hedley PE, Huitema E (2013) Identification and characterisation CRN effectors in *Phytophthora capsici* shows modularity and functional diversity. PLoS ONE 8:e59517. doi:10.1371/journal.pone.0059517

Tooley PW, Kyde KL (2007) Susceptibility of some eastern forest species to *Phytophthora ramorum*. Plant Dis 91:435–438

Tooley PW, Kyde KL, Englander L (2004) Susceptibility of selected Ericaceous ornamental host species to *Phytophthora ramorum*. Plant Dis 88:993–999

Torto TA, Li S, Styer A, Huitema E, Testa A, Gow NAR, Van West P, Kamoun S (2003) EST mining and functional expression assays identify extracellular effector proteins from the plant pathogen *Phytophthora*. Genome Res 13:1675–1685

Tyler BM (2009) Entering and breaking: virulence effector proteins of oomycete plant pathogens. Cell Microbiol 11:13–20

Tyler BM, Tripathy S, Zhang X, Dehal P, Jiang RHY, Aerts A, Arredondo FD, Baxter L, Bensasson D, Beynon JL, Chapman J, Damasceno CMB, Dorrance AE, Dou D, Dickerman AW, DubchakI L, Garbelotto M, Gijzen M, Gordon SG, Govers F, Grünwald NJ, Huang W, Ivors KL, Jones RW, Kamoun S, Krampis K, Lamour Kurt H, Lee MK, McDonald WH, Medina M, Meijer HJG, Nordberg EK, Maclean DJ, Ospina-Giraldo MD, Morris PF, Phuntumart V, Putnam NH, Rash S, Rose JKC, Sakihama Y, Salamov A, Savidor A, Scheuring CF, Smith BM, Sobral BWS, Terry A, Torto-Alalibo TA, Win J, Xu Z, Zhang H, Grigoriev IV, Rokhsar DS, Boore JL (2006) *Phytophthora* genome sequences uncover evolutionary origins and mechanisms of pathogenesis. Science 313:1261–1266

Van West P, Shepherd SJ, Walker CA, Li S, Appiah AA, Grenville-Briggs LJ, Govers F, Gow NAR (2008) Internuclear gene silencing in *Phytophthora infestans* is established through chromatin remodelling. Microbiology 154:1482–1490

Vercauteren A, Boutet X, D'hondt L, Bockstaele EV, Maes M, Leus L, Chandelier A, Heungens K (2011) Aberrant genome size and instability of *Phytophthora ramorum* oospore progenies. Fungal Genet Biol 48:537–543

Vercauteren A, De Dobbelaere I, Grünwald NJ, Bonants P, Van Boackstaele E, Maes M, Heungens K (2010) Clonal expansion of the Belgian *Phytophthora ramorum* populations based on new microsatellite markers. Mol Ecol 19:92–107

Walters K, Sansford C, Slawson D (2009) *Phytophthora ramorum* and *Phytophthora kernoviae* in England and Wales—Public Consultation and New Programme. In: Frankel SJ, Kliejunas JT, Palmieri KM (tech. coords) Proceedings of the sudden oak death fourth science symposium, Gen.Tech. Rep. PSW-GTR-229. USDA Forest Service, Pacific Southwest Research Station, Albany, CA, pp 6–14

Werres S, Marwitz R, Veld W, De Cock A, Bonants PJM, De Weerdt M, Themann K, Ilieva E, Baayen RP (2001) *Phytophthora ramorum* sp. nov., a new pathogen on *Rhododendron* and *Viburnum*. Mycol Res 105:1155–1165

Whisson SC, Vetukuri RR, Avrova AO, Dixelius C (2012) Can silencing of transposons contribute to variation in effector gene expression in *Phytophthora infestans*? Fungal Biol 2(2):110–114

Zeh DW, Zeh JA, Ishida Y (2009) Transposable elements and an epigenetic basis for punctuated equilibria. BioEssays 31:715–726

# Phytophthora infestans

# 9

Howard S. Judelson

## 9.1 Introduction

The oomycete *Phytophthora infestans* is one of most devastating plant pathogens known to mankind, affecting potato and tomato and feared by farmers worldwide. *P. infestans* significantly impacted human history, gave birth to the modern science of plant pathology, and contributed to our understanding of all infectious diseases. Scientists referred to *P. infestans* as a "fungus" during its first century of study, before recognizing that oomycetes belong to a distinct kingdom, the Stramenopiles. *P. infestans* has led oomycetes into the genomic era as the first for which large-scale transcriptome sequencing and expression profiling was performed, and for which reliable methods were developed for DNA-mediated transformation and functional genomics. This chapter focuses on those structural and functional genomics resources and how they have helped to resolve issues in oomycete biology, pathology, and evolution.

H. S. Judelson (✉)
Department of Plant Pathology and Microbiology, University of California, Riverside, CA 92521, USA
e-mail: howard.judelson@ucr.edu

### 9.1.1 Discovery of *P. infestans* and Its Impact on Plant Pathology

The first recorded reports of the disease of potato caused by *P. infestans* were in 1843 in the United States near Philadelphia and New York (Peterson et al. 1992). Its ability to decimate entire crops in only a few weeks caused major concern. In 1844, the disease was also found in Belgium and England, before its infamous march across Europe in 1845 (Bourke 1964). In Ireland where peasants were dependent on potatoes for nourishment, the disease caused over a million deaths from starvation and disease, and large losses also occurred on continental Europe (Zadoks 2008). The disease was called the "potato murrain" at the time, but is now named late blight which distinguishes it from early blight caused by *Alternaria solani.*

Several scientists of the day (often gentleman amateurs) such as Teschemacher in the United States, and Berkeley and Montagne in Europe, suggested that a fungus was responsible for the potato ailment and in 1845 assigned it to the genus *Botrytis* (Peterson et al. 1992; Bourke 1964). At the time, speculation that any disease was caused by a microbe was unorthodox and possibly heretical. That the "Botrytis" caused late blight was not accepted widely until the work of Anton de Bary in 1876, who showed that potato plants displayed disease symptoms only after being treated with its spores (DeBary 1876). De Bary renamed the pathogen *P. infestans,* which means "infectious plant destroyer."

R. A. Dean et al. (eds.), *Genomics of Plant-Associated Fungi and Oomycetes: Dicot Pathogens,*
DOI: 10.1007/978-3-662-44056-8_9, © Springer-Verlag Berlin Heidelberg 2014

Several factors contributed to the severity of late blight in nineteenth century Europe. Conquistadors brought potatoes from Latin America to Europe by the mid-1600s, and while initially grown as a botanical curiosity, potato was soon recognized as a valuable crop (McNeill 1999). Tubers vegetatively propagated from a narrow genetic base resulted in a dangerous situation whereby every plant in most fields was genetically near-identical and susceptible to *P. infestans*. It is assumed that the source of the epidemic was blighted tubers transported from the New World during the mid-1840s, when Europe experienced a prolonged period of unusually wet, cool weather conducive to the spread of *P. infestans*.

From the tragedy of the Great Famine grew the new science of plant pathology, and a new perspective of the natural world. Following de Bary's work, others began to report fungi capable of infecting plants, and Pasteur's new germ theory received broader acceptance. Crop protection shifted from superstitious rituals to scientifically founded approaches.

### 9.1.2 Contemporary Impact on Agriculture

Late blight continues to be a major problem on potato, and is also a significant pathogen of tomato (Fry et al. 2013). Infection occurs on leaves, stems, tubers, and tomato fruit, but not roots. Infections of petunia are reported but this has not emerged as a major disease (Rathbone et al. 2002). Damage to foliage reduces photosynthetic assimilates and decreases the quality and quantity of the harvest. Plant-to-plant transmission and lesion expansion can be especially rapid, resulting in destruction of the canopy within a week. In addition, affected tomato fruit and potato tubers can quickly rot. In potato production where harvests are often stored for months, a few blighted tubers can spread disease through an entire warehouse.

Recent estimates place the cost of potato late blight between 7 and 16 billion dollars per year, equal to about 15 % of the farm price (Haverkort et al. 2008, 2009). Control on potato and tomato relies heavily on fungicides, as plant-based resistance is limited. The past century has witnessed the appearance of new *P. infestans* strains capable of overcoming a major fungicide and defeating many once-promising resistance (*R*) genes (Fry 2008). As discussed later, analyses of *P. infestans* genomes have helped to reveal the mechanisms used by *P. infestans* to beat *R* genes, and spur strategies for developing more durable resistance.

### 9.1.3 Taxonomy and Origins of *P. infestans*

Most oomycetes resemble fungi as both display filamentous growth and heterotrophic absorptive nutrition, and occupy similar ecological niches. Nevertheless, phylogenies based on molecular markers place oomycetes in the Kingdom Stramenopila (Beakes et al. 2011). Although most *Phytophthora* spp. infect terrestrial hosts, more basal clades within the oomycetes primarily inhabit aquatic environments, as do other Stramenopiles such as brown and golden-brown algae, and diatoms. Several features distinguish oomycetes from true fungi. Oomycetes are diploid for most of their life cycles, unlike fungi which are typically haploid (Shaw and Khaki 1971). While most fungi have septated hyphae with chitin-rich cell walls, oomycetes form coenocytic cells with walls comprised mainly of the $\beta$-1,4 glucan cellulose plus $\beta$-1,3 and $\beta$-1,6 glucans (Bartnicki-Garcia and Wang 1983). Finally, oomycetes produce a unique multi-walled spherical sexual spore (oospore) that enables survival through adverse environmental conditions (Judelson and Blanco 2005).

The taxonomy of *Phytophthora* has evolved from schemes founded on morphology to phylogenetic clades based on molecular markers, with many of the markers identified from sequencing projects. Ten major clades of *Phytophthora* were proposed based on the sequences of internal transcribed spacer regions of the ribosomal DNA repeat, and later validated by multilocus phylogenies based on nuclear and mitochondrial loci (Blair et al. 2008; Cooke et al. 2000). Such analyses split some groups and grew the number of *Phytophthora* species to 120 or more, but *P. infestans* remained intact as a member of clade 1. Interestingly, that clade includes pathogens with both narrow and broad host ranges on diverse plants (Erwin and Ribeiro 1996). Economically significant members of the clade besides *P. infestans* include *P. cactorum* (pathogen of woody ornamentals and fruit trees), *P. nicotianae* (pathogen of herbaceous and woody plants), and *P. phaseoli* (pathogen of lima bean).

Gene-based phylogenetics, combined with enhanced interest in conducting field surveys, have helped identify close relatives of *P. infestans* which are grouped in clade 1, subsection c. Other members of clade 1c include *P. phaseoli* as described above, *P. mirabilis* (pathogen of four-o-clock, *Mirabilis jalapa*), *P. ipomoeae* (pathogen of morning glory), and *P. andina* (pathogen of several *Solanum* spp.). Like *P. infestans,* they are mainly foliar pathogens. While they have unique host ranges, nucleotide diversity within coding sequences of orthologs is reported to average less than a few percent (Blair et al. 2012). *P. andina* is believed to be a hybrid of *P. infestans* and an unknown species similar to *P. mirabilis* or *P. ipomoeae* (Goss et al. 2011).

The close similarity of the clade 1c species raises questions about where *P. infestans* originated, and how it moved to the sites of the nineteenth century epidemics. Several theories were proposed (Ristaino 2002; Fry et al. 2009). One placed the origin of *P. infestans* in the highlands of central Mexico, where nuclear markers show the greatest diversity. A Mexican origin is also supported by the observation that *P. infestans, P. mirabilis* and *P. ipomoeae* and their hosts coexist there (Flier et al. 2002; Oliva et al. 2010); divergence of *P. infestans* from a shared ancestor may have been driven by changes in host specificity followed by reproductive isolation (Raffaele et al. 2010a, b). A second theory proposed Peru as the center of origin of *P. infestans*, and the source of inoculum for the historical epidemics (Ristaino 2002). This is consistent with the first appearance of cultivated potato in the Andes, where wild hosts of *P. infestans* still exist, and the fact that trade in potato between Peru and Europe in the 1840s was more common than between Mexico and Europe. A third theory acknowledged Mexico as the center of origin, with isolates migrating from Mexico to the Andes and then to the United States and Europe in the nineteenth century (Ristaino 2002).

Once genome sequences became available, studies of the population genetics of *P. infestans* advanced from relying on restriction fragment and amplified fragment length polymorphisms (RFLPs, AFLPs) to more robust and informative markers. Currently popular are microsatellites, which were identified by mining expressed sequence tag (cDNA) databases (Lees et al. 2006), mitochondrial loci (Gavino and Fry 2002; Griffith and Shaw 1998; Avila-Adame et al. 2006), and polymorphisms within specific genes, including some with roles in pathobiology such as avirulence genes (Cardenas et al. 2011; Li et al. 2012a, b). One group pioneered the analysis of DNA from herbarium samples from the historical famines (Ristaino et al. 2001; Gomez-Alpizar et al. 2007). Their initial success with mitochondrial DNA from herbarium samples led recently to the generation by two groups of whole-genome shotgun sequences of nineteenth century isolates of *P. infestans* from Austria, Belgium, Denmark, Germany, Sweden, and the United Kingdom (Yoshida et al. 2013; Martin et al. 2013). The results confirmed indications from prior analyses of mitochondrial DNA (Ristaino et al. 2001) that the US-1 genotype that predominated worldwide until the 1970s did not cause the nineteenth century epidemics. Responsible instead was a lineage which one group of researchers named HERB-1, which persisted in Europe for at least 50 years. Some heterogeneity was noted between herbarium samples, suggesting either that mutations occurred within the lineage after being introduced to Europe, or that closely related strains were imported (Yoshida et al. 2013; Martin et al. 2013). HERB-1 has close affinity in phylogenetic analyses to US-1, and it seems probable that both spread to Europe from a site outside the center of diversity, perhaps in the Andes.

### 9.1.4 Life and Disease Cycles

*P. infestans* exhibits both sexual and asexual life cycles, with reproduction during the growing season predominately involving asexual growth including the formation of sporangia. In nature, growth begins when sporangia move by wind or water to a new host (Fig. 9.1). Sporangia may extend germ tubes directly, but more common is indirect germination of these multinucleate spores, during which each releases about eight mononucleate zoospores (Judelson and Blanco 2005). The liberation of zoospores typically occurs within 1 or 2 h after sporangia are placed on moist cool surfaces, a situation common early in the day. Zoospores are biflagellated and exhibit chemotaxis (Latijnhouwers et al. 2004). After a period of motility, zoospores lose their flagella and form a walled cyst that adheres to the plant. Each extends a germ tube which usually culminates in an appressorium on the plant cuticle, but occasionally a germ tube accesses the apoplast through a stomata. In most foliar infections, it is believed that colonization occurs from appressoria, which use enzymes and physical forces to enable penetration of the underlying epidermal cell. Assuming that *P. infestans* can defeat host defenses, hyphae then ramify through living plant tissue. Most hyphae are intercellular, but form haustoria within plant cells which are used to deliver to the host many pathogenesis-related proteins, which are collectively termed effectors (Whisson et al. 2007). After 4–6 days, depending on environment and pathogen genotype, hyphae emerge through stomata, and develop sporangiophores which become decorated with new asexual sporangia.

Late blight is a polycyclic disease, with epidemics enabled by the rapid asexual reproduction of the pathogen. *P. infestans* is considered to be a hemibiotroph, as infection occurs only on living tissue with necrosis taking place late in infection. With some isolates and especially on tomato, sporulation on above-ground tissues may be the first macroscopic symptom of infection. While most research into *P. infestans* biology has focused on foliage infections, tuber blight is also important. This often begins when spores from foliage are washed into soil, or contact potatoes at harvest (Lapwood 1977). Entry into tubers occurs mainly through wounds or lenticels and does not require appressoria. Infected tubers are a major source of early-season inoculum in potato production, and aid survival over the winter; persistence of *P. infestans* as a saprophyte is considered insignificant. How *P. infestans* survives between years in tomato production is unknown. Survival in seed is not described, but may occur in bridge plants such as wild solanaceous species.

*P. infestans* is heterothallic and normally requires both mating types (A1 and A2) for

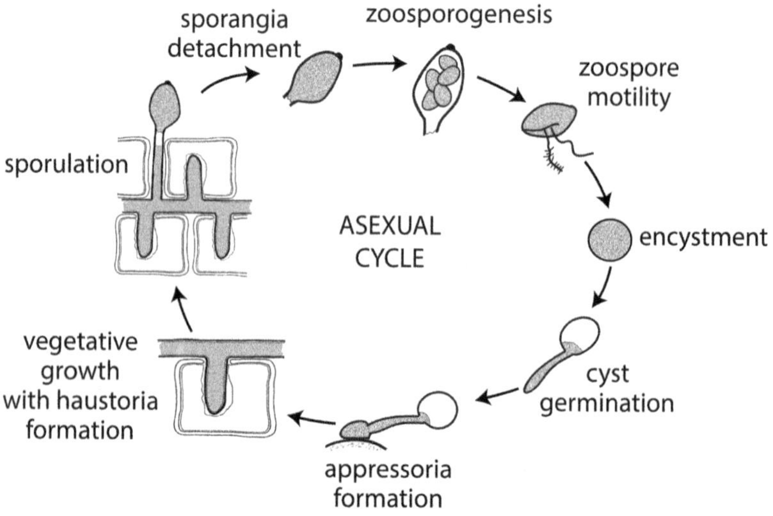


**Fig. 9.1** Asexual life cycle of *P. infestans*

sexual reproduction (Judelson 2007). Each form is bisexual and produces hormones that induce antheridia and/or oogonia in the opposite type, which fuse to form oospores. Oospores germinate to form germ tubes and then germ sporangia, which behave like asexual sporangia. The two mating hormones ($\alpha 1$, $\alpha 2$) have recently been characterized (Qi et al. 2005; Ojika et al. 2011). Both are acyclic diterpene alcohols related to phytol, a plant sterol. It is proposed that phytol is the precursor of $\alpha 2$, which is converted within A1 strains to the $\alpha 1$ form. Genetic analyses have defined a single nuclear locus for mating type, with A1 being heterozygous or hemizygous, and A2 homozygous (Fabritius and Judelson 1997; Judelson et al. 1995). The underlying genes have not been identified.

### 9.1.5 Current Structure of *P. infestans* Populations

Substantial changes have occurred over the past 30 years in most geographic areas due to migrations and the establishment of sexual populations (Fry et al. 2009; Yuen and Andersson 2012). Most known isolates of *P. infestans* until recent decades were of the A1 mating type; the A2 was discovered only in the mid-1950s during a survey in Mexico. At the time, most isolates outside of central Mexico were relatively uniform, belonging to a genotype named US-1. Populations in Africa, North America, South America, northern Mexico, and some parts of Asia have changed since the mid-twentieth century but remained mostly clonal, with epidemics usually involving a predominant genotype (Fry et al. 2013; Pule et al. 2013). Deeper sampling and the use of DNA markers with increased resolution revealed some diversity within "clonal" lineages in some geographic regions (Delgado et al. 2013). It is unclear if this resulted from spontaneous variation, limited sexual recombination, or new introductions. In contrast, *P. infestans* populations in parts of Europe and in China have become more sexual, with diverse genotypes found in some areas (Yuen and Andersson 2012; Li et al. 2012a). The sexual cycle is significant due to the potential of forming genotypes that are more aggressive or able to overcome *R* genes; a notable example is the "blue 13" A2 strain which was first found in The Netherlands around 2005 and has spread throughout most of Europe and into Asia (Cooke et al. 2012). The sexual spores (oospores) themselves are also worrisome as they can remain dormant for years, and survive harsh conditions such as freezing or fumigation. In some Scandinavian countries, oospore formation within potato fields is now common and believed responsible for the earlier occurrence of epidemics during the growing season (Widmark et al. 2011).

### 9.1.6 *P. infestans* as an Experimental System

*P. infestans* is normally maintained in the laboratory on complex media, such as rye-sucrose or pea broth. Defined media are also available. Hyphae or sporangia from such cultures are easily used for studies of biochemistry, developmental biology, or pathology. Infections can be performed on young plants or seedlings, with a complete disease cycle taking less than a week. Controlled crosses are also possible and have been used to study the inheritance of mating type, fungicide resistance, and virulence (Fabritius et al. 1997; Lee et al. 1999; Judelson et al. 1995; Al-Kherb et al. 1995), and to construct a linkage map based on AFLP markers (Van der Lee et al. 1997).

DNA transformation is a routine technique in many laboratories. It has been used for gene silencing, ectopic expression, subcellular localization, and promoter function studies (Whisson et al. 2007; Hua et al. 2013; Ah Fong and Judelson 2003; Roy et al. 2013a). A set of vectors that facilitates the fusion of proteins to a range of fluorescent markers is also available (Ah-Fong and Judelson 2011). Some groups have reported single studies involving more than one thousand independent transformants, although skill and practice is required for routine success.

Transformation was first achieved using selectable marker genes regulated by promoters and transcriptional terminators from another oomycete, *Bremia lactucae,* which causes downy mildew of lettuce (Judelson et al. 1991, 1992). Those regulatory sequences, taken from the *ham34* and *hsp70* genes, had been isolated specifically to construct oomycete transformation vectors; it was assumed that sequences from non-oomycetes would function poorly. Functional and bioinformatics studies later proved this supposition to be true, due in part to divergence in core promoter structure (Judelson et al. 1992; Roy et al. 2013b). An early report of *Phytophthora* spp. being transformed using markers driven by fungal promoters proved to be irreproducible. Reliable transformation was achieved in *P. infestans* using the *B. lactucae* promoters to drive genes for resistance to G418, hygromycin, or streptomycin, and then successfully transferred to other *Phytophthora* spp., *Pythium*, and *Saprolegnia* (Judelson et al. 1993a; Weiland 2003; Mort-Bontemps and Fevre 1997). The first successful reports of transformation entailed treating protoplasts with DNA mixed with calcium-polyethylene glycol, combined with cationic liposomes. Transformation was subsequently also achieved using electroporation, microprojectile bombardment, or *Agrobacterium tumefaciens*-based methods (Cvitanich and Judelson 2003; Vijn and Govers 2003; Ah-Fong et al. 2008). The protoplast and electroporation methods have proved to be the most popular.

Most transformation studies continue to rely on the *ham34* and *hsp70* promoters. To date, no homologous promoter has been found that out-performs the *B. lactucae* sequences for driving transgenes in *P. infestans.* Indeed, the selection of the *B. lactucae* sequences has proved to be a fortuitous choice. Comparisons of *P. infestans* and *P. capsici* promoters in *P. infestans* have suggested that homologous promoters function suboptimally since they are more likely to trigger silencing (Andreeva et al. unpublished results). *P. infestans* contains orthologs of *ham34* and *hsp70,* but their promoters lack similarity to the *B. lactucae* sequences and thus should not induce an RNAi response.

Regardless of promoter, the strength of transgene expression varies between transformants due to position effects and copy number variation (Judelson et al. 1993b). Chromosomal integration of transforming DNA is the rule in *P. infestans,* but there is no evidence for homologous recombination between plasmid and chromosome. Integration sites are random, and transformants typically contain tandem copies of the transforming plasmid, which complicates quantitative analyses of gene or promoter function (Ah Fong et al. 2007). Extrachromosomal persistence of plasmids as reported in *P. nicotianae* (Gaulin et al. 2007) has not been described in *P. infestans.*

Gene silencing has allowed the function of *P. infestans* genes in growth, development, and pathogenesis to be tested. To date silencing has been reported for over 20 genes (Table 9.1). Following approaches established in other organisms, several *P. infestans* researchers showed that inserting sense, antisense, or inverted repeat copies of a native gene into stable transformants often triggers silencing of both the native locus and transgene. Silencing frequently eliminates detectable expression of the target gene, but partial knock-downs also occur (Ah-Fong et al. 2008). At least initially, silencing involves the formation of 21-nt double stranded RNA derived from the targeted genes. It appears that transcriptional silencing then is imposed, based on data from nuclear run-on assays and evidence that silenced loci become heterchromatinized (Judelson and Tani 2007; van West et al. 2008). Systematic comparisons indicated that silencing occurs more often in transformants obtained using the protoplast approach compared to other methods. Also, 500-nt inverted repeat transgenes triggered silencing more than 21-nt hairpins, and inverted repeat constructs out-performed sense or anti-sense transgenes (Judelson and Tani 2007; Ah-Fong et al. 2008). Treatments with in vitro synthesized short double-stranded RNAs have been shown to transiently induce partial silencing (Walker et al. 2008; Whisson et al. 2005).

**Table 9.1** Gene silencing studies in *P. infestans*

| Target gene(s) | Phenotype | Reference |
|---|---|---|
| Argonaute[a] | Reduced silencing | Vetukuri et al. (2012) |
| Avr3a (RXLR effector) | Impaired virulence | Bos et al. (2010) |
| bZIP transcription factor family | Defects in zoospore motility, appressorium formation, ROS hypersensitivity | Blanco and Judelson (2005); Gamboa-Melendez et al. (2013) |
| Cdc14 phosphatase | Impaired sporangium formation | Ah Fong and Judelson (2003) |
| Cellulose synthase CesA[a] | Appressorium defects | Grenville-Briggs et al. (2008) |
| Dicer-like[a] | Reduced silencing | Vetukuri et al. (2012) |
| GPCR-PIPK protein GK4 | Reduced spore germination and virulence | Hua et al. (2013) |
| G$\alpha$ protein Gpa1 | Impaired zoospore motility and virulence | Latijnhouwers et al. (2004) |
| G$\beta$ protein Gpb1 | Impaired sporangium formation | Latijnhouwers and Govers (2003) |
| G$\beta$-like protein | Reduced virulence | Kim (2006) |
| Haustorial protein hmp1[a] | Reduced host penetration | Avrova et al. (2008) |
| Histone deacetylase[a] | Lower gene silencing efficiency | Vetukuri et al. (2012) |
| Inf1 elicitin | Larger lesions on *N. benthamiana* | Kamoun et al. (1998) |
| NifC phosphatase | Reduced zoospore cyst germination | Judelson and Tani (2007) |
| RNA helicase Rnh1 | Aberrant zoospores | Walker et al. (2008) |

[a] Studies reporting transient silencing

## 9.2 Genome Structure, Composition, and Evolution

### 9.2.1 Nuclear Genome Size and Topography

An annotated assembly of the genome of *P. infestans* was released in 2009, which was developed from eightfold coverage by Sanger sequencing of a strain from The Netherlands (Haas et al. 2009). This joined data from expressed sequence tag projects that identified 95,000 sequences from libraries based on 20 developmental stages and growth conditions (Randall et al. 2005; Kamoun et al. 1999). DNA and RNA from other isolates from the United States and Europe have subsequently been sequenced by Illumina technology (Yoshida et al. 2013; Martin et al. 2013; Raffaele et al. 2010a).

*P. infestans* has one of the largest sequenced oomycete genomes, with 190 Mb of DNA assembled into contigs and an extrapolated total genome size of about 230 Mb. This is close to a prior estimate of 237 Mb based on microfluorometry, and close to values for other clade 1c species such as *P. mirabilis* and *P. phaseoli* (Tooley and Therrien 1987). Other oomycete genomes range from 37 to 280 Mb, with differences in repetitive DNA responsible for most variation (Kemen et al. 2011; Baxter et al. 2010; Levesque et al. 2010; Raffaele and Kamoun 2012). *P. infestans* has the most repetitive DNA (74 %) and *Py. ultimum* the least (7 % repetitive, 43 Mb) of any oomycete assembly released to date. *Phytophthora* genomes are larger than those of fungal phytopathogens such as *Magnaporthe oryzae* (40 Mb) or diatoms such as *Thalassiosira pseudonana* (34 Mb).

*P. infestans* is believed to contain about 9–13 chromosomes, compared to the assembly which has a scaffold N50 of 1.59 Mb and contig N50 of 44 Mb (Haas et al. 2009; Sansome and Brasier 1973). While a genetic map of *P. infestans* exists, it has not been aligned to the assembly (Van der Lee et al. 1997). It should be noted that there could be significant variation between isolates, as researchers have reported variation in chromosome number, trisomy, and aneuploidy (Ritch and Daggett 1995; Catal et al. 2010; Van der Lee et al. 2004).

The current estimate of the nuclear content of *P. infestans* is 17,797 protein-coding genes (Haas et al. 2009). By comparison, 13,000–20,000 genes are predicted for other oomycetes. It is important to recognize that gene counts are estimates that change over time, and comparing values from different sequencing technologies can be problematic. Interspecific comparisons with *P. capsici, P. ramorum,* and *P. sojae* indicate that *Phytophthora* spp. contain a core set of about 9,000 genes (Haas et al. 2009; Seidl et al. 2012a). Nearly half of each proteome is therefore variable, with differences attributable to the growth of gene families and/or gene loss.

Contributing to the fluidity of *Phytophthora* genomes are their novel organization. While genes in most eukaryotes are also not distributed evenly, this is taken to an extreme in *P. infestans* and relatives. As illustrated in Fig. 9.2 for supercontig1 of the *P. infestans* genome assembly, gene density varies dramatically along chromosomes (Fig. 9.2a). About three-quarters of genes are closely spaced, residing within 1 kb of each other (e.g., Fig. 9.2d). In such regions, gene order is conserved to a large degree between *Phytophthora* spp., *Pythium,* and downy mildews.

The remaining genes are found in gene-sparse regions in which intergenic distances typically exceed several kilobases, and genes are interspersed with transposable elements especially *gypsy* and *copia*-like sequences (e.g., Fig. 9.2b). These zones exhibit little microsynteny, and are enriched for genes with roles in pathogenesis. In the example shown in Fig. 9.2b, three of the five genes encode potential plant cell wall degrading enzymes including pectin lyase. Other gene-sparse regions contain additional types of pathogenesis-related proteins such as the CRN and RXLR effectors (Haas et al. 2009). Interspecific and intraspecific comparisons indicate that such genes vary more than average in sequence and copy number. The CRN and RXLR families of *P. infestans,* for example, are dramatically expanded compared to other *Phytophthora* spp., with 196 and 563 members in *P. infestans* compared to 335 and 100 in *P. sojae,* respectively (Table 9.2). Likely contributors to their rapid evolution are the nearby repeated sequences, which may foster illegitimate recombination, and the fact that such regions contain few essential genes, which minimizes the frequency of lethal recombination events.

Repeated genes also occur within the gene-dense regions. Examples include genes encoding metabolic proteins such as glucokinases and invertases, proteins involved in sexual spore development, and protein kinases. In the case of the latter, *P. infestans* encodes 354 proteins with classic eukaryotic protein kinase domains (ePKs; Judelson and Ah-Fong 2010). Unequal crossing over appears to have contributed to its expansion in *Phytophthora,* as about 20 % of ePK genes reside in clusters having 2–13 members. Many of these clusters are shared with *H. arabidopsidis,* which contains 207 predicted ePKs, but usually not with *Py. ultimum* which encodes only about 100 ePKs. Expansion of ePKs appears to have slowed after the last common ancestor of *Phytophthora,* as the ePK content of *P. infestans, P. ramorum,* and *P. sojae* are nearly identical.

A genome-wide analysis of the evolutionary history of ten stramenopile genomes (three diatoms, one brown algae, and six oomycetes including *Phytophthora*, *Pythium*, *Saprolegnia, and H. arabidopsidis*) indicated that the rate of gene duplications was highest in the branch leading to *Phytophthora* (Seidl et al. 2012a, b). This was speculated to have contributed to the pace of speciation within the 100-plus member genus. Whether an ancestor of *Phytophthora* experienced a whole-genome duplication has also been discussed (Martens and Van de Peer 2010). Related to the pace of genome evolution in oomycetes may also be the absence of cytosine-5-methyltransferases (C5-MTases; Judelson

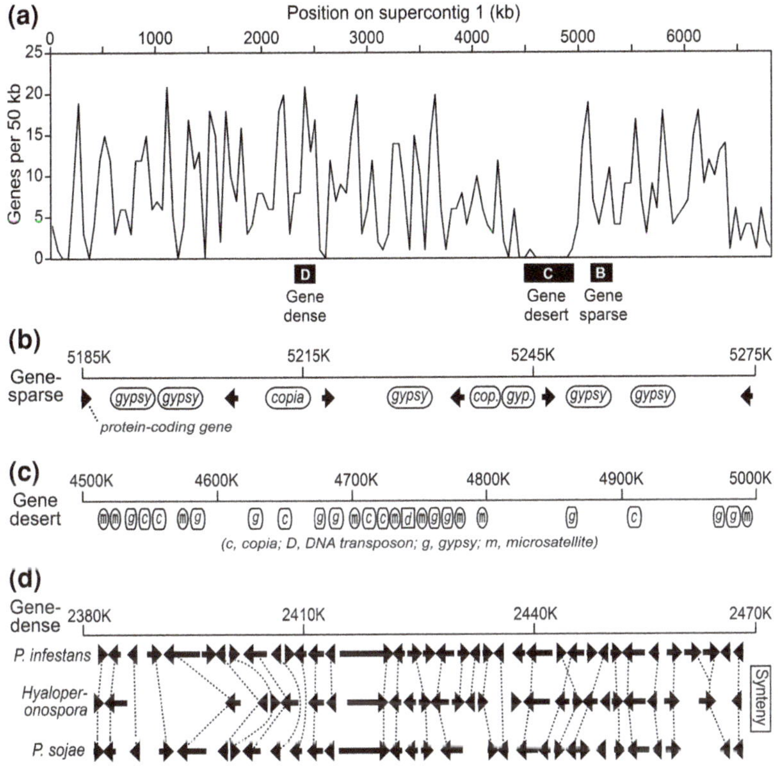


**Fig. 9.2** Genome organization in *P. infestans.* **a** Gene density along supercontig 1. Bars labeled **b**, **c**, and **d** denote representative gene-sparse, gene desert, and gene-dense regions, respectively. **b** Gene-sparse region from *panel a*, showing locations of genes (*arrows*) and *copia* or *gypsy* retroelement-like sequences. **c** Gene desert from *panel a*, marking locations of *copia* (*c*), *gypsy* (*g*), DNA transposon (*d*), and microsatellite (*m*) sequences. **d** Gene-dense, transposon-sparse region from panel a. *Arrows* indicate *P. infestans* genes and orthologs on syntenous contigs from *H. arabidopsidis* and *P. sojae,* with *dotted lines* connecting orthologs

and Tani 2007), which may help restrict the activity of transposable elements which may drive genome change. In contrast, C5-MTases are present in most other stramenopiles including diatoms and brown algae, which contain C5-MTase families 2, 3, and 5 (Maumus et al. 2011). Also facilitating gene duplications in oomycetes may be DNA transposon families; as described next, several of these are specific to oomycete or *Phytophthora* lineages.

### 9.2.2 Transposable Elements

*P. infestans* contains diverse families of retrotransposon and DNA transposon-like sequences, which comprise the bulk of the 74 % repeat content of the genome. As in many eukaryotes, the most abundant retrotransposons belong to the long terminal repeat (LTR)-containing *gypsy* and *copia* families. Non-LTR LINE elements are also present. *P. infestans* contains about 10,000 *gypsy* and 4,000 *copia* like elements which comprise 37 % of the genome, compared to *P. ramorum* and *P. sojae* which contain only 2,400 and 3,000 total retroelement-like sequences, respectively (Haas et al. 2009). In contrast, *gypsy* and *copia* make up only 2 and 9 % of the genomes of *Py. ultimum* and *Albugo laibachii* (a white rust), respectively, with *copia* outnumbering *gypsy* (Levesque et al. 2010; Kemen et al. 2011). Expansion of *gypsy* families therefore appears to have been the major force underlying expansion of the *P. infestans* genome. Most retrotransposon-like sequences are degenerate and incapable of movement.

**Table 9.2** Pathogenesis-related protein families in oomycetes

| Function | *P. infestans* | *P. sojae* | *A. laibachii* | *H. arabidopsidis* | *Py. ultimum* |
|---|---|---|---|---|---|
| Aspartyl protease | 12 | 13 | 10 | 9 | 22 |
| Crinkler (CRN) | 196 | 100 | 3 | 20 | 26 |
| Cutinase | 4 | 14 | 2 | 2 | 0 |
| Cysteine protease | 33 | 29 | 16 | 7 | 32 |
| Glucanase inhibitor | 4 | 7 | 0 | 0 | 0 |
| Glycosyl hydrolase | 157 | 190 | 44 | 66 | 85 |
| Lectin-like | 10 | 13 | 6 | 6 | 20 |
| Lipase | 19 | 27 | 12 | 10 | 19 |
| NEP1-like (NLP) | 27 | 39 | 0 | 24 | 7 |
| PcF/Scr | 16 | 8 | 0 | 0 | 3 |
| Pectate lyase | 30 | 30 | 1 | 8 | 15 |
| Pectin esterase | 11 | 19 | 0 | 4 | 0 |
| Phospholipase | 36 | 31 | 13 | 13 | 6 |
| Protease inhibitor | 38 | 26 | 0 | 3 | 1 |
| RXLR | 563 | 335 | 49 | 115 | 0 |
| Serine carboxypeptidase | 24 | 21 | 32 | 6 | 1 |

Values include data from new analyses and published studies (Kemen et al. 2011; Haas et al. 2009)

DNA transposons comprise about 10 % of the *P. infestans* genome, and include members of families known to replicate by cut and paste, rolling circle, and self-synthesizing processes. Listed in declining order of copy number are the piggyBac, helitron, hAT, crypton, Tc1/mariner/pogo, MuDR/foldback, Sola, Maverick, and PIF/harbinger families (Haas et al. 2009; Vadnagara 2010; Feschotte and Pritham 2007). Helitrons are of particular interest since in other species they have demonstrated a propensity to form novel genes by combining exons from unrelated loci. Most appear nonfunctional, but 13 of the 273 helitron-like sequences within *P. infestans* seem intact, which is more than any other eukaryote. In contrast, all helitrons in *P. ramorum* and *P. sojae* appear to be defective. Most transposon families are ancient members of the oomycete lineage as they have representatives in *A. laibachii, H. arabidopsidis,* and *Py. ultimum*. The exception is cyptonF, which seems to be a relatively recent acquisition as it is present in *Phytophthora* but not *A. laibachii, H. arabidopsidis,* or *Py. ultimum.* CryptonF has only been found in ascomycetes and *Phytophthora,* which suggests horizontal transfer (Kojima and Jurka 2011). HAT and Tc1/mariner elements are absent from the diatom *T. pseudonana,* so may have been acquired after oomycetes diverged from other stramenopiles.

There is evidence for limited movement of transposable elements in *P. infestans.* One gene was found to be interrupted by a *gypsy* element, and to contain a host target site duplication (Haas et al. 2009). Inverted terminal repeats of hAT, helitron, and PIF elements with target site duplications flanking transposase/replicase and host genes have also been reported (Vadnagara 2010). The captured host genes include SET domain proteins and AdoMet-dependent methyltransferases, which interestingly are both involved in epigenetic regulation, plus an ABC transporter, cysteine protease, and transglutaminase.

While most transposable elements are interspersed throughout the gene-sparse regions of the genome, others cluster in large gene deserts. Illustrated in Fig. 9.2 is one such site which spans 500 kb. In this example, the region is composed primarily of degenerate *gypsy* and *copia* elements, with a smaller number of DNA transposons. Simple sequence repeats (microsatellites) of CAG, TA, CTG, and CCG are also unusually common in this region.

### 9.2.3 Horizontal Gene Transfer

Interspecific transfer of DNA during the evolution of *P. infestans* is not limited to transposable elements. Early studies of polygalacturonases, which degrade plant cell walls, identified similarities between *Phytophthora* and fungal genes suggestive of horizontal gene transfer (HGT) or convergent evolution (Torto et al. 2002). Careful phylogenetic analyses using whole-genome data now provide strong support for HGT. Transfers from fungi to oomycetes may have involved about 20 types of genes. Two-thirds of these encode proteins that are secreted and/or have potential roles in pathogenesis including lipases, carbohydrate depolymerizing enzymes, sugar and nitrogenous base transporters, enzymes that degrade plant defense compounds, and phytotoxins (Richards et al. 2011; Gijzen and Nurnberger 2006). HGT seems to have been less common in *A. laibachii* and *S. parasitica* (Richards et al. 2011), which suggests that the transposon-rich *Phytophthora* genomes are more tolerant of insertions by foreign DNA.

HGT has not been limited to genes related to pathogenesis, as other candidates for HGT include several classes of metabolic enzymes. These include enzymes involved in cofactor metabolism, amino acid biosynthesis, and lipopolysaccharide formation (Whitaker et al. 2009). Several enzymes in glycolysis also appear to have been obtained from fungi, plants, and bacteria (Whitaker et al. 2009; Judelson et al. 2009b).

### 9.2.4 Endosymbiotic Gene Transfer

The acquisition of genes through endosymbiosis has also been proposed as a feature of the ancestry of *P. infestans* and relatives. In one model for eukaryotic evolution, *P. infestans* is placed in the Chromalveolata which includes cryptophytes, alveolates, haptophytes, and stramenopiles and comprises about half of known protists and algae (Keeling 2009). The group is proposed to be derived from a biflagellated cell with a red algal endosymbiont; the alga is suggested to have been retained as a photosynthetic plastid in lineages such as diatoms, but lost from oomycetes after transferring some genes to the nucleus (Martens et al. 2008).

The chromalveolate hypothesis has proved to be controversial. Support for endosymbiosis in the diatom lineage appears solid. Considerable molecular data have been invoked in support of the chromalveolate hypothesis, including a study of *P. ramorum* and *P. sojae* that reported 855 genes of likely red algal origin based on sequence similarity (Tyler et al. 2006). However, the occurrence of endosymbiotic transfer prior to the emergence of oomycetes was challenged by a study that argued that other events, such as HGT or stepwise mutation, more plausibly explain the apparent presence of red algal-like genes (Stiller et al. 2009).

### 9.2.5 Gene Structure and Transcriptional Landscape

As noted earlier, most *P. infestans* genes reside in dense blocks. About 40 % of adjacent genes are transcribed divergently and 60 % in the same direction, and in both situations the median intergenic distance is about 430 nt. Despite this close spacing, the likelihood of adjacent genes being co-expressed during the life cycle is only slightly more than expected by random chance, if one excludes tandemly repeated genes which often show similar transcription patterns due to promoter duplication (Roy et al. 2013a). This differs from the situation in fungi and animals, where many adjoining genes are co-expressed as a consequence of chromatin-level regulation or employ bidirectional promoters with common transcription factor binding sites (Tsai et al. 2009; Kruglyak and Tang 2000). Although not very common in *P. infestans*, there are interesting cases where genes transcribed from a shared promoter region have related functions and similar patterns of expression. Examples include genes for ribosomal proteins and nitrogen assimilation. Some represent ancient gene pairs, with similar organizations found in metazoans, plants, and fungi (Davila-Lopez et al. 2010).

Slightly more than half of *P. infestans* genes contain introns, with an mean of 2.7 introns per gene, a median intron size of 73 nt, and 1.5 kb per mature transcript (Haas et al. 2009). Intron retention is common, and usually results in premature termination of translation. To date there are no known examples of alternative splicing leading to proteins with distinct biological functions in *P. infestans*. There have been reports that intron retention is more common in sporangia, perhaps due to slower processing of primary transcripts.

*P. infestans* genes are GC-rich (58 %), compared to the entire genome which is 51 % G+C. This results from a strong bias at the third position of each codon, and proved useful for distinguishing host from pathogen sequences in infection cDNA libraries (Sierra et al. 2010). TRNAs with anticodons matching 47 of the 61 sense codons are present, with the rest addressable by wobble binding as in other eukaryotes. Interestingly, the anticodons missing from *P. infestans* only partially overlap those absent from *P. ramorum* and *P. sojae*. For example, *P. infestans* tRNAs include those with the AUU anticodon which are absent from the other two species, but lack a CCG anticodon which are present in *P. ramorum* and *P. sojae* (Tripathy and Tyler 2006). Analyses of codon usage reveal adaptation to these differences (Fig. 9.3). The frequencies of the corresponding AAU and CGG codons in *P. infestans* are much higher and reduced, respectively.

### 9.2.6 Novel Gene Fusions

*P. Infestans* encodes proteins with an unusually high number of novel domain combinations, compared to 66 other eukaryotes (Seidl et al. 2011; Morris et al. 2009). The tight spacing of genes in *P. infestans* and the presence of active transposons both likely contribute to such innovations. Fusions can evolve readily from the loss of a stop codon in an upstream gene, for example. Since oomycetes are diploid, fusions in one haplotype might show only limited lethality, and persist for long periods during which their benefits could be tested.

Many of the novel proteins in *P. infestans* have roles in cellular signaling, and appear conserved throughout *Phytophthora*. One notable example are phosphatidylinositol-4-phosphate kinases that contain transmembrane domains characteristic of G protein-coupled receptors (GPCRs; Meijer and Govers 2006). The evolution of this novel mechanism for transducing intracellular or extracellular signals may be related to the fact that *Phytophthora* spp. express only single G$\alpha$ and G$\beta$ proteins, compared to most eukaryotes that diversify signaling through several $\alpha$ and $\beta$ subunits. A related role may be played by another family of proteins in which aspartic proteases are fused to GPCR-like transmembrane domains (Kay et al. 2011; Meijer and Govers 2006). *P. infestans* also contains a family of phospholipase D proteins with the unusual feature of having a signal peptide (Meijer et al. 2011), which may affect extracellular signaling. A fourth class of potential signaling protein with an innovative structure is a MAP kinase with a N-terminal PAS domain, which is a signal sensing fold (Judelson and Ah Fong 2010). While present throughout oomycetes, this does not occur in other eukaryotes or stramenopiles. Another novel protein joins an N-terminal cyclin-dependent kinase with a C-terminal cyclin, which in most other organisms are on separate proteins. This unusual configuration is found in most oomycetes but otherwise appears only in apicomplexans and ciliates, which have moderate taxonomic affinity to oomycetes.

### 9.2.7 Mitochondrial Genome

Four different mitochondrial types (Ia, Ib, IIa, IIb) have been identified in *P. infestans* and sequenced (Avila-Adame et al. 2006). These range in size from 37.9 to 39.9 kb and are 22 %

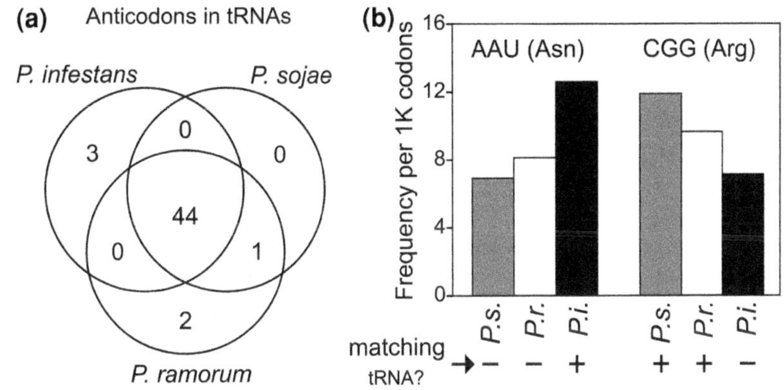


**Fig. 9.3** Transfer RNAs and codon adaptation in *Phytophthora*. **a** Number of anticodons represented in tRNAs from *P. infestans, P. ramorum*, and *P. sojae*. **b** Occurrence in gene space of two codons for which matching tRNAs are lacking in some *Phytophthora* spp., illustrating shift in *P. infestans* (*P. i.*) towards greater use of a codon for which a tRNA is present only in that species (5′-AAU), and away from a codon for which tRNA is lacking (5′-CGG)

G+C. The four types lack the large inverted repeat seen in the mitochondrial DNA of *Pythium*, but contain two inversions compared to *P. ramorum* and *P. sojae*. Genes span about 90 % of the mitochondrial genome and include 18 that encode proteins involved in electron transport, 16 ribosomal protein genes, 25 tRNA genes, and two encoding ribosomal RNA. Between 6 and 13 other open reading frames of unknown function are also predicted in the different genotypes.

The mitochondrial genome of *P. infestans* exhibits maternal inheritance, and its polymorphisms have been used extensively in population genetic studies (Ristaino et al. 2012; Ristaino 2002). Types I and II are proposed to have evolved independently from a common ancestor, followed by the development of Ia and Ib, and IIa and IIb. Analyses of herbarium specimens showed that the Ia haplotype was predominant in the nineteenth century famine populations (Ristaino et al. 2012). In contrast, Ib was dominant worldwide until the mid 1970s, indicating that new exports from the Americas had replaced the famine strains. After disappearing from most of Britain and Ireland by the 1990s, the Ib type has recently reappeared (Kildea et al. 2013). However, it now occurs with new nuclear genotypes, which implicates either a new migration or sexual recombination.

About 10 kb or more of sequences in the nuclear genome assemblies of *P. infestans* and other *Phytophthora* spp. such as *P. sojae* show similarity to mitochondrial DNA, excluding regions of very high identity that may represent assembly errors. These likely represent *numts* (nuclear sequence of mitochondrial origin), which are common in eukaryotic genomes (Hazkani-Covo et al. 2010). Such sequences are believed to represent recent transfers from the mitochondrion to the nucleus. In humans, *numts* have been associated with five genetic diseases. Although not yet studied in detail in any oomycete, intraspecific comparisons of *numts* may provide insight into the plasticity of their genomes, phenotypic variation, and reproductive isolation.

### 9.2.8 Viruses

Four *P. infestans* viruses have been characterized from isolates found to contain double-stranded RNA, although evidence for encapsulated forms of the RNA genomes are lacking (Cai and Hillman 2013; Cai et al. 2009, 2012). Based on sequence analysis, PiRV-4 belongs to the Narnaviridae, which was previously only known to have members in fungi. The other three viruses (PiRV-1, PiRV-2, PiRV-3) probably represent novel eukaryotic families. At 3.0 kb, PiRV-4 is the smallest *P. infestans* virus and only encodes an RNA-dependent RNA polymerase (RdRp). PiRV-3 has a 8.0 kb genome that is predicted to encode an RdRp and an additional protein of unknown function, and may affect colony morphology. PiRV-1 consists of separate 3.2 and 2.8 kb mRNAs, one encoding an RdRp and the other a protease and additional reading frame. PiRV-2 has the largest genome at 11.2 kb, which includes a long open reading frame. The possibility of using these viruses as a new transformation system has been suggested. A viroid-like sequence has also been observed in some isolates, but probably does not represent a potential vector for transgenes as it induces a stress response (Judelson et al. 2010).

## 9.3 Genomic Insights into *P. infestans* Biology

### 9.3.1 An Unusually Dynamic Transcriptome

The first large-scale study of gene expression in any oomycete was performed in *P. infestans* using microarrays, and examined hyphal growth, spore formation, and spore germination. That analysis revealed striking shifts in the transcriptome throughout the life cycle (Fig. 9.4). Levels of mRNA for at least one-half of genes changed > twofold during life-stage transitions, with 14 % being expressed exclusively in one stage or another (Judelson et al. 2008). Other microarray studies identified genes that are upregulated during growth *in planta*, or during sexual development (Prakob and Judelson 2007; Haas et al. 2009).

Genes that show dramatic changes during development include many that are logically required at only certain life stages. For example, of the 259 flagella-associated proteins predicted from the *P. infestans* genome, two-thirds are induced fivefold or more during sporulation (Judelson et al. 2009a, 2012). This is consistent with inhibitor studies that showed that most flagella components are preformed in sporangia (Clark et al. 1978). Conversely, during the transition from growing hyphae to sporangia there is a strong down-regulation of mRNAs encoding cell cycle proteins required for mitotic progression, such as G1 cyclins and the cdc2 kinase (Judelson et al. 2008). Such genes are reactivated in germinated cysts, coincident with the resumption of mitosis. Also induced in germinated cysts are many secreted virulence factors, which presumably are needed in the subsequent appressorium and plant colonization stages (Raffaele et al. 2010b). These include plant cell wall degrading enzymes such as polygalacturonases and cutinases, proteases, RXLR effectors, and inhibitors of plant apoplastic proteases. Interestingly, many of the predicted virulence factors are expressed with distinct kinetics. For example, while about 85 % of expressed RXLRs are induced in germinating cysts, 15 % show little mRNA until *in planta* growth begins. In addition, about three-quarters have abundant mRNA *in planta* only in the early

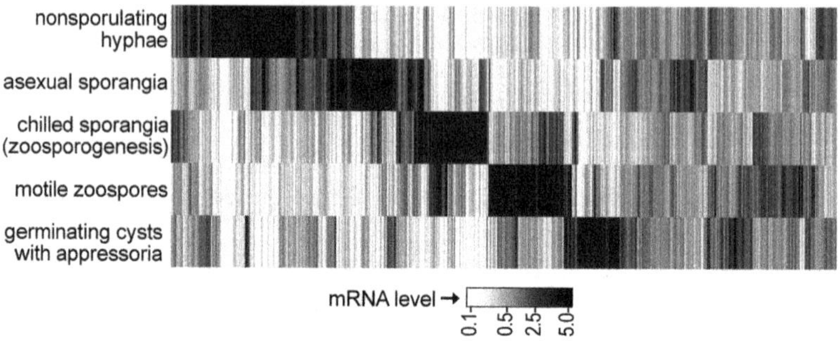


**Fig. 9.4** Transcript patterns during *P. infestans* development. Shown are per gene normalized values of loci annotated as genes in the T30-4 assembly (Judelson 2007)

biotrophic stage, while others are expressed predominantly in late stages or throughout infection. Many RXLRs are also expressed at higher levels *in planta* compared to artificial media (Raffaele et al. 2010b; Haas et al. 2009).

Such observations raise questions about what triggers changes in gene expression throughout the life and disease cycles. To what extent are they regulated by environmental cues, such as hydrophobic surfaces during appressoria formation or plant metabolites during host colonization? Do plant defense compounds influence the expression of effectors that act to suppress defense? Are some patterns hard-wired into development? For example, does zoospore encystment trigger an irreversible transcription factor cascade that induces genes in cysts and then appressoria? This would represent a form of developmental commitment conceptually similar to that occurring during *Bacillus* sporulation and T-cell differentiation; once a pathway is stimulated, development proceeds without continued stimulation (Dworkin and Losick 2005). Finally, do the transcriptional changes result from de novo synthesis of transcription factors, or post-translational modifications of existing factors?

### 9.3.2 Transcription Factors as Engines of Development

*P. infestans* encode approximately 315 DNA-binding transcription factors (Rayko et al. 2010; Seidl et al. 2013; Iyer et al. 2008). This value excludes basal factors such as TFIID, proteins that bind histones rather than DNA, and proteins that have putative DNA-binding domains but are not believed to regulate transcription. Most common are zinc finger proteins in the CCHC and C2H2 classes, and the bZIP, Myb, and HSF groups. Providing evidence that oomycetes have a unique evolutionary history compared to plants and fungi, *P. infestans* lack proteins resembling the plant-dominant WRKY or fungal-dominant C2H6 types. When calculating the number of transcription factors in *P. infestans,* it is important to note that many of its DNA-binding domains have diverged from those modeled in databases. For example, large subgroups of *P. infestans* bZIP and Myb proteins contain novelties that require the use of custom position-specific scoring matrices to be detected (Xiang and Judelson 2010; Blanco and Judelson 2005; Gamboa-Melendez et al. 2013). How often similar challenges occur when matching other oomycete proteins against generic domain databases has not been addressed systematically.

Many transcription factors show strong differential expression, especially those in the bZIP and Myb families (Xiang and Judelson 2010; Judelson et al. 2008; Gamboa-Melendez et al. 2013). For example, about half of bZIPs are expressed primarily in sporangia, zoospores, or germinating cysts, while about half of Myb factors are expressed preferentially during sporulation or zoosporogenesis. In contrast, other families show little differential expression during the life cycle, such as most C2H2 TFs, which are expressed constitutively or mainly in hyphae and germinated cysts. Differential expression during the life cycle does not necessarily indicate a role in development, however. For example, when eight bZIP genes that are induced in the zoospore or cyst germination stages of *P. infestans* were silenced using DNA-directed RNAi, no defects were observed in life cycle progression or pathogenicity (Gamboa-Melendez et al. 2013). Instead, defects were observed in the strains' ability to defend against peroxide stress. A *P. infestans* bZIP that is more constitutively expressed was shown to be needed for appressorium development (Blanco and Judelson 2005).

Several approaches have been used to identify transcription factor binding sites (TFBSs) in stage-specific promoters, as part of understanding the regulatory networks that drive development. Using promoter mutagenesis, several DNA motifs that determine expression specific to sporulation and zoosporogenesis were identified (Xiang et al. 2009; Ah Fong et al. 2007; Tani and Judelson 2006). Bioinformatics was used more recently to discover 120 stage-specific TFBS candidates by searching for motifs over-represented in co-regulated genes (Roy et al. 2013a;

Seidl et al. 2012b). Predictions of TFBSs linked to transcription during sporulation, zoosporogenesis, or cyst germination were shown to be robust as they included motifs identified previously by promoter mutagenesis. Many were also shown to specifically bind proteins from nuclear extracts, and confer stage-specific transcription to a reporter gene (Roy et al. 2013a). At least half of the predicted TFBSs are conserved in *P. ramorum* and *P. sojae.*

Some features of the TFBSs provide a better understanding of the organization of oomycete promoters. Most TFBSs occur within 150 nt of the transcription start site, which suggests that *P. infestans* promoters are compact and is consistent with the small intergenic distances in most of the genome. Over three-quarters of motifs occur preferentially in the forward versus reverse direction, compared to only about 50 % of yeast and mouse TFBSs (Roy et al. 2013a). This helps explain why *P. infestans* genes transcribed in divergent directions from a shared promoter region typically show unrelated patterns of expression.

### 9.3.3 Divergence of the Oomycete Core Promoter

Prior to the availability of oomycete promoters, several groups attempted but failed to transform *P. infestans* using marker genes driven by fungal promoters. The availability of genome sequences has helped to reveal that such failures were at least partly due to differences between the core promoters of oomycetes and other eukaryotes. The core promoter is normally defined as the 50 bases on either side of the transcription start site (TSS), which contain sites that position RNA polymerase during establishment of the preinitiation complex (Smale and Kadonaga 2003). Researchers examining the few oomycete genes available in the pre-genomics era reported that many contained a 16–19 nt region near the transcription start site, which contained a 7 nt element resembling the eukaryotic Initiator (INR) core element at its 5′ end followed by an approximately 9 nt sequence unique to oomycetes named Flanking Promoter Region or FPR (McLeod et al. 2004; Pieterse et al. 1994). Some genes contained TATA-like motifs, but these were not functionally tested and could be spurious matches (Yan and Liou 2005). Once whole genomes could be studied, statistical analyses suggested that TATA box-like elements were indeed absent from *P. infestans* as they were not over-represented within promoter space. Instead it was found that the predominant core motifs are the INR, FPR, an INR + FPR supramotif, and a novel element called DPEP (Roy et al. 2013b). While INR, FPR, and the supramotif are found in all oomycetes, DPEP is uncommon or absent in *H. arabidopsidis* and *S. parasitica.* Genes with core promoters closely matching the INR + FPR consensus display much higher mRNA levels than average and are more likely to be developmentally regulated and have roles related to pathogenesis. In contrast, DPEP-bearing genes are linked to housekeeping functions. Such results may have value for predicting the start sites of genes and optimizing the expression of transgenes.

### 9.3.4 Understanding Pathogenesis: Genes for Offense

*P. infestans* secretes diverse types of proteins that contribute to its ability to colonize a host. Used for offensive purposes are many classes of extracellular degradative enzymes including aspartyl, cysteine, and serine proteases, glycoside hydrolases, pectate lyases, cutinases, and lipases. Some insight into the pathogenic strategies of *P. infestans,* a hemibiotroph, comes from genome-wide comparisons with the downy mildew *H. arabidopsidis* and white rust *A. laibachii* (both biotrophs) and the broad host-range necrotroph *Py. ultimum* (Table 9.2). In general, both *P. infestans* and *Py. ultimum* encode more degradative enzymes than the biotrophs. For example, *P. infestans* and *Py. ultimum* each make about 30 cysteine proteases compared to only 7 for the downy mildew, and the same trend is seen for pectate lyases. Similar reductions also occur in the biotrophic white rust *A. laibachii.*

Cases where *P. infestans* has more genes than *Py. ultimum* include glycoside hydrolases (157 vs. 85), phospholipases (36 vs. 6), and cutinases (4 vs. none). An explanation for the latter may be that *P. infestans* normally colonizes plant tissues that have thick cuticles, while *Py. ultimum* mainly infects nonsuberized young tissue. When considering the significance of the number of glycoside hydrolases, it must be noted that such enzymes also act on the pathogen's own walls to enable growth.

Several classes of *P. infestans* proteins encode demonstrated or suspected phytotoxins (Table 9.2). Having 16 members in *P. infestans* is the small cysteine-rich PcF toxin family, which was first discovered in *Phytophthora cactorum* (Orsomando et al. 2001). A second group are named Nep1-like proteins or NLPs after a class of phytotoxins also found in bacteria and fungi (Gijzen and Nurnberger 2006), and have 27 representatives in *P. infestans*. Some NLPs act as toxins during the necrotrophic phase of infection, but studies in *P. sojae* have shown that only a subset of NLPs induce necrosis and many are expressed early in infection (Dong et al. 2012). Consistent with this is the observation that *P. infestans* and *H. arabidopsidis* express similar numbers of NLPs, although the gene is absent from *A. laibachii*. Although one might predict that a necrotroph such as *Py. ultimum* would encode more toxins than *P. infestans,* the opposite is true as *Py. ultimum* encodes only 7 NLPs and 3 PcFs.

*P. infestans* also encodes several lectin-like proteins that may help it attach to the host, a property required both during early infection and haustoria development. Such proteins include glue-like proteins released by zoospores upon encystment, mucins, jacalin domain proteins, cadherins, and the cellulose-binding protein CBEL. Only CBEL has a confirmed role in pathogenesis, based on studies in *P. nicotianae* (Gaulin et al. 2002). CBEL contains a PAN/Apple domain that helps the pathogen adhere to cellulosic surfaces, and interestingly the same domain is used by parasitic apicomplexans to bind their hosts (Gong et al. 2012).

### 9.3.5 Understanding Pathogenesis: Genes for Defense

Proteins secreted by *P. infestans* also help protect it from the host immune response during plant colonization. As in fungal and bacterial pathosystems, these proteins are collectively termed effectors based on their abilities to alter host physiology, and are classified as apoplastic if they act in the extracellular space between host and pathogen, or cytoplasmic if translocated into the plant cell (Stassen and Van den Ackerveken 2011). Some researchers use a broader definition of effectors by including the degradative enzymes described in the previous section. Excluding such enzymes, all oomycete effectors characterized to date inhibit plant defenses but this may be an artifact of how they have been identified. In *P. infestans* and relatives, this usually involves mining genome or cDNA databases for secreted proteins, which are tested *in planta* for their ability to suppress defense or induce necrosis (Oh et al. 2009). There is no reason to restrict effectors to these roles, however. Others might aid the pathogen by translocating host nutrients to sites of infection, for example. It is also difficult to interpret if toxins such as NLPs and PcFs are employed for defense or offense. They may be used to release nutrients from plant cells to the pathogen, or poison plant cells to hinder their defenses. Indeed, some of the NLPs in *P. sojae* that fail to induce necrosis have been shown to suppress defense responses in soybean (Dong et al. 2012).

Apoplastic effectors synthesized by *P. infestans* include inhibitors of plant glucanases and proteases. The former bind and inhibit plant apoplastic endo-$\beta$-1,3-glucanases that are believed to target the pathogen cell wall. Such glucanase inhibitor proteins (GIPs) are encoded by four *P. infestans* genes; two are expressed both in artificial culture and *in planta,* and two mainly *in planta* (Damasceno et al. 2008). These GIPs have been shown to complex with the plant enzymes in infected tomato leaves, but their role has not yet been tested by gene silencing. Studies of codon evolution in GIPs and their plant targets in *Phytophthora* pathosystems identified pathogen and host sites that exhibit positive (adaptive) selection, which is a hallmark of a coevolving molecular arms race. *P. infestans* also produces about 38 proteinase inhibitors classified as Kazal or cystatin-like, which bind and inhibit serine and cysteine proteases, respectively (Tian et al. 2005, 2007). Most are expressed predominantly at early stages of infection, such during cyst germination or the biotrophic growth phase (Raffaele et al. 2010b). The proteins are absent from *A. laibachii, H. arabidopsidis,* and *Py. ultimum* (Table 9.2).

The most abundant and most studied class of cytoplasmic effectors from *P. infestans* are the RXLRs, with 563 predicted from the reference genome (Haas et al. 2009). The name denotes an N-terminal sequence found downstream of a conventional signal peptide, which frequently is followed by a dEER motif (Fig. 9.5a). The RXLR motif has been proposed to participate in host uptake by binding phosphoinositol-3-phosphate, but this has been controversial (Tyler et al. 2013; Wawra et al. 2013; Yaeno and Shirasu 2013). RXLRs are typically small, having a median size of 144 amino acids but ranging up to 800 amino acids, and bear one or more copies of a conserved fold thought to participate in protein-protein interactions (Win et al. 2012). Many RXLRs suppress plant immunity, such as Avr3a and AvrBlb2 which respectively stabilize an E3 ubiquitin-ligase used in plant defense and inhibit transport of a plant protease (Bozkurt et al. 2011; Bos et al. 2010). The names of these effectors indicate that besides functioning as virulence factors, they are recognized as avirulence factors by plant *R* genes; all known avirulence proteins from oomycetes are RXLRs. Others are not known to be recognized by a *R* gene, such as some shown to inhibit small RNA pathways in the host that are needed for the defense response (Qiao et al. 2013).

RXLRs exhibit diverse patterns of expression, with some induced in germinated cysts and the biotrophic stage, and others expressed mainly late in infection (Raffaele et al. 2010b; Haas et al. 2009). The 563-member RXLR family of *P. infestans* is larger than that of other oomycetes (Table 9.2), as 350 and 335 are predicted in the genomes of *P. ramorum* and *P. sojae,* and the biotrophs *A. laibachii* and *H. arabidopsidis* encode 49 and 115, respectively (Baxter et al. 2010; Kemen et al. 2011). There is convincing evidence for the expression of only about 200 RXLRs from *P. infestans,* so the

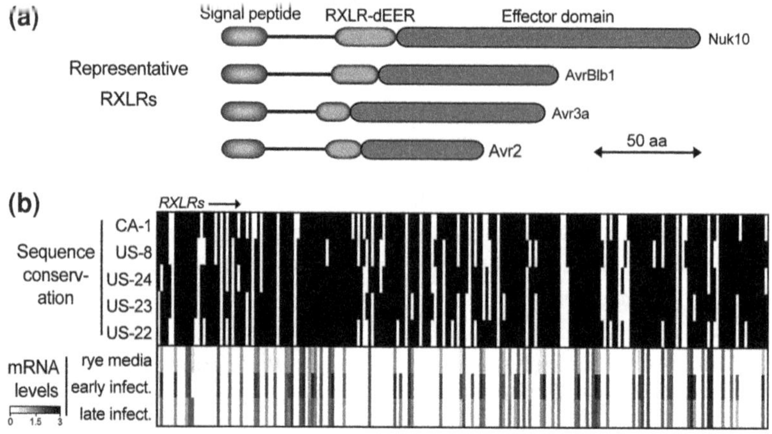


**Fig. 9.5** Structure, expression, and conservation of RXLRs in *P. infestans.* **a** Composition of four typical RXLRs indicating N-terminal signal peptide, RXLR-dEER region, and variable C-terminal effector domain. **b** Conservation in different isolates and expression pattern of representative RXLRs. Genes from the T30-4 assembly were matched to the sequenced genomes of isolates of genotypes CA-1, US-8, US-22, US-23, and US-24; *black* or *white* shading represent presence or absence of a matching sequence, respectively. mRNA levels were determined by RNA-seq analysis of a rye media culture, and early and late-stage infections of tomato leaflets; genes lacking matching sequence reads in all three stages are shown in *white*

number of active RXLRs in these species may be closer. Most reside within repeat-rich, gene-sparse regions of the genome, which may have facilitated their expansion in *P. infestans.* As with GIPs, the C-terminal regions of RXLR effectors show evidence of positive selection consistent with their coevolution with host factors. Interestingly, *Py. ultimum* does not encode RXLR-like proteins, which may be related to its necrotrophic life style.

The second major family of cytoplasmic effectors made by *P. infestans* are the Crinklers (CRNs). These are named after a "crinkling" or necrotic phenotype that many cause when expressed ectopically at high levels in *N. benthamiana* (Schornack et al. 2010). Some CRNs are required for virulence, suggesting that the cell death phenotype caused by ectopic expression may not reveal their true biological roles. CRNs are relatively large, with a median size of 428 amino acids. They contain a signal peptide plus adjacent LFLAK and DWL motifs required for translocation into the host cell, and some also contain a nuclear localization signal. At least 33 types of C-terminal domains have been identified in CRNs, including some with protein kinase activity (Haas et al. 2009). CRNs tend to be expressed constitutively, unlike most other virulence proteins. Like RXLRs, CRNs are found predominantly in repeat-rich regions of the *P. infestans* genome where they have expanded in copy number. While *P. infestans* encodes 196 CRNs, *P. ramorum* and *P. sojae* are predicted to make 19 and 100 CRNs. CRNs are also present but at relatively low numbers in the white rust, downy mildew, and *Py. ultimum* (Table 9.2).

Effectors not only differ in number between species, they also show diversity within *P. infestans.* Both CRNs and RXLRs vary in copy number between isolates, and exhibit higher rates of nonsynonymous substitutions than non-effector genes. One particularly aggressive isolate of *P. infestans* was found to express several RXLRs at higher levels than other strains (Cooke et al. 2012). The rapid evolution of RXLRs and CRNs through codon changes and recombination within their C-terminal domains was also observed in a comparison of *P. infestans* with other clade 1c species such as *P. andina, P. ipomoeae, P. mirabilis,* and *P. phaseoli* (Raffaele et al. 2010a). This was taken as being consistent with host jumps within the clade, in which an ancestral species evolved its effector repertoire to adapt to new hosts.

Several themes have emerged from studies of effectors. One is that their architectures are modular, with C-terminal regions conferring effector activity and N-terminal regions directing localization. A second theme is that their genes are more likely to reside in the gene-sparse and plastic regions of the genome. A third but related theme is that most are members of families, in which sequences are evolving rapidly and diversifying selection is common. A final theme is that the expression profile of each family is not monolithic; some members are expressed in artificial media and *in planta* while others are plant-specific, some are induced in preinfection stages (germinated cysts and appressoria), and some are expressed predominantly at early or late (biotrophic or necrotrophic) stages of infection. These patterns suggest that expression of the effector repertoire is carefully orchestrated, with some genes hard-wired to the developmental program of *P. infestans* and others expressed in response to plant signals.

### 9.3.6 Effector Genomics and Agricultural Solutions

Genome resources enable several strategies for improving the use of *R* genes against late blight. Since all known *R* genes active against *P. infestans* detect RXLRs, these can be used to screen potential sources of resistance as part of "effectoromics" approaches (Vleeshouwers et al. 2011). In addition, information about effector diversity in *P. infestans* can help predict which *R* genes will be most durable and manage their use.

Effectoromics entails using effectors to probe plant germplasm for new *R* genes, or identify the pathogen gene reacting with a known *R* gene. In the late blight system, Potato Virus X or *A. tumefaciens* rather than *P. infestans* is most commonly used to deliver the effector. Plants are checked for cell death which signals effector-*R* gene recognition. In early studies, this was done using libraries of effector candidates (mostly but not exclusively RXLRs) selected from cDNA libraries, and later using sequences mined from the genome sequence; in one study, 63 effectors were tested against 13 lines of potato (Torto et al. 2003). If the *R* gene is cloned, a screen of potential avirulence genes can also be performed by expressing those candidates with the *R* gene in a surrogate plant such as *N. benthamiana.* When hunting for new *R* genes, subsequent analyses of segregating plant populations can indicate whether one or more plant genes are responsible for recognition; this was used to isolate *Rpi-sto1* and *Rpi-pta1* from *S. stoloniferum,* for example (Vleeshouwers et al. 2008). The approach of testing effectors against known *R* genes was used to identify *Avrblb1* and *Avrblb2,* which interact with *Rpi-blb1* and *Rpi-blb2* from the Mexican wild potato *S. bulbocastanum* (Oh et al. 2009).

In a search for a new *R* gene, it would be ideal if effectors used to probe germplasm resources were widespread throughout *P. infestans.* Most *R* genes identified in the past failed in the field due to diversity within *P. infestans* populations. Consequently, an effort coordinated among several European groups has focused on the subset of RXLRs that are well-conserved, such as certain members of the Avr3a family (Whisson et al. 2011). Tests for intraspecific variation in RXLRs by polymerase chain reaction (PCR) have started to shift towards whole-genome sequencing. Of the more than 500 RXLRs predicted in a typical genome, less than 50 are both highly conserved between strains and expressed during infection (Fig. 9.5b), and transient and stable gene silencing experiments have identified a subset that contribute to virulence. These genes may be essential to *P. infestans* and therefore represent good targets for *R* gene screens.

The deployment of *R* genes can also be improved by monitoring *P. infestans* populations. Once the *P. infestans* protein that is recognized by an *R* gene is known, isolates can be checked by PCR instead of plant-based phenotyping which would allow many isolates to be tested in a short time-frame. This could help detect *P. infestans* strains that evade *R* gene recognition before they reach epidemic proportions, and provide advance information to allow growers to shift to new *R* genes in the next growing period (Champouret et al. 2009; Halterman et al. 2010). Another application of effector monitoring is to provide growers with data about the effectiveness of *R* genes in their fields against local isolates of *P. infestans.* This may allow growers to reduce fungicide use.

### 9.3.7 Genomics and Crop Protection Chemicals

Genome resources also have applications in developing oomyceticides (commonly but improperly referred to as fungicides) and understanding their modes of action. Besides the potential of using genome data to identify targets for new chemicals, expression profiling may help reveal modes of action for agents discovered by other routes. Fluopicolide, for example, was found to misregulate genes for vesicle transport in *Phytophthora,* which was consistent with cytological data that suggested it affected the cytoskeleton (Toquin et al. 2009). In the case of mandipropamid, biochemical and cell biology data suggested an effect on cell wall biosynthesis; using the *P. infestans* genome sequence, it was possible to identify the genes encoding cellulose synthase, sequence them from sensitive and resistant strains, and confirm that they were the target (Blum et al. 2010). A similar approach helped demonstrate that a subunit of RNA polymerase I is the target of phenylamides (Whisson 2013).

Genomics and transcriptomics data can also inform business decisions and facilitate approvals of oomyceticides by regulatory bodies (Cools and Hammond-Kosack 2013). For example, suggestions about a chemical's effect on *P. infestans* based on transcriptome responses can allow a company to ensure that chemical leads being pursued do not target the same pathway, thus reducing business risk. It should be noted that this approach is challenged by the difficulty of distinguishing changes in the primary biochemical target from secondary responses. Nevertheless, regulatory approvals could also be aided by assessing the risk of resistance by profiling the genomic response of the pathogen. Even if resistance occurs, if its genetic basis is known then DNA assays can be used to predict isolate sensitivity and inform growers in time for management decisions.

### 9.3.8 Insights into Metabolism

Metabolism is an under-studied topic in *P. infestans,* but important due to the potential of finding inhibitors for crop protection applications and yielding insight into evolution. Metabolism and nutrient acquisition are also central to the pathobiology of *P. infestans,* which must utilize host compounds for growth. It is generally assumed that *Phytophthora* spp. species acquire nutrients primarily through haustoria, which are specialized hyphae that insert into living plant cells, as do fungi such as rusts and powdery mildews (Staples 2001). The role of oomycete haustoria in nutrient uptake remains to be proved, in contrast to their role in delivering effectors to plant cells (Whisson et al. 2007).

Metabolic enzymes discovered from the genome sequence indicate that *P. infestans* is largely autotrophic and can use diverse molecules as substrates for growth. *P. infestans* however grows rather poorly on most defined media. Only limited studies have been performed of the capacity of *P. infestans* to modulate its gene expression to different growth substrates. This may be restricted in *P. infestans* due to its adaptation to growth on living plant tissue as opposed to survival as a saprophyte. Nevertheless, some pathways vary depending on the host tissue (leaf versus tuber; Judelson et al. 2009b). In vitro, changes have been reported in nitrate reductase and fatty acid metabolism enzymes depending on the media employed (Pieterse et al. 1995; Chen et al. 2013).

*P. infestans* apparently encodes enzymes capable of utilizing many of the mono- and disaccharides abundant in plant tissues. Interestingly, as indicated in Fig. 9.6, four of the glycolytic enzymes seem to have been acquired through horizontal or endosymbiotic gene transfer (glucokinase, 6-phosphofructokinase, pyruvate kinase, phosphoglycerate kinase, and glucose-6-phosphate isomerase; Whitaker et al. 2009). For example, five glucokinases resemble red algal or cyanobacterial enzymes, and one is most similar to a bacterial ROK family glucokinase. The first five are found in the same region of the genome, which is consistent with the trend of localized expansion of gene families seen throughout *Phytophthora.* Also present are three invertases for breaking down sucrose into glucose and fructose, which have the closest affinity with fungal invertases (Fig. 9.6). Each invertase contains signal peptides which suggests that they are secreted into the plant apoplast or reside in the cell wall, and their genes are clustered in the genome. Specialized fructokinases are absent, although this function could be provided by the glucokinases. Also present are enzymes for each step of glycolysis, the citric acid cycle, and for shunting carbon from five-carbon sugars into glycolysis through the pentose phosphate pathway. A large family of enzymes for breaking down fatty acids into acetyl-CoA are also present in *P. infestans.*

Certain deviations from conventional glycolysis occur in *P. infestans.* Several enzymes utilize pyrophosphate ($PP_i$) instead of ATP, for example. This may free up ATP for other processes, and be an adaptation to the relatively high levels of $PP_i$ common in potato tissue (Farre et al. 2000). One such enzyme is a pyruvate phosphate dikinase, which converts phosphoenolpyruvate, AMP and $PP_i$ into pyruvate and ATP as a bypass to pyruvate kinase in glycolysis, and in the reverse direction forms phosphoenolpyruvate for gluconeogenesis. The presence of this enzyme in an oomycete is of interest since prior studies showed it was largely limited to plants and bacteria (Marshall et al. 2001; Slamovits and Keeling 2006), and since it

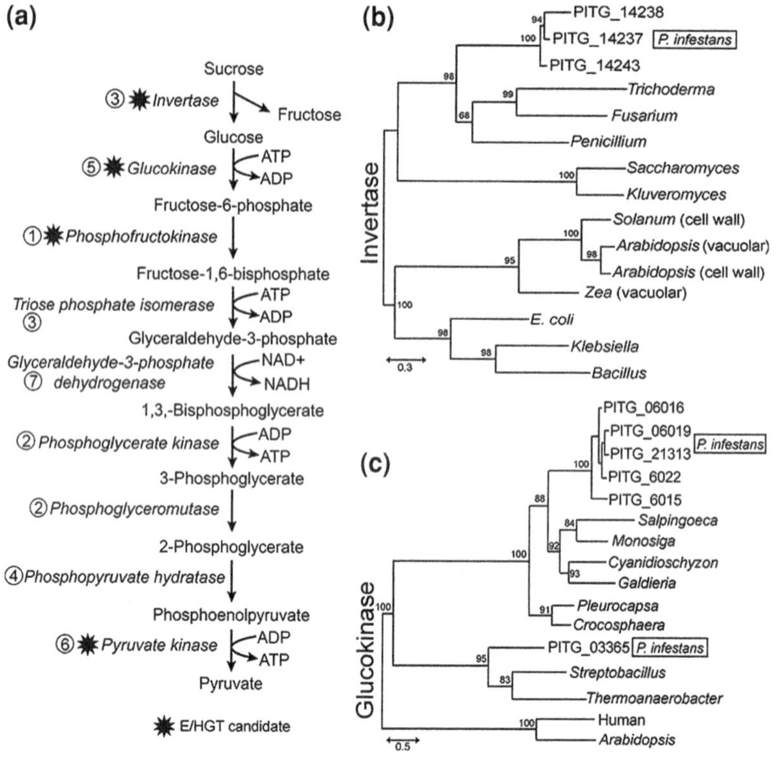


**Fig. 9.6** Enzymes for utilization of sucrose and glucose in *P. infestans*. **a** Steps through glycolysis. *Circled numbers* indicate the gene count for each activity in the reference genome, and *stars* indicate genes proposed to have been acquired by horizontal or endosymbiotic gene transfer (Whitaker et al. 2009). **b** Phylogenetic tree of invertases from *P. infestans* (PITG nomenclature), filamentous fungi *(Trichoderma, Fusarium, Penicillium)*, yeasts *(Saccharomyces, Kluveromyces)*, plants (*Solanum esculentum, Arabidopsis, Zea mays*), and bacteria *(E. coli, Klebsiella, Bacillus)*. **c** Tree of glucokinases from *P. infestans,* choanoflagellates *(Salpingoeca, Monosiga)*, *red* algae *(Cyanidioschyzon, Galdieria)*, cyanobacteria *(Pleurocapsa, Crocosphaera)*, bacteria *(Streptobacillus, Thermoanaerobacter)*, and human and *Arabidopsis* (both predicted hexokinases)

coexists with the more typical ATP-dependent pyruvate kinase. *P. infestans* also encodes a $PP_i$-dependent 6-phosphofructose-1-kinase, but not the ATP-dependent form which is typical of animals and bacteria.

At least part of glycolysis may occur in *P. infestans* mitochondria, compared to its textbook location in the cytosol. While *P. infestans* encodes several canonical cytosolic triose phosphate isomerases (TPI) and glyceraldehyde-3-phosphate dehydrogenases (GAPDH), it also makes TPI-GAPDH fusion proteins that bear predicted mitochondrial localization sequences (Liaud et al. 2000; Nakayama et al. 2012). Also predicted as mitochondrial are enzymes from the entire last half of glycolysis, namely phosphoglycerate kinase, phosphoglycerate mutase, enolase, and pyruvate kinase. The same enzymes are also predicted to be mitochondrial in other stramenopiles, suggesting they were an ancient acquisition. Whether glycolysis in *P. infestans* truly involves both cytosolic and mitochondrial forms of these enzymes requires confirmation, but could explain a prior finding that different genes encoding the same activity showed divergent patterns of expression during growth and pathogenesis (Judelson et al. 2009a, b).

Pathways used by *P. infestans* for storing carbon are also of interest, since its spores must rely on reserves until a feeding relationship is established with the host. Unlike fungi, glycogen and sugar alcohols do not accumulate significantly in *P. infestans* and relatives (Pfyffer et al. 1986). Carbon storage instead involves mainly lipids and $\beta$-1,3-glucans (Bartnicki-Garcia and Wang 1983). Enzymes for their synthesis and breakdown ($\beta$-1,3-glucan synthase and $\beta$-1,3-glucosidase) can be identified in the genome. Interestingly, a gene encoding a chitin synthase-like protein is also present, even though *P. infestans* has no measurable chitin or chitin synthase activity (Grenville-Briggs et al. 2013).

Another interesting feature of metabolism, which was first discovered in a *P. infestans* expressed sequence tag project, are phosphagen kinases. Such enzymes are well-described in animals where they play roles in energy homeostasis by transferring a high energy phosphoryl group from ATP to an acceptor molecule, and vice versa. The *Phytophthora* genes may encode taurocyamine kinases (Uda et al. 2013). Since some are induced at the same time as structural genes for flagella, the phosphagen kinases have been proposed to help maintain ATP levels in the motile stage (Kim and Judelson 2003). Consistent with this is the observation that *H. arabidopsidis,* which lost the ability to produce flagella, lacks phosphagen kinases (Judelson et al. 2012).

The *P. infestans* genome indicates that nitrogen sources usable by *P. infestans* include nitrate, nitrite, ammonium, and amino acids. Most amino acid biosynthesis pathways appear standard, except for lysine. While true fungi make lysine using the $\alpha$-aminodipic acid pathway, *P. infestans* and other oomycetes make this amino acid through the diaminopimelic pathway. Several of the *P. infestans* genes in this pathway have been proposed as candidates for horizontal gene transfer from archaea (Richards et al. 2011).

The number of bioactive secondary metabolites produced by *Phytophthora* is limited compared to that of phytopathogenic fungi. It is therefore not surprising that *P. infestans* expresses only about 30 cytochrome P450s, which are heme-containing monoxygenases used to produce and degrade diverse metabolites. By comparison, phytopathogenic fungi addressed by a recent study were each found to encode an average of 116 P450s (Park et al. 2008). Such enzymes are thought to be important for inactivating phytoalexins, including those produced by potato and tomato. In *P. infestans,* degrading phytoalexins may be a low priority compared to suppressing their biosynthesis using effectors. *P. infestans* also lacks the CYP51-type cytochrome P450 enzymes needed to make sterols, which must be obtained from an external source to support maximum growth and sexual reproduction (Gaulin et al. 2010).

Like other haustoria-forming oomycetes, *P. infestans* lacks the gene for synthesizing thiamine, which is essential for growth and must be obtained from a host or artificial media. This contrasts with fungi that produce haustoria, which not only make thiamine but generate the highest levels in haustoria (Sohn et al. 2000). It is possible that haustorial oomycetes lost this ability as part of its stealthy infection strategies, as thiamine is known to induce the synthesis of phytoalexins through the phenylpropanoid pathway in some plants (Boubakri et al. 2013).

### 9.3.9 Transporters for Growth and Pathogenesis

*P. infestans* is predicted to encode approximately 770 proteins involved in cross-membrane transport of ions ($Ca^{2+}$, $K^+$, $Na^+$, and $Cl^-$), nutrients and other metabolites (glucose, fructose, amino acids, ATP), and water. Of specific interest for plant pathogens are those involved in the acquisition of host nutrients and toxicant efflux.

About 210 transporters are predicted to play roles in the transport of carbonaceous growth substrates (mostly sugars and organic acids) and potential nitrogen sources (ammonium, nitrate, nucleosides). This value excludes predicted mitochondrial transporters and ABC transporters (which are mostly involved in efflux), and thus represents an estimate of the number of proteins that may be used to acquire nutrients from the plant. The ten largest families in *P. infestans* are shown in Fig. 9.7a. The largest group in *P. infestans* is the major facilitator superfamily (MFS), which contains 102 members predicted to transport sugars, amino acids, organic acids,

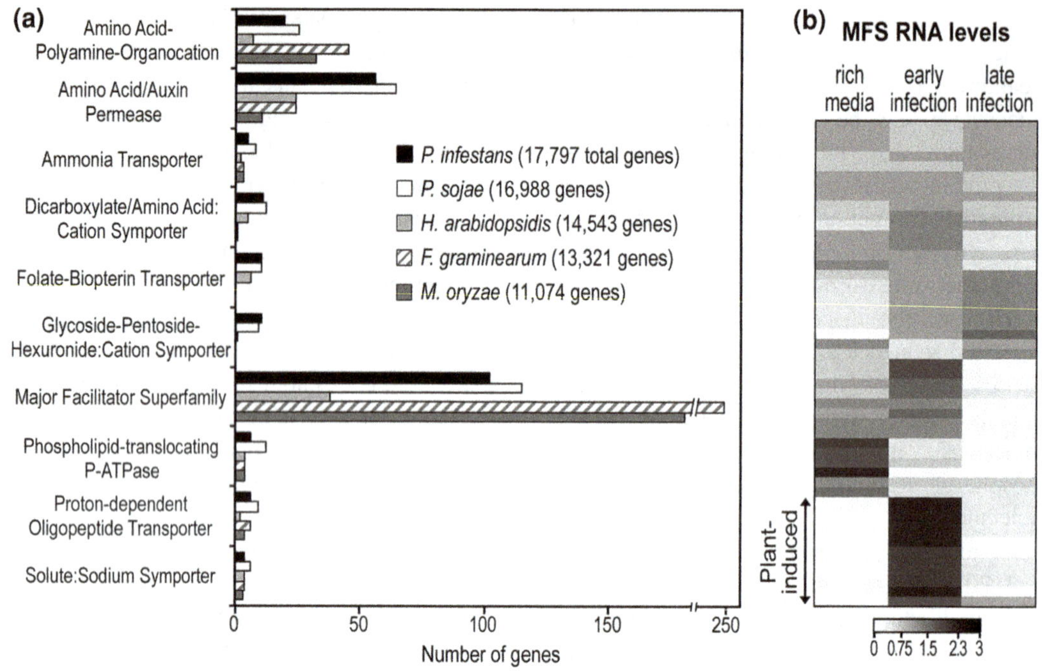


**Fig. 9.7** Transporter families with potential roles in *P. infestans* nutrition. **a** Major groups implicated in nutrient uptake, showing gene numbers in three oomycetes and two fungi. **b** Expression pattern of major facilitator superfamily (MFS) from *P. infestans* grown on rich media, or from early and late stages of infection

nitrate and phosphate. *P. infestans* encodes more transporters than fungi in each category except the MFS group, which is twice as large in *F. graminearum* and *M. oryzae*. The number of genes for transporters in *P. infestans* and *P. sojae* are similar, about 1.5 % of total genes, but twice that of *H. arabidopsidis*. Several transporters are up-regulated during leaf infection by *P. infestans,* particularly during the biotrophic stage, as illustrated for MFS proteins in Fig. 9.7b. These might play roles related to transporters of fungal rusts that are found primarily in haustoria (Hahn et al. 1997).

ABC (ATP-binding cassette) transporters represent another large gene family in *P. infestans.* Such proteins are best known for their efflux activities, which in plant pathogens may involve exporting toxic phytoalexins produced by plants, antimicrobials produced by competing microflora and fungicides (Del Sorbo et al. 2000). The proteins generally have two transmembrane and two nucleotide binding domains (TMD and NBD, respectively), and are part of a slightly larger ABC superfamily that includes some TMD-lacking nontransporters. *P. infestans* encodes 137 ABC proteins, similar to *P. ramorum* and *P. sojae* which have 135 and 136, respectively (Fig. 9.8). These are classified into seven main families (A to G) based on sequence and domain orientation (Paumi et al. 2009). The transporter genes encode either full or half-transporters, depending on whether they encode a $TMD_2$-$NBD_2$ protein or TMD-NBD monomers that must dimerize for activity. *P. infestans* makes 11 full and 0 half-transporters in the ABCA family, as well as 8 full and 8 half ABCB, 25 full ABCC, 1 half ABCD, and 45 full and 23 half ABCG transporters. *P. infestans* also encodes 1 and 7 members of the nontransporter ABCE and ABCF families, respectively. The transporters most likely to participate in toxicant efflux are the ABCB family, also known from studies in animal cells as the multidrug resistance group, the ABCC family or multi-drug resistance-associated protein group, and the ABCG family or pleiotropic drug resistance

(a)

| Species | # of ABC transporters |
|---|---|
| *P. infestans* | 137 |
| *P. ramorum* | 135 |
| *P. sojae* | 136 |
| *H. arabidopsidis* | 53 |
| *A. laibachii* | 41 |
| *Py. ultimum* | 169 |
| *T. pseudonana* | 57 |
| *F. graminearum* | 76 |
| *M. oryzae* | 60 |

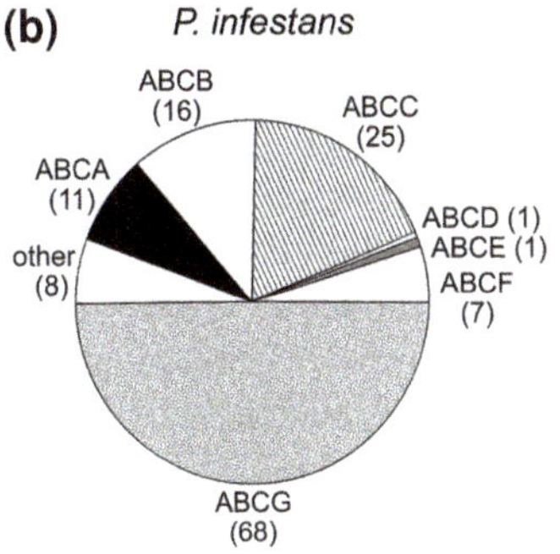


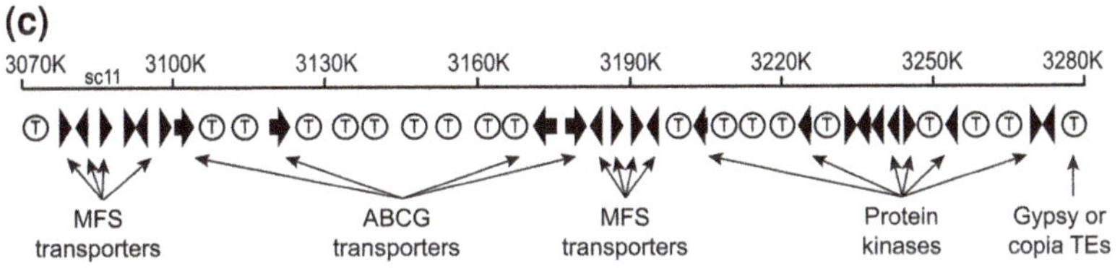


**Fig. 9.8** ABC transporter superfamilies. **a** Size of ABC families in five oomycetes, the diatom *T. pseudonana,* and the fungi *F. graminearum* and *M. oryzae.* **b** Number of *P. infestans* proteins in ABC subgroups. **c** Region of *P. infestans* genome containing expanded arrays of ABC transporters from subgroup G, MFS transporters, and protein kinases

group. Several are induced by treatment with oomyceticides, possibly as part of a general stress response (Judelson and Senthil 2006).

About three times fewer ABC transporters are in *A. laibachii* and *H. arabidopsidis* (Morris and Phuntumart 2009). This reduction may reflect the biotrophic life styles of the two latter species, which may cause fewer host toxins to be produced. *Py. ultimum* in contrast has 20 % more ABC transporters than *P. infestans,* which might be explained by the fact that like true fungi but unlike *Phytophthora, Pythium* can exist saprophytically in soil where defenses against microbes or environmental toxicants are more important. Nevertheless, plant pathogenic fungi such as *F. graminearum* and *M. oryzae* have far fewer ABC proteins than *P. infestans.* The largest difference exists in the ABCG family, with *M. oryzae* and *P. infestans* containing 11 and 68, respectively.

The genomic organization of the ABCG family helps explain its large size in *P. infestans.* At least 48 % of genes encoding ABCGs reside in clusters of 2 to 5 loci, suggestive of expansion through unequal crossing over. This is illustrated in Fig. 9.8c, which portrays a tandem array of four genes encoding full transporters. This array is not conserved in *P. sojae* where only 1 ABC transporter gene resides at the syntenous region. Despite this, the total number of ABC transporters in the two species has remarkably remained nearly identical, suggesting that expansions at one region are compensated by contractions elsewhere. Another remarkable observation is that this array of ABC transporters is embedded within an cluster of seven predicted sugar transporters from the unrelated MFS group, and flanks an array of four protein kinases. This region of the genome has apparently been especially prone to expansion.

### 9.3.10 Epigenetics and Phenotypic Diversity

Eukaryotes exhibit several forms of epigenetic regulation which are manifest as cytosine methylation, histone modifications, and small regulatory RNAs. Evidence exists only for the latter two in *P. infestans.* While genes predicted to encode histone methylases and acetylases are in the genome, cytosine methyltransferases seem absent. Components of canonical gene silencing pathways have also been detected including genes encoding five Argonaute, one RNA-directed RNA polymerase, and one Dicer-like protein (Vetukuri et al. 2011).

That *P. infestans* has the ability to regulate genes through small RNA (sRNA) pathways was suggested by the observation that inverted repeat transgenes can silence native genes, and confirmed when 21-nt RNAs matching the target genes were detected in such transformants (Ah-Fong et al. 2008). This was not surprising considering the prevalence of RNA interference in most eukaryotes, and the assumption that a mechanism must exist for taming the many transposable elements in the genome.

Recently, *P. infestans* was shown to produce sRNAs that range from 19–40 nt with three predominant size classes of 21, 25/26, and 32 nt (Vetukuri et al. 2012). By mapping sRNA reads to the genome assembly, all size classes were found to include sequences homologous to RXLR genes, while 21-nt sRNAs contained CRN-derived sequences. Also common were sRNAs matching DNA transposons and retrotransposons. An inverse relationship was observed between levels of sRNAs and mRNA from the corresponding RXLR and CRN genes. That RXLRs and CRNs might be subject to epigenetic control is consistent with their presence as repeat families, which are known to trigger RNA interference. Moreover, many *P. infestans* RXLR genes lack detectable transcripts based on microarray and RNA-seq studies (Fig. 9.5), and epigenetic processes have been invoked to explain the silence of certain RXLR loci in *P. sojae* (Qutob et al. 2013). Most effector genes are also within 2-kb of a transposable element, which may help to initiate silencing. It is interesting to place these findings in context with reports from a half century ago of reversible changes of virulence in strains of *P. infestans* (Watson 1970).

## 9.4 Future Perspectives

Genomics has provided valuable insight into the biology and evolution of *P. infestans* and relatives. Their complex genomes have been influenced by horizontal gene transfer from multiple kingdoms, transposable elements, and the formation of novel genes and expansion of families. The influence of epigenetics is apparent but just starting to be understood. Advances in tools for functional analysis have helped test the role of many genes, although more effort and technological improvements are needed.

Perhaps the largest impact of genomics has been on illuminating the role of effectors and their diversity between isolates. Such knowledge is being used to help understand population dynamics and phenotypic variation, and improve strategies for identifying and deploying R genes. Studies of effector targets in the host are also revealing pathways that could be engineered for improved disease resistance. Such translational studies should increase over the coming decade.

While much has been learned about effectors, most studies have targeted those that cause necrosis in a natural host or a surrogate such as *N. benthamiana*. This will identify effectors that activate or misregulate defense pathways, but other functions may be missed. Are there effectors that mobilize nutrients to infection sites? Do effectors influence the behavior of stomata, which remain open at the end of the disease cycle which allows hyphae to exit and sporulate? Another question concerns effectors that cause necrosis late in infection; does a necrotic stage truly benefit *P. infestans*? Whether effectors determine host-species specificity also remains to be learned. Many researchers once favored RXLRs for this role, but since these are absent from *Py. ultimum* and often little-conserved within a species, effectors that are more widespread such as CRNs may be responsible. Undoubtedly, the pathogen's ability to perceive and exploit host metabolites also plays a role in host specificity.

Our understanding of the biochemistry and nutrition of *P. infestans* has also been enhanced by genomics. Most metabolic pathways conform to traditional canons, but there are exceptions including novel enzymes and steps that may occur in multiple cellular compartments. The genome data should also make it easier to assess the role of haustoria in metabolite uptake, through studies of *P. infestans* transporters. Studies of metabolism should also contribute to translational research on oomyceticides, for example by helping to identify promising targets. Another application would be to understand the basis of variation in oomyceticide sensitivity between isolates, which may result from sequence or expression level differences in the target site, efflux pathways, or degradative enzymes. When new insensitive strains are found, the new sequencing technologies could be used to characterize them within days of their discovery. The resulting data could be translated into revised chemistries, and DNA-based diagnostics to help growers assess how to control their local isolates of *P. infestans.*

Genomics also has the power to dissect many traits in *P. infestans* besides fungicide sensitivity, such as general or host-specific aggressiveness and latent period. These are likely complex traits but could be dissected through SNP analysis of genetic crosses or genome-wide association studies. This could translate into disease management tools, allowing growers to refine their application schedules of crop protection chemicals. Pursuing such approaches in an oomycete would have been unthinkable in the pre-genome era.

**Acknowledgments** This work was supported by the United States Department of Agriculture-National Institute of Food and Agriculture under grant numbers 2008-35600-04686 and 2011-68004-30154 and the National Science Foundation under grant number 0949426.

## References

Ah Fong A, Xiang Q, Judelson HS (2007) Architecture of the sporulation-specific Cdc14 promoter from the oomycete Phytophthora infestans. Eukaryot Cell 6:2222–2230

Ah Fong AM, Judelson HS (2003) Cell cycle regulator Cdc14 is expressed during sporulation but not hyphal growth in the fungus-like oomycete Phytophthora infestans. Molec Microbiol 50:487–494

Ah-Fong AM, Bormann-Chung CA, Judelson HS (2008) Optimization of transgene-mediated silencing in Phytophthora infestans and its association with small-interfering RNAs. Fungal Genet Biol 45:1197–1205

Ah-Fong AM, Judelson HS (2011) Vectors for fluorescent protein tagging in Phytophthora: tools for functional genomics and cell biology. Fungal Biol 115:882–890

Al-Kherb SM, Fininsa C, Shattock RC, Shaw DS (1995) The inheritance of virulence of Phytophthora infestans to potato. Plant Pathol 44:552–562

Avila-Adame C, Gomez-Alpizar L, Zismann V, Jones KM, Buell CR, Ristaino JB (2006) Mitochondrial genome sequences and molecular evolution of the Irish potato famine pathogen, Phytophthora infestans. Curr Genet 49:39–46

Avrova AO, Boevink PC, Young V, Grenville-Briggs LJ, van West P, Birch PR, Whisson SC (2008) A novel Phytophthora infestans haustorium-specific membrane protein is required for infection of potato. Cell Microbiol 10:2271–2284

Bartnicki-Garcia S, Wang MC (1983) Biochemical aspects of morphogenesis in Phytophthora. In: Erwin DC, Bartnicki-Garcia S, Tsao PH (eds) Phytophthora, its biology, taxonomy, ecology, and pathology. APS Press, St. Paul, pp 121–137

Baxter L, Tripathy S, Ishaque N, Boot N, Cabral A, Kemen E, Thines M, Ah-Fong A, Anderson R, Badejoko W, Bittner-Eddy P, Boore JL, Chibucos MC, Coates M, Dehal P, Delehaunty K, Dong S, Downton P, Dumas B, Fabro G, Fronick C, Fuerstenberg SI, Fulton L, Gaulin E, Govers F, Hughes L, Humphray S, Jiang RH, Judelson H, Kamoun S, Kyung K, Meijer H, Minx P, Morris P, Nelson J, Phuntumart V, Qutob D, Rehmany A, Rougon-Cardoso A, Ryden P, Torto-Alalibo T, Studholme D, Wang Y, Win J, Wood J, Clifton SW, Rogers J, Van den Ackerveken G, Jones JD, McDowell JM, Beynon J, Tyler BM (2010) Signatures of adaptation to obligate biotrophy in the Hyaloperonospora arabidopsidis genome. Science 330:1549–1551

Beakes GW, Glockling SL, Sekimoto S (2011) The evolutionary phylogeny of the oomycete fungi. Protoplasma 249:3–19

Blair JE, Coffey MD, Martin FN (2012) Species tree estimation for the late blight pathogen, Phytophthora infestans, and close relatives. PLoS ONE 7:e37003

Blair JE, Coffey MD, Park SY, Geiser DM, Kang S (2008) A multi-locus phylogeny for Phytophthora utilizing markers derived from complete genome sequences. Fungal Genet Biol 45:266–277

Blanco FA, Judelson HS (2005) A bZIP transcription factor from Phytophthora interacts with a protein kinase and is required for zoospore motility and plant infection. Molec Microbiol 56:638–648

Blum M, Boehler M, Randall E, Young V, Csukai M, Kraus S, Moulin F, Scalliet G, Avrova AO, Whisson SC, Fonne-Pfister R (2010) Mandipropamid targets the cellulose synthase-like PiCesA3 to inhibit cell wall biosynthesis in the oomycete plant pathogen, Phytophthora infestans. Molec Plant Pathol 11:227–243

Bos JIB, Armstrong MR, Gilroy EM, Boevink PC, Hein I, Taylor RM, Tian ZD, Engelhardt S, Vetukuri RR, Harrower B, Dixelius C, Bryan G, Sadanandom A,

Whisson SC, Kamoun S, Birch PRJ (2010) Phytophthora infestans effector AVR3a is essential for virulence and manipulates plant immunity by stabilizing host E3 ligase CMPG1. Proc Natl Acad Sci USA 107:9909–9914

Boubakri H, Poutaraud A, Wahab MA, Clayeux C, Baltenweck-Guyot R, Steyer D, Marcic C, Mliki A, Soustre-Gacougnolle I (2013) Thiamine modulates metabolism of the phenylpropanoid pathway leading to enhanced resistance to Plasmopara viticola in grapevine. BMC Plant Biol 13:31

Bourke PMA (1964) Emergence of potato blight 1843-46. Nature 203:805–806

Bozkurt TO, Schornack S, Win J, Shindo T, Ilyas M, Oliva R, Cano LM, Jones AME, Huitema E, van der Hoorn RAL, Kamoun S (2011) Phytophthora infestans effector AVRblb2 prevents secretion of a plant immune protease at the haustorial interface. Proc Natl Acad Sci USA 108:20832–20837

Cai G, Hillman BI (2013) Phytophthora viruses. Adv Virus Res 86:327–350

Cai G, Myers K, Hillman BI, Fry WE (2009) A novel virus of the late blight pathogen, Phytophthora infestans, with two RNA segments and a supergroup 1 RNA-dependent RNA polymerase. Virology 392:52–61

Cai GH, Myers K, Fry WE, Hillman BI (2012) A member of the virus family Narnaviridae from the plant pathogenic oomycete Phytophthora infestans. Arch Virol 157:165–169

Cardenas M, Grajales A, Sierra R, Rojas A, Gonzalez-Almario A, Vargas A, Marin M, Fermin G, Lagos LE, Grunwald NJ, Bernal A, Salazar C, Restrepo S (2011) Genetic diversity of Phytophthora infestans in the Northern Andean region. BMC Genet 12:23

Catal M, King L, Tumbalam P, Wiriyajitsomboon P, Kirk WW, Adams GC (2010) Heterokaryotic nuclear conditions and a heterogeneous nuclear population are observed by flow cytometry in Phytophthora infestans. Cytometry 77:769–775

Chen Y, Chi HY, Meesapyodsuk D, Qiu X (2013) Phytophthora infestans cholinephosphotransferase with substrate specificity for very-long-chain polyunsaturated fatty acids. Appl Environ Microbiol 79:1573–1579

Clark MC, Melanson DL, Page OT (1978) Purine metabolism and differential inhibition of spore germination in Phytophthora infestans. Can J Microbiol 24:1032–1038

Champouret N, Bouwmeester K, Rietman H, van der Lee T, Maliepaard C, Heupink A, van de Vondervoort PJ, Jacobsen E, Visser RG, van der Vossen EA, Govers F, Vleeshouwers VG (2009) Phytophthora infestans isolates lacking class I ipiO variants are virulent on Rpi-blb1 potato. Molec Plant-Microbe Interact 22:1535–1545

Cooke DE, Cano LM, Raffaele S, Bain RA, Cooke LR, Etherington GJ, Deahl KL, Farrer RA, Gilroy EM, Goss EM, Grunwald NJ, Hein I, MacLean D, McNicol JW, Randall E, Oliva RF, Pel MA, Shaw DS, Squires JN, Taylor MC, Vleeshouwers VG, Birch PR, Lees AK, Kamoun S (2012) Genome analyses of an aggressive and invasive lineage of the Irish potato famine pathogen. PLoS Pathog 8:e1002940

Cooke DEL, Drenth A, Duncan JM, Wagels G, Brasier CM (2000) A molecular phylogeny of Phytophthora and related oomycetes. Fungal Genet Biol 30:17–32

Cools HJ, Hammond-Kosack KE (2013) Exploitation of genomics in fungicide research: current status and future perspectives. Mol Plant Pathol 14:197–210

Cvitanich C, Judelson H (2003) Stable transformation of the oomycete, Phytophthora infestans, using microprojectile bombardment. Curr Genet 42:228–235

Damasceno CM, Bishop JG, Ripoll DR, Win J, Kamoun S, Rose JK (2008) Structure of the glucanase inhibitor protein (GIP) family from phytophthora species suggests coevolution with plant endo-beta-1,3-glucanases. Molec Plant-microbe Interact 21:820–830

Davila-Lopez M, Martinez-Guerra JJ, Samuelsson T (2010) Analysis of gene order conservation in eukaryotes identifies transcriptionally and functionally linked genes. PLoS ONE 5:e10654

DeBary A (1876) Researches into the nature of the potato fungus Phytophthora infestans. J Roy Agr Soc 12:9–34

Delgado RA, Monteros-Altamirano AR, Li Y, Visser RGF, van der Lee TAJ, Vosman B (2013) Large subclonal variation in Phytophthora infestans populations associated with Ecuadorian potato landraces. Plant Pathol 62:1081–1088

Del Sorbo G, Schoonbeek H-J, De Waard MA (2000) Fungal transporters involved in efflux of natural toxic compounds and fungicides. Fungal Genet Biol 30:1–15

Dong S, Kong G, Qutob D, Yu X, Tang J, Kang J, Dai T, Wang H, Gijzen M, Wang Y (2012) The NLP toxin family in Phytophthora sojae includes rapidly evolving groups that lack necrosis-inducing activity. Molec Plant-Microbe Interact 25:896–909

Dworkin J, Losick R (2005) Developmental commitment in a bacterium. Cell 121:401–409

Erwin DC, Ribeiro OK (1996) Phytophthora diseases worldwide. APS Press, St. Paul, Minn

Fabritius A-L, Judelson HS (1997) Mating-type loci segregate aberrantly in Phytophthora infestans but normally in Phytophthora parasitica: implications for models of mating-type determination. Curr Genet 32:60–65

Fabritius A-L, Shattock RC, Judelson HS (1997) Genetic analysis of metalaxyl insensitivity loci in Phytophthora infestans using linked DNA markers. Phytopathology 87:1034–1040

Farre EM, Geigenberger P, Willmitzer L, Trethewey RN (2000) A possible role for pyrophosphate in the coordination of cytosolic and plastidial carbon metabolism within the potato tuber. Plant Physiol 123:681–688

Feschotte C, Pritham EJ (2007) DNA transposons and the evolution of eukaryotic genomes. Annu Rev Genet 41:331–368

Flier WG, Grunwald NJ, Kroon LPNM, van den Bosch TBM, Garay-Serrano E, Lozoya-Saldana H, Bonants

PJM, Turkensteen LJ (2002) Phytophthora ipomoeae sp nov., a new homothallic species causing leaf blight on Ipomoea longipedunculata in the Toluca Valley of central Mexico. Mycol Res 106:848–856

Fry WE (2008) Phytophthora infestans: the plant (and R gene) destroyer. Molecular Plant Pathol 9:385–402

Fry WE, Grunwald NJ, Cooke DEL, McLeod A, Forbes GA, Cao K (2009) Population genetics and population diversity of Phytophthora infestans. In: Lamour K, Kamoun S (eds) Wiley, Hoboken, pp 139–164

Fry WE, McGrath MT, Seaman A, Zitter TA, McLeod A, Danies G, Small IM, Myers K, Everts K, Gevens AJ, Gugino BK, Johnson SB, Judelson H, Ristaino J, Roberts R, Secor G, Seebold K, Snover-Clift K, Wyenandt A, Grunwald NJ, Smart CD (2013) The 2009 late blight pandemic in the eastern United States—causes and results. Plant Dis 97:296–306

Gamboa-Melendez H, Huerta AI, Judelson HS (2013) bZIP transcription factors in the oomycete Phytophthora infestans with novel DNA-binding domains are Involved in defense against oxidative stress. Eukaryot Cell 12:1403–1412

Gaulin E, Bottin A, Dumas B (2010) Sterol biosynthesis in oomycete pathogens. Plant Signal Behav 5:258–260

Gaulin E, Haget N, Khatib M, Herbert C, Rickauer M, Bottin A (2007) Transgenic sequences are frequently lost in Phytophthora parasitica transformants without reversion of the transgene-induced silenced state. Can J Microbiol 53:152–157

Gaulin E, Jauneau A, Villalba F, Rickauer M, Esquerre-Tugaye MT, Bottin A (2002) The CBEL glycoprotein of Phytophthora parasitica var. nicotianae is involved in cell wall deposition and adhesion to cellulosic substrates. J Cell Sci 115:4565–4575

Gavino PD, Fry WE (2002) Diversity in and evidence for selection on the mitochondrial genome of Phytophthora infestans. Mycologia 94:781–793

Gijzen M, Nurnberger T (2006) Nep1-like proteins from plant pathogens: recruitment and diversification of the NPP1 domain across taxa. Phytochemistry 67: 1800–1807

Gomez-Alpizar L, Carbone I, Ristaino JB (2007) An Andean origin of Phytophthora infestans inferred from mitochondrial and nuclear gene genealogies. Proc Natl Acad Sci USA 104:3306–3311

Gong H, Kobayashi K, Sugi T, Takemae H, Kurokawa H, Horimoto T, Akashi H, Kato K (2012) A novel PAN/apple domain-containing protein from Toxoplasma gondii: characterization and receptor identification. PLoS ONE 7:e30169

Goss EM, Cardenas ME, Myers K, Forbes GA, Fry WE, Restrepo S, Grunwald NJ (2011) The plant pathogen Phytophthora andina emerged via hybridization of an unknown Phytophthora species and the Irish potato famine pathogen P. infestans. PloS One 6:e24543

Grenville-Briggs LJ, Anderson VL, Fugelstad J, Avrova AO, Bouzenzana J, Williams A, Wawra S, Whisson SC, Birch PR, Bulone V, van West P (2008) Cellulose synthesis in Phytophthora infestans is required for normal appressorium formation and successful infection of potato. Plant Cell 20:720–738

Grenville-Briggs LJ, Klinter S, Vilaplana F, Inman A, Melida H, Bulone V (2013) Cell wall biology to illuminate mechanisms of pathogenicity in Phytophthora infestans. In: Abstracts of the Oomycete molecular genetics network meeting, Pacific Grove, California, 10–12 March 2013

Griffith GW, Shaw DS (1998) Polymorphisms in Phytophthora infestans: four mitochondrial haplotypes are detected after PCR amplification of DNA from pure cultures or from host lesions. Appl Environ Microb 64:4007–4014

Haas BJ, Kamoun S, Zody MC, Jiang RH, Handsaker RE, Cano LM, Grabherr M, Kodira CD, Raffaele S, Torto-Alalibo T, Bozkurt TO, Ah-Fong AM, Alvarado L, Anderson VL, Armstrong MR, Avrova A, Baxter L, Beynon J, Boevink PC, Bollmann SR, Bos JI, Bulone V, Cai G, Cakir C, Carrington JC, Chawner M, Conti L, Costanzo S, Ewan R, Fahlgren N, Fischbach MA, Fugelstad J, Gilroy EM, Gnerre S, Green PJ, Grenville-Briggs LJ, Griffith J, Grunwald NJ, Horn K, Horner NR, Hu CH, Huitema E, Jeong DH, Jones AM, Jones JD, Jones RW, Karlsson EK, Kunjeti SG, Lamour K, Liu Z, Ma L, Maclean D, Chibucos MC, McDonald H, McWalters J, Meijer HJ, Morgan W, Morris PF, Munro CA, O'Neill K, Ospina-Giraldo M, Pinzon A, Pritchard L, Ramsahoye B, Ren Q, Restrepo S, Roy S, Sadanandom A, Savidor A, Schornack S, Schwartz DC, Schumann UD, Schwessinger B, Seyer L, Sharpe T, Silvar C, Song J, Studholme DJ, Sykes S, Thines M, van de Vondervoort PJ, Phuntumart V, Wawra S, Weide R, Win J, Young C, Zhou S, Fry W, Meyers BC, van West P, Ristaino J, Govers F, Birch PR, Whisson SC, Judelson HS, Nusbaum C (2009) Genome sequence and analysis of the Irish potato famine pathogen Phytophthora infestans. Nature 461:393–398

Hahn M, Neef U, Struck C, Gottfert M, Mendgen K (1997) A putative amino acid transporter is specifically expressed in haustoria of the rust fungus Uromyces fabae. Molec Plant-Microbe Interact 10:438–445

Halterman DA, Chen Y, Sopee J, Berduo-Sandoval J, Sanchez-Perez A (2010) Competition between Phytophthora infestans effectors leads to increased aggressiveness on plants containing broad-spectrum late blight resistance. PLoS ONE 5:e10536

Haverkort AJ, Boonekamp PM, Hutten R, Jacobsen E, Lotz LAP, Kessel GJT, Visser RGF, van der Vossen EAG (2008) Societal costs of late blight in potato and prospects of durable resistance through cisgenic modification. Potato Res 51:47–57

Haverkort AJ, Struik PC, Visser RGF, Jacobsen E (2009) Applied biotechnology to combat late blight in potato caused by Phytophthora infestans. Potato Res 52:249–264

Hazkani-Covo E, Zeller RM, Martin W (2010) Molecular poltergeists: mitochondrial DNA copies (numts) in sequenced nuclear genomes. PLoS Genet 6:e1000834

Hua C, Meijer HJ, de Keijzer J, Zhao W, Wang Y, Govers F (2013) GK4, a G-protein-coupled receptor with a phosphatidylinositol phosphate kinase domain in Phytophthora infestans, is involved in sporangia development and virulence. Molec Microbiol 88:352–370
Iyer LM, Anantharaman V, Wolf MY, Aravind L (2008) Comparative genomics of transcription factors and chromatin proteins in parasitic protists and other eukaryotes. Int J Parasitol 38:1–31
Judelson HS (2007) Sexual reproduction in plant pathogenic oomycetes. In: Heitman J, Kronstad J, Taylor J, Casselton L (eds) Sex in fungi, molecular determination and evolutionary implications. ASM press, Washington, DC, pp 445–458
Judelson HS, Ah-Fong AM (2010) The kinome of Phytophthora infestans reveals oomycete-specific innovations and links to other taxonomic groups. BMC Genom 11:700
Judelson HS, Ah-Fong AM, Aux G, Avrova AO, Bruce C, Cakir C, da Cunha L, Grenville-Briggs L, Latijnhouwers M, Ligterink W, Meijer HJ, Roberts S, Thurber CS, Whisson SC, Birch PR, Govers F, Kamoun S, van West P, Windass J (2008) Gene expression profiling during asexual development of the late blight pathogen Phytophthora infestans reveals a highly dynamic transcriptome. Molec Plant-Microbe Interact 21:433–447
Judelson HS, Ah-Fong AM, Fabritius AL (2010) An RNA symbiont enhances heat tolerance and secondary homothallism in the oomycete Phytophthora infestans. Microbiology 156:2026–2034
Judelson HS, Blanco FA (2005) The spores of Phytophthora: weapons of the plant destroyer. Nature Microbiol Rev 3:47–58
Judelson HS, Coffey MD, Arredondo FR, Tyler BM (1993a) Transformation of the oomycete pathogen Phytophthora megasperma f. sp. glycinea occurs by DNA integration into single or multiple chromosomes. Curr Genet 23:211–218
Judelson HS, Dudler R, Pieterse CMJ, Unkles SE, Michelmore RW (1993b) Expression and antisense inhibition of transgenes in Phytophthora infestans is modulated by choice of promoter and position effects. Gene 133:63–69
Judelson HS, Narayan RD, Ah-Fong AM, Kim KS (2009a) Gene expression changes during asexual sporulation by the late blight agent Phytophthora infestans occur in discrete temporal stages. Mol Genet Genom 281:193–206
Judelson HS, Narayan RD, Tani S (2009b) Metabolic adaptation of Phytophthora infestans during growth on leaves, tubers, and artificial media. Molec Plant Pathol 10:843–855
Judelson HS, Senthil GS (2006) Investigating the role of ABC transporters in multifungicide insensitivity in Phytophthora infestans. Molec Plant Pathol 7:17–29
Judelson HS, Shrivastava J, Manson J (2012) Decay of genes encoding the oomycete flagellar proteome in the downy mildew Hyaloperonospora arabidopsidis. PLoS ONE 7:e47624
Judelson HS, Spielman LJ, Shattock RC (1995) Genetic mapping and non-Mendelian segregation of mating type loci in the oomycete, Phytophthora infestans. Genetics 141:503–512
Judelson HS, Tani S (2007) Transgene-induced silencing of the zoosporogenesis-specific PiNIFC gene cluster of Phytophthora infestans involves chromatin alterations. Eukaryot Cell 6:1200–1209
Judelson HS, Tyler BM, Michelmore RW (1991) Transformation of the oomycete pathogen, Phytophthora infestans. Mol. Plant Microbe Interact 4:602–607
Judelson HS, Tyler BM, Michelmore RW (1992) Regulatory sequences for expressing genes in oomycete fungi. Molec Gen Genet 234:138–146
Kamoun S, Hraber P, Sobral B, Nuss D, Govers F (1999) Initial assessment of gene diversity for the oomycete pathogen Phytophthora infestans based on expressed sequences. Fungal Genet Biol 28:94–106
Kamoun S, van West P, Vleeshouwers VG, de Groot KE, Govers F (1998) Resistance of Nicotiana benthamiana to Phytophthora infestans is mediated by the recognition of the elicitor protein INF1. Plant Cell 10:1413–1426
Kay J, Meijer HJ, ten Have A, van Kan JA (2011) The aspartic proteinase family of three Phytophthora species. BMC Genom 12:254
Keeling PJ (2009) Chromalveolates and the evolution of plastids by secondary endosymbiosis. J Eukaryot Microbiol 56:1–8
Kemen E, Gardiner A, Schultz-Larsen T, Kemen AC, Balmuth AL, Robert-Seilaniantz A, Bailey K, Holub E, Studholme DJ, Maclean D, Jones JD (2011) Gene gain and loss during evolution of obligate parasitism in the white rust pathogen of Arabidopsis thaliana. PLoS Biol 9:e1001094
Kildea S, Quinn L, Mehenni-Ciz J, Cooke DEL, Perez FM, Deahl KL, Griffin D, Cooke LR (2013) Re-emergence of the Ib mitochondrial haplotype within the British and Irish Phytophthora infestans populations. Eur J Plant Pathol 135:237–242
Kim KS (2006) Molecular mechanisms governing the development of sporangia in the oomycete phytopathogen Phytophthora infestans. Ph.D. thesis, University of California-Riverside
Kim KS, Judelson HS (2003) Sporangia-specific gene expression in the oomyceteous phytopathogen Phytophthora infestans. Eukaryot Cell 2:1376–1385
Kojima KK, Jurka J (2011) Crypton transposons: identification of new diverse families and ancient domestication events. Mob DNA 2:12
Kruglyak S, Tang H (2000) Regulation of adjacent yeast genes. Trends Genet 16:109–111
Lapwood DH (1977) Factors affecting the field infection of potato tubers of different cultivars by blight (Phytophthora infestans). Ann Appl Biol 85:23–42
Latijnhouwers M, Govers F (2003) A Phytophthora infestans G-protein & $\beta$-subunit is involved in sporangium formation. Eukaryot Cell 2:971–977
Latijnhouwers M, Ligterink W, Vleeshouwers VGAA, Van West P, Govers F (2004) A G-alpha subunit

controls zoospore motility and virulence in the potato late blight pathogen Phytophthora infestans. Molec Microbiol 51:925–936

Lee TY, Mizubuti E, Fry WE (1999) Genetics of metalaxyl resistance in Phytophthora infestans. Fungal Genet Biol 26:118–130

Lees AK, Wattier R, Shaw DS, Sullivan L, Wn A, Cooke DEL (2006) Novel microsatellite markers for the analysis of Phytophthora infestans populations. Plant Pathol 55:311–319

Levesque CA, Brouwer H, Cano L, Hamilton JP, Holt C, Huitema E, Raffaele S, Robideau GP, Thines M, Win J, Zerillo MM, Beakes GW, Boore JL, Busam D, Dumas B, Ferriera S, Fuerstenberg SI, Gachon CM, Gaulin E, Govers F, Grenville-Briggs L, Horner N, Hostetler J, Jiang RH, Johnson J, Krajaejun T, Lin H, Meijer HJ, Moore B, Morris P, Phuntmart V, Puiu D, Shetty J, Stajich JE, Tripathy S, Wawra S, van West P, Whitty BR, Coutinho PM, Henrissat B, Martin F, Thomas PD, Tyler BM, De Vries RP, Kamoun S, Yandell M, Tisserat N, Buell CR (2010) Genome sequence of the necrotrophic plant pathogen Pythium ultimum reveals original pathogenicity mechanisms and effector repertoire. Genome Biol 11:R73

Li Y, van der Lee T, Zhu JH, Jin GH, Lan CZ, Zhu SX, Zhang RF, Liu BW, Zhao ZJ, Kessel G, Huang SW, Jacobsen E (2012a) Population structure of Phytophthora infestans in China—geographic clusters and presence of the EU genotype Blue_13. Plant Pathol 62:932–942

Li Y, van der Lee TA, Evenhuis A, van den Bosch GB, van Bekkum PJ, Forch MG, van Gent-Pelzer MP, van Raaij HM, Jacobsen E, Huang SW, Govers F, Vleeshouwers VG, Kessel GJ (2012b) Population dynamics of Phytophthora infestans in the Netherlands reveals expansion and spread of dominant clonal lineages and virulence in sexual offspring. Genes Genom Genet 2:1529–1540

Liaud MF, Lichtle C, Apt K, Martin W, Cerff R (2000) Compartment specific isoforms of TPI and GAPDH are imported into diatom mitochondria as a fusion protein: evidence in favor of a mitochondrial origin of the eukaryotic glycolytic pathway. Molec Biol Evol 17:213–223

Marshall JS, Ashton AR, Govers F, Hardham AR (2001) Isolation and characterization of four genes encoding pyruvate, phosphate dikinase in the oomycete plant pathogen Phytophthora cinnamomi. Curr Genet 40:73–81

Martens C, Van de Peer Y (2010) The hidden duplication past of the plant pathogen Phytophthora and its consequences for infection. BMC Genom 11:353

Martens C, Vandepoele K, Van de Peer Y (2008) Whole-genome analysis reveals molecular innovations and evolutionary transitions in chromalveolate species. Proc Natl Acad Sci USA 105:3427–3432

Martin MD, Cappellini E, Samaniego JA, Zepeda ML, Campos PF, Seguin-Orlando A, Wales N, Orlando L, Ho SY, Dietrich FS, Mieczkowski PA, Heitman J, Willerslev E, Krogh A, Ristaino JB, Gilbert MT (2013) Reconstructing genome evolution in historic samples of the Irish potato famine pathogen. Nature Comm 4:2172

Maumus F, Rabinowicz P, Bowler C, Rivarola M (2011) Stemming epigenetics in marine stramenopiles. Curr Genom 12:357–370

McLeod A, Smart CD, Fry WE (2004) Core promoter structure in the oomycete Phytophthora infestans. Eukaryot Cell 3:91–99

McNeill WH (1999) How the potato changed the world's history. Soc Res 66:67–83

Meijer HJ, Govers F (2006) Genomewide analysis of phospholipid signaling genes in Phytophthora spp.: novelties and a missing link. Molec Plant-Microbe Interact 19:1337–1347

Meijer HJ, Hassen HH, Govers F (2011) Phytophthora infestans has a plethora of phospholipase D enzymes including a subclass that has extracellular activity. PLoS ONE 6:e17767

Morris PF, Phuntumart V (2009) Inventory and comparative evolution of the ABC superfamily in the genomes of Phytophthora ramorum and Phytophthora sojae. J Molec Evol 68:563–575

Morris PF, Schlosser LR, Onasch KD, Wittenschlaeger T, Austin R, Provart N (2009) Multiple horizontal gene transfer events and domain fusions have created novel regulatory and metabolic networks in the oomycete genome. PLoS ONE 4:e6133

Mort-Bontemps M, Fevre M (1997) Transformation of the oomycete Saprolegnia monoica to hygromycin B resistance. Curr Genet 31:272–275

Nakayama T, Ishida K, Archibald JM (2012) Broad distribution of TPI-GAPDH fusion proteins among eukaryotes: evidence for glycolytic reactions in the mitochondrion? PLoS ONE 7:e52340

Oh SK, Young C, Lee M, Oliva R, Bozkurt TO, Cano LM, Win J, Bos JI, Liu HY, van Damme M, Morgan W, Choi D, Van der Vossen EA, Vleeshouwers VG, Kamoun S (2009) In planta expression screens of Phytophthora infestans RXLR effectors reveal diverse phenotypes, including activation of the Solanum bulbocastanum disease resistance protein Rpi-blb2. Plant Cell 21:2928–2947

Ojika M, Molli SD, Kanazawa H, Yajima A, Toda K, Nukada T, Mao H, Murata R, Asano T, Qi J, Sakagami Y (2011) The second Phytophthora mating hormone defines interspecies biosynthetic crosstalk. Nature Chem Biol 7:591–593

Oliva RF, Kroon LPNM, Chacon G, Flier WG, Ristaino JB, Forbes GA (2010) Phytophthora andina sp nov., a newly identified heterothallic pathogen of solanaceous hosts in the Andean highlands. Plant Pathol 59:613–625

Orsomando G, Lorenzi M, Raffaelli N, Dalla Rizza M, Mezzetti B, Ruggieri S (2001) Phytotoxic protein PcF, purification, characterization, and cDNA sequencing of a novel hydroxyproline-containing factor secreted by the strawberry pathogen Phytophthora cactorum. J Biol Chem 276:21578–21584

Park J, Lee S, Choi J, Ahn K, Park B, Park J, Kang S, Lee YH (2008) Fungal cytochrome P450 database. BMC Genom 9:402

Paumi CM, Chuk M, Snider J, Stagljar I, Michaelis S (2009) ABC transporters in Saccharomyces cerevisiae and their interactors: new technology advances the biology of the ABCC (MRP) subfamily. Microbiol Molec Biol Rev 3:577–593

Peterson PD, Campbell CL, Griffith CS, James E (1992) Teschemacher and the cause and management of potato-blight in the United-States. Plant Dis 76:754–756

Pfyffer GE, Pfyffer BU, Rast DM (1986) The polyol pattern, chemotaxonomy, and phylogeny of the fungi. Sydowia 39:160–201

Pieterse CMJ, Van West P, Verbakel HM, Brasse PWHM, Van Den Berg-Velthuis GCM, Govers F (1994) Structure and genomic organization of the ipiB and ipiO gene clusters of Phytophthora infestans. Gene 138:67–77

Pieterse CMJ, Van't Klooster J, Van Den Berg-Velthuis GCM, Govers F (1995) NiaA, the structural nitrate reductase gene of Phytophthora infestans: isolation, characterization and expression analysis in Aspergillus nidulans. Curr Genet 27:359–366

Prakob W, Judelson HS (2007) Gene expression during oosporogenesis in heterothallic and homothallic Phytophthora. Fungal Genet Biol 44:726–739

Pule BB, Meitz JC, Thompson AH, Linde CC, Fry WE, Langenhoven SD, Meyers KL, Kandolo DS, van Rij NC, McLeod A (2013) Phytophthora infestans populations in central, eastern and southern African countries consist of two major clonal lineages. Plant Pathol 62:154–165

Qi J, Asano T, Jinno M, Matsui K, Atsumi K, Sakagami Y, Ojika M (2005) Characterization of a Phytophthora mating hormone. Science 309:1828

Qiao Y, Liu L, Xiong Q, Flores C, Wong J, Shi J, Wang X, Liu X, Xiang Q, Jiang S, Zhang F, Wang Y, Judelson HS, Chen X, Ma W (2013) Oomycete pathogens encode RNA silencing suppressors. Nature Genet 45:330–333

Qutob D, Patrick Chapman B, Gijzen M (2013) Transgenerational gene silencing causes gain of virulence in a plant pathogen. Nature Comm 4:1349

Raffaele S, Farrer RA, Cano LM, Studholme DJ, MacLean D, Thines M, Jiang RH, Zody MC, Kunjeti SG, Donofrio NM, Meyers BC, Nusbaum C, Kamoun S (2010a) Genome evolution following host jumps in the Irish potato famine pathogen lineage. Science 330:1540–1543

Raffaele S, Kamoun S (2012) Genome evolution in filamentous plant pathogens: why bigger can be better. Nature Rev Microbiol 10:417–430

Raffaele S, Win J, Cano LM, Kamoun S (2010b) Analyses of genome architecture and gene expression reveal novel candidate virulence factors in the secretome of Phytophthora infestans. BMC Genom 11:637

Randall TA, Dwyer RA, Huitema E, Beyer K, Cvitanich C, Kelkar H, Ah Fong AMV, Gates K, Roberts S, Yatzkan E, Gaffney T, Law M, Testa A, Torto T, Zhang M, Zheng L, Mueller E, Windass J, Binder A, Birch PRJ, Gisi U, Govers F, Gow NAR, Mauch F, van West P, Waugh ME, Yu J, Boller T, Kamoun S, Lam ST, Judelson HS (2005) Large-scale gene discovery in the oomycete Phytophthora infestans reveals likely components of phytopathogenicity shared with true fungi. Molec Plant-Microbe Interact 18:229–243

Rathbone MC, Smart CD, Fry WE (2002) Isolates of Phytophthora infestans that infect Petunia x hybrida and Nicotiana benthamiana also produce INF1. Phytopathology 92:S145

Rayko E, Maumus F, Maheswari U, Jabbari K, Bowler C (2010) Transcription factor families inferred from genome sequences of photosynthetic stramenopiles. New Phytol 188:52–66

Richards TA, Soanes DM, Jones MD, Vasieva O, Leonard G, Paszkiewicz K, Foster PG, Hall N, Talbot NJ (2011) Horizontal gene transfer facilitated the evolution of plant parasitic mechanisms in the oomycetes. Proc Natl Acad Sci USA 108:15258–15263

Ritch DL, Daggett SS (1995) Nuclear DNA content and chromosome number in german isolates of Phytophthora infestans. Mycologia 87:579–581

Ristaino JB (2002) Tracking historic migrations of the Irish potato famine pathogen, Phytophthora infestans. Microbes Infect 4:1369–1377

Ristaino JB, Groves CT, Parra GR (2001) PCR amplification of the Irish potato famine pathogen from historic specimens. Nature 411:695–697

Ristaino JB, Hu CH, Fitt BDL (2012) Evidence for presence of the founder Ia mtDNA haplotype of Phytophthora infestans in 19th century potato tubers from the Rothamsted archives. Plant Pathol 62:492–500

Roy S, Kagda M, Judelson HS (2013a) Genome-wide prediction and functional validation of promoter motifs regulating gene expression in spore and infection stages of Phytophthora infestans. PLoS Pathog 9:e1003182

Roy S, Poidevin L, Jiang T, Judelson HS (2013b) Novel core promoter elements in the oomycete pathogen Phytophthora infestans and their influence on expression detected by genome-wide analysis. BMC Genom 14:106

Sansome E, Brasier CM (1973) Diploidy and chromosomal structural hybridity in Phytophthora infestans. Nature 241:344–345

Schornack S, van Damme M, Bozkurt TO, Cano LM, Smoker M, Thines M, Gaulin E, Kamoun S, Huitema E (2010) Ancient class of translocated oomycete effectors targets the host nucleus. Proc Natl Acad Sci USA 107:17421–17426

Seidl MF, Van den Ackerveken G, Govers F, Snel B (2011) A domain-centric analysis of oomycete plant pathogen genomes reveals unique protein organization. Plant Physiol 155:628–644

Seidl MF, Van den Ackerveken G, Govers F, Snel B (2012a) Reconstruction of oomycete genome evolution identifies differences in evolutionary trajectories leading to present-day large gene families. Genome Biol Evol 4:199–211

Seidl MF, Wang R-P, Van den Ackerveken G, Govers F, Snel B (2013) Bioinformatic inference of specific and general transcription factor binding sites in the plant pathogen Phytophthora infestans. PLoS ONE 7:e51295

Seidl MF, Wang RP, Van den Ackerveken G, Govers F, Snel B (2012b) Bioinformatic inference of specific and general transcription factor binding sites in the plant pathogen Phytophthora infestans. PLoS ONE 7:e51295

Shaw DS, Khaki IA (1971) Genetical evidence for diploidy in Phytophthora. Genet Res 17:165–167

Slamovits CH, Keeling PJ (2006) Pyruvate-phosphate dikinase of Oxymonads and Parabasalia and the evolution of pyrophosphate-dependent glycolysis in anaerobic eukaryotes. Eukaryot Cell 5:148–154

Smale ST, Kadonaga JT (2003) The RNA polymerase II core promoter. Ann Rev Biochem 72:449–479

Sohn J, Voegele RT, Mendgen K, Hahn M (2000) High level activation of vitamin B1 biosynthesis genes in haustoria of the rust fungus Uromyces fabae. Molec Plant-Microbe Interact 13:629–636

Sierra R, Rodríguez LM, Chaves D, Pinzón A, Grajales A, Rojas A, Mutis G, Cardenas M, Burbano D, Jimenez P, Bernal A, Restrepo S (2010) Discovery of Phytophthora infestans genes expressed in planta through mining of cDNA libraries. PLoS ONE 5:e9847

Staples RC (2001) Nutrients for a rust fungus: the role of haustoria. Trends Plant Sci 6:496–498

Stassen J, Van den Ackerveken G (2011) How do oomycete effectors interfere with plant life? Curr Opin Plant Biol 14:407–414

Stiller JW, Huang J, Ding Q, Tian J, Goodwillie C (2009) Are algal genes in nonphotosynthetic protists evidence of historical plastid endosymbioses? BMC Genom 10:484

Tani S, Judelson HS (2006) Activation of zoosporogenesis-specific genes in Phytophthora infestans involves a 7-nucleotide promoter motif and cold-induced membrane rigidity. Eukaryot Cell 5:745–752

Tian M, Benedetti B, Kamoun S (2005) A Second Kazal-like protease inhibitor from Phytophthora infestans inhibits and interacts with the apoplastic pathogenesis-related protease P69B of tomato. Plant Physiol 138:1785–1793

Tian M, Win J, Song J, van der Hoorn R, van der Knaap E, Kamoun S (2007) A Phytophthora infestans cystatin-like protein targets a novel tomato papain-like apoplastic protease. Plant Physiol 143:364–377

Tooley PW, Therrien CD (1987) Cytophotometric determination of the nuclear DNA content of 23 Mexican and 18 non-Mexican isolates of Phytophthora infestans. Exp Mycol 11:19–26

Toquin V, Barja F, Sirven C, Gamet S, Mauprivez L, Peret P, Latorse M-P, Zundel J-L, Schmitt F, Lebrun M-H, Beffa R (2009) Novel tools to identify the mode of action of fungicides as exemplified with fluopicolide. In: Gisi U, Chet I, Gullino M (eds) Recent developments in management of plant Diseases. Springer, New York, pp 19–36

Torto TA, Li S, Styer A, Huitema E, Testa A, Gow NA, van West P, Kamoun S (2003) EST mining and functional expression assays identify extracellular effector proteins from the plant pathogen Phytophthora. Genome Res 13:1675–1685

Torto TA, Rauser L, Kamoun S (2002) The pipg1 gene of the oomycete Phytophthora infestans encodes a fungal-like endopolygalacturonase. Curr Genet 40:385–390

Tripathy S, Tyler BM (2006) The repertoire of transfer RNA genes is tuned to codon usage bias in the genomes of Phytophthora sojae and Phytophthora ramorum. Molec Plant-Microbe Interact 19:1322–1328

Tsai HK, Huang PY, Kao CY, Wang D (2009) Co-Expression of neighboring genes in the Zebrafish (Danio rerio) genome. Int J Mol Sci 10:3658–3670

Tyler BM, Kale SD, Wang Q, Tao K, Clark HR, Drews K, Antignani V, Rumore A, Hayes T, Plett JM, Fudal I, Gu B, Chen Q, Affeldt KJ, Berthier E, Fischer GJ, Dou D, Shan W, Keller NP, Martin FM, Rouxel T, Lawrence C (2013) Microbe-independent entry of oomycete RxLR effectors and fungal RxLR-like effectors into plant and animal cells is specific and reproducible. Molec Plant-Microbe Interact 26:611–616

Tyler BM, Tripathy S, Zhang X, Dehal P, Jiang RH, Aerts A, Arredondo FD, Baxter L, Bensasson D, Beynon JL, Chapman J, Damasceno CM, Dorrance AE, Dou D, Dickerman AW, Dubchak IL, Garbelotto M, Gijzen M, Gordon SG, Govers F, Grunwald NJ, Huang W, Ivors KL, Jones RW, Kamoun S, Krampis K, Lamour KH, Lee MK, McDonald WH, Medina M, Meijer HJ, Nordberg EK, Maclean DJ, Ospina-Giraldo MD, Morris PF, Phuntumart V, Putnam NH, Rash S, Rose JK, Sakihama Y, Salamov AA, Savidor A, Scheuring CF, Smith BM, Sobral BW, Terry A, Torto-Alalibo TA, Win J, Xu Z, Zhang H, Grigoriev IV, Rokhsar DS, Boore JL (2006) Phytophthora genome sequences uncover evolutionary origins and mechanisms of pathogenesis. Science 313:1261–1266

Uda K, Hoshijima M, Suzuki T (2013) A novel taurocyamine kinase found in the protist Phytophthora infestans. Comp Biochem Phys Part B 165:42–48

Vadnagara K (2010) A tale of three phytopathogens: impact of transposable elements on genome evolution. Dissertation, University of Texas, Arlington

Van der Lee T, De Witte I, Drenth A, Alfonso C, Govers F (1997) AFLP linkage map of the oomycete Phytophthora infestans. Fungal Genet Biol 21:278–291

Van der Lee T, Testa A, Robold A, van't Klooster JW, Govers F (2004) High density genetic linkage maps *of Phytophthora infestans re*veal trisomic progeny and chromosomal rearrangements. Genetics 157:949–956

Van West P, Shepherd SJ, Walker CA, Li S, Appiah AA, Grenville-Briggs LJ, Govers F, Gow NA (2008) Internuclear gene silencing in Phytophthora infestans is established through chromatin remodelling. Microbiology 154:1482–1490

Vetukuri RR, Asman AK, Tellgren-Roth C, Jahan SN, Reimegard J, Fogelqvist J, Savenkov E, Soderbom F, Avrova AO, Whisson SC, Dixelius C (2012) Evidence for small RNAs homologous to effector-encoding genes and transposable elements in the oomycete Phytophthora infestans. PLoS ONE 7:e51399

Vetukuri RR, Avrova AO, Grenville-Briggs LJ, Van West P, Soderbom F, Savenkov EI, Whisson SC, Dixelius C (2011) Evidence for involvement of dicer-like, argonaute and histone deacetylase proteins in gene silencing in Phytophthora infestans. Molec Plant Pathol 12:772–785

Vijn I, Govers F (2003) Agrobacterium tumefaciens mediated transformation of the oomycete plant pathogen Phytophthora infestans. Molec Plant Pathol 4:456–467

Vleeshouwers VG, Raffaele S, Vossen JH, Champouret N, Oliva R, Segretin ME, Rietman H, Cano LM, Lokossou A, Kessel G, Pel MA, Kamoun S (2011) Understanding and exploiting late blight resistance in the age of effectors. Ann Rev Phytopathol 49: 507–531

Vleeshouwers VG, Rietman H, Krenek P, Champouret N, Young C, Oh SK, Wang M, Bouwmeester K, Vosman B, Visser RG, Jacobsen E, Govers F, Kamoun S, Van der Vossen EA (2008) Effector genomics accelerates discovery and functional profiling of potato disease resistance and phytophthora infestans avirulence genes. PLoS ONE 3:e2875

Walker CA, Koppe M, Grenville-Briggs LJ, Avrova AO, Horner NR, McKinnon AD, Whisson SC, Birch PR, van West P (2008) A putative DEAD-box RNA-helicase is required for normal zoospore development in the late blight pathogen Phytophthora infestans. Fungal Genet Biol 45:954–962

Watson IA (1970) Changes in virulence and population shifts in plant pathogens. Ann Rev Phytopathol 8:209–230

Wawra S, Djamei A, Albert I, Nurnberger T, Kahmann R, van West P (2013) In vitro translocation experiments with rxlr-reporter fusion proteins of avr1b from Phytophthora sojae and avr3a from Phytophthora infestans fail to demonstrate specific autonomous uptake in plant and animal cells. Molec Plant-microbe Interact 26:528–536

Weiland JJ (2003) Transformation of Pythium aphanidermatum to geneticin resistance. Curr Genet 42:344–352

Whisson SC, Avrova AO, Boevink PC, Armstrong MR, Seman ZA, Hein I, Birch PRJ (2011) Exploiting knowledge of pathogen effectors to enhance late blight resistance in potato. Potato Res 54:325–340

Whisson SC, Avrova AO, van West P, Jones JT (2005) A method for double-stranded RNA-mediated transient gene silencing in Phytophthora infestans. Molec Plant Pathol 6:153–163

Whisson SC, Boevink PC, Moleleki L, Avrova AO, Morales JG, Gilroy EM, Armstrong MR, Grouffaud S, van West P, Chapman S, Hein I, Toth IK, Pritchard L, Birch PR (2007) A translocation signal for delivery of oomycete effector proteins into host plant cells. Nature 450:115–118

Whisson SC, Randall E, Young, V, Birch PRJ, Cooke DEL, Csukai M (2013) Involvement of RNA polymerase I in mefenoxam insensitivity in Phytophthora infestans. In: Abstracts of the Oomycete Molecular Genetics Network meeting, Pacific Grove, California, 10–12 March 2013

Whitaker JW, McConkey GA, Westhead DR (2009) The transferome of metabolic genes explored: analysis of the horizontal transfer of enzyme encoding genes in unicellular eukaryotes. Genome Biol 10:R36

Widmark AK, Andersson B, Sandstrom M, Yuen JE (2011) Tracking Phytophthora infestans with SSR markers within and between seasons-a field study in Sweden. Plant Pathol 60:938–945

Win J, Krasileva KV, Kamoun S, Shirasu K, Staskawicz BJ, Banfield MJ (2012) Sequence divergent RXLR effectors share a structural fold conserved across plant pathogenic oomycete species. PLoS Pathog 8: e1002400

Xiang Q, Judelson HS (2010) Myb transcription factors in the oomycete Phytophthora with novel diversified DNA-binding domains and developmental stage-specific expression. Gene 453:1–8

Xiang Q, Kim KS, Roy S, Judelson HS (2009) A motif within a complex promoter from the oomycete Phytophthora infestans determines transcription during an intermediate stage of sporulation. Fungal Genet Biol 46:400–409

Yaeno T, Shirasu K (2013) The RXLR motif of oomycete effectors is not a sufficient element for binding to phosphatidylinositol monophosphates. Plant Signal Behav 8:e23865

Yan HZ, Liou RF (2005) Cloning and analysis of pppg1, an inducible endopolygalacturonase gene from the oomycete plant pathogen Phytophthora parasitica. Fungal Genet Biol 42:339–350

Yoshida K, Schuenemann VJ, Cano LM, Pais M, Mishra B, Sharma R, Lanz C, Martin FN, Kamoun S, Krause J, Thines M, Weigel D, Burbano HA (2013) The rise and fall of the Phytophthora infestans lineage that triggered the Irish potato famine. eLife 2:e00731

Yuen JE, Andersson B (2012) What is the evidence for sexual reproduction of Phytophthora infestans in Europe? Plant Pathol 62:485–491

Zadoks JC (2008) The potato murrain on the european continent and the revolutions of 1848. Potato Res 51:5–45

# 10 Hyaloperonospora arabidopsidis: A Model Pathogen of Arabidopsis

John M. McDowell

## 10.1 Introduction

Downy mildew disease of *Arabidopsis thaliana* made its debut in the literature in 1901 (Lindau 1901), appeared again in 1918 (Gäumann 1918), but was largely absent during the following seven decades. This situation changed dramatically in the 1980s with the emergence of *Arabidopsis* as a genetic and molecular model for many aspects of plant biology, including interactions with pathogens (Dangl 1993; Crute et al. 1994). As part of searches for "model pathogens" that could be used for *Arabidopsis* experiments, several investigators independently rediscovered *Hyaloperonospora arabidopsidis* in field populations of *Arabidopsis* (Koch and Slusarenko 1990; Slusarenko and Mauch-Mani 1991; Crute et al. 1993; Holub et al. 1994b; Holub 2008) or in one case, from a fortuitous infection of *Arabidopsis* grown in a glasshouse (Parker et al. 1996; Reignault et al. 1996).

The identification of *Arabidopsis*-compatible isolates of *H. arabidopsidis* was a seminal event that led to the development of this species as a very important model pathogen of *Arabidopsis*. At that time, *H. arabidopsidis* was referred to as *Peronospora parasitica* and was taxonomically lumped together with related strains that cause downy mildew on cultivated brassica species (Holub 2008). DNA evidence prompted two subsequent taxonomic reclassifications: First to *Hyaloperonospora parasitica* (Constantinescu and Fatehi 2002) and soon afterwards as *H. arabidopsidis* (Voglmayr et al. 2004). *Hyaloperonospora arabidopsidis* isolates were named according to a standardized system comprised of two letters referencing the geographic location from which the isolate was obtained, followed by two letters designating a laboratory accession of *Arabidopsis* upon which compatibility was observed. For example, the reference genome sequence was obtained from an *H. arabidopsidis* isolate named Emoy2 ("Em" for East Malling, England, and "oy" for susceptible *Arabidopsis* accession *Oystese*, (Koch and Slusarenko 1990; Holub and Beynon 1996)).

The early history of the *Arabidopsis-H. arabidopsidis* pathosystem is described in several excellent reviews (Crute et al. 1994; Holub and Beynon 1996; Coates and Beynon 2010; Holub 2008). This chapter lays out key attributes of this reference organism in relation to its agronomically important relatives, summarizes insights from the reference genome of *H. arabidopsidis*, and describes how this new information has been (and will continue to be) applied for fundamental and practical advances in plant-pathogen interactions.

J. M. McDowell (✉)
Department of Plant Pathology, Physiology, and Weed Science, Virginia Tech, Blacksburg, VA 24061-0329, USA
e-mail: johnmcd@vt.edu

R. A. Dean et al. (eds.), *Genomics of Plant-Associated Fungi and Oomycetes: Dicot Pathogens*,
DOI: 10.1007/978-3-662-44056-8_10, © Springer-Verlag Berlin Heidelberg 2014

### 10.1.1 Taxonomic Position of *H. arabidopsidis*

*Hyaloperonospora arabidopsidis* resides within a diverse group of eukaryotic microbes known as oomycetes (Phylum oomycota, (Dick 2001)). This phylum contains many important pathogens of plants, animals, and marine plants. Oomycetes are often found in aquatic environments and are therefore referred to informally as "water molds" (Beakes et al. 2012).

Oomycetes were originally classified alongside fungi (ascomycetes, basidiomycetes, deuteromycetes) (Dick 2001). At that time, this was perfectly logical, because oomycetes and "true" fungi share many superficial similarities, including assimilative metabolism, a filamentous growth habit, intracellular feeding structures (e.g., haustoria, see below), and reproduction via sporulation (Latijnhouwers et al. 2003). However, it became evident late in the twentieth century that oomycetes are actually quite distant from fungi (Beakes et al. 2012). The most definitive evidence came from DNA-based classifications that unambiguously aligned oomycetes with brown algae, diatoms, and kelp, in the kingdom *Stramenopila* (Baldauf et al. 2000). Thus, oomycetes and fungi are quite divergent. In fact, you, the reader, are more closely related to fungi than are oomycetes. The evolutionary distance between oomycetes and fungi holds fundamental and practical implications: From a fundamental standpoint, oomycetes and fungi have independently evolved to colonize plants via structures that appear very similar at a morphological level, and thus represent a clear case of convergent evolution. As is discussed in Sect. 10.4.1, this convergence extends to the molecular and genomic level (McDowell 2011). From a practical standpoint, chemistries developed for control of fungal diseases can be ineffective against oomycetes, because of divergence in the physiological targets of the chemicals (Kamoun 2003; Latijnhouwers et al. 2003). Thus, the divergence between oomycetes and fungi complicates disease control strategies, but also provides a fascinating context for which to apply comparative genomics to understand convergent evolution.

Over 800 species of oomycetes can cause downy mildew disease (Clark and Spencer-Phillips 2000). Molecular studies indicate that all downy mildew species are monophyletic, forming a single clade with a common ancestor (Goker et al. 2007). The sister clade of the downy mildews is comprised of species in the genus *Phytophthora* ("plant destroyer"), which cause a number of devastating diseases (Voglmayr 2003; Blair et al. 2008). The placement of downy mildews relative to *Phytophthora* species is somewhat controversial (see Sect. 10.4.4 and (Runge et al. 2011)), but there is no doubt that *Phytophthoras* and downy mildew species are each other's closest relatives. Other phytopathogenic genera of oomycetes include *Pythium*, which causes damping off-diseases, and the white blister rust pathogens in the genus *Albugo* (Cooke et al. 2000). These genera will be referenced in discussions of comparative genomics in Sect. 10.4.1.

### 10.1.2 Importance of *H. arabidopsidis*

Downy mildew species are distributed worldwide and are estimated to parasitize ~15 % of flowering plant families, including a wide range of monocot and dicot crops (Clark and Spencer-Phillips 2000). For example, *H. arabidopsidis* and related species collectively infect more than 140 species in the family Cruciferae, including crops such as cabbage, oilseed rape, broccoli, and cauliflower (Lucas et al. 1995). An accidental introduction of *Plasmopara viticola* from North America to Europe nearly wiped out the European wine industry in the late 1800s, and *P. viticola* remains a significant problem for grape growers today (Lucas et al. 1995). *Peronospora* species also cause disease on soybean, onions, and tobacco. *Bremia lactucae* infects more than 230 members of the Compositae family, including lettuce and globe artichoke (Lucas et al. 1995). This pathogen was estimated to cause ~1.2 million (pounds sterling) loss to the

UK lettuce crop in 1989 (Clark and Spencer-Phillips 2000). Downy mildew of pea in the UK can cause up to 55 % yield loss (Clark and Spencer-Phillips 2000). Outbreaks of cucumber downy mildew during the 1980s in former Czechoslovakia decreased yield by up to 85 % (Clark and Spencer-Phillips 2000). Downy mildews from the *Peronosclerospora*, *Sclerospora*, and *Sclerophthora* genera can also cause major problems on monocot crops, including maize, rice, sorghum, millets, and sugarcane (Clark and Spencer-Phillips 2000; Jeger et al. 1998). Their effect can be particularly devastating in developing nations. For example, losses of up to 100 % of the maize crop in Zaire have been reported, and overall losses of grains to downy mildews in Africa are estimated at 20 %. *Peronosclerospora philippinesis* and *Sclerophthora rayssiae* can cause 40–60 % loss of maize harvests in India and Southeast Asia. *P. philippinesis* and *S. rayssiae* are not endemic to the US and are considered as significant potential threats to US corn production. These two downy mildews are listed among the top ten potential plant bioterror threats (Agricultural Bioterrorism Protection Act of 2002). The importance of downy mildews to agriculture is underscored by estimates that ~20 % of the $4.7 billion world fungicide market goes to downy mildew control (Gisi 2002; Clark and Spencer-Phillips 2000).

As mentioned above, downy mildew species are closely related to extremely destructive oomycete pathogens in the *Phytophthora* and *Pythium* genera. *Phytophthora* species cause rots of roots, stems, leaves, and fruits of a huge range of agriculturally and ornamentally important plants. The economic damage to crops in the United States by *Phytophthora* species is estimated in the tens of billions of dollars, including the costs of control measures, and worldwide it is many times this figure (Erwin and Ribiero 1996). Because downy mildews and other oomycetes are evolutionarily and physiologically distinct from true fungi, these organisms are unaffected by the majority of fungicides and therefore are very difficult to control. Additionally, many oomycetes have displayed an extraordinary ability to overcome chemical control measures and genetic resistance bred into plant hosts (Kamoun 2003; Latijnhouwers et al. 2003). This phenotypic adaptability implies unusual genetic flexibility, which is an important justification for detailed comparative genomic examinations of oomycete genomes (Jiang and Tyler 2012).

*Hyaloperonospora arabidopsidis* is a natural pathogen of *Arabidopsis* (Holub 2008). The laboratory isolates of *H. arabidopsidis* were all collected from naturally occurring populations of *Arabidopsis*. Based on laboratory tests, the host range of *H. arabidopsidis* appears to be restricted to *Arabidopsis*; no *H. arabidopsidis* isolate has been found which can cause disease on cultivated relatives of *Arabidopsis* (Holub and Beynon 1996). Thus, *H. arabidopsidis* is not a crop pathogen. Rather, its value lies in its ability to efficiently colonize *Arabidopsis*. Indeed, *H. arabidopsidis* is one of only a few eukaryotic microbes that can efficiently colonize wild-type *Arabidopsis*. Although a couple of *Phytophthora* species can complete their life cycle on *Arabidopsis* under laboratory conditions (Wang et al. 2013; Schlaeppi et al. 2010; Attard et al. 2010; Rookes et al. 2008; Roetschi et al. 2001), no other oomycete has been identified as a natural pathogen of *Arabidopsis*. Thus, *H. arabidopsidis* is a very important model pathogen for exploiting the experimental tools of *Arabidopsis* in the context of plant-oomycete molecular interactions and coevolution, for understanding basic aspects of downy mildew pathogenesis and evolution, for studying the molecular basis and evolution of obligate biotrophy, and as a test bed for new strategies aimed at control of oomycete diseases.

### 10.1.3 Life Cycle of *H. arabidopsidis*

*Hyaloperonospora arabidopsidis* reproduces sexually through conidiospores and asexually through oospores (Koch and Slusarenko 1990). *Hyaloperonospora arabidopsidis* isolates are homothallic (i.e., self-compatible), but can also outcross. Sexual reproduction occurs when hyphal tips differentiate into antheridia (male) and oogonia (female) and then fuse. The resultant spore is thick walled and serves as a durable propagule that can survive in the soil following decomposition of the plant, and then infect new compatible hosts (Lucas et al. 1995). Asexual sporulation occurs on tree-like conidiophores, which are produced as hyphae emerge to the outside of infected organs through stomata. Such sporulation typically occurs under darkness, and is promoted by cool temperatures (i.e., 10–18 °C) and high humidity (Koch and Slusarenko 1990). Spores are released via hygroscopic twisting and can be carried by wind or water to new hosts (Slusarenko and Schlaich 2003). Heavily colonized plants can produce thousands of conidiospores under favorable conditions; this comprises the epidemic phase of downy mildew disease.

The infection cycle is initiated when a spore germinates on the surface of a host organ. An infection peg is produced and forms an appressorium for attachment to the host. This is followed by emergence of a penetration hypha that typically grows through the anticlinal junction between epidermal cells (Soylu and Soylu 2003). The hypha then enters the mesophyll layer, wherein it initiates a program of dichotomous branching, leading to the formation of a web-like mycelium (Soylu and Soylu 2003). Hyphae remain in the intracellular spaces and do not grow directly through host cells. However, feeding structures, called haustoria, emerge from hyphae to penetrate the cell wall and form an interface with the host cell (Mims et al. 2004). Haustoria have traditionally been regarded as feeding structures, through which water and nutrients are extracted from the host (Spencer-Phillips 1997; Szabo and Bushnell 2001). However, nutrient uptake through hyphae has also been documented, and the relative importance of hyphae and haustoria for nutrient uptake is unclear. Another issue to be addressed is the role of hyphae and haustoria in delivery of effectors. Based on studies of *Phytophthora infestans*, the haustoria appear to be the primary site of effector delivery (Whisson et al. 2007), but this remains to be generalized to *H. arabidopsidis* and other oomycetes.

The hallmark symptom of downy mildew disease is the formation of asexual fruiting bodies on the exterior of colonized leaves, lending a downy appearance. Downy mildew disease of *Arabidopsis* typically has a latent phase of at least 4 days under laboratory conditions that favor colonization. In many compatible *Arabidopsis* ecotype-*H. arabidopsidis* isolate combinations, the first evidence of disease is sporulation. However, some *Arabidopsis* accessions display chlorosis prior to sporulation (Holub 2008). Typically, infected plants can recover if they are grown under optimal conditions. However, infection can negatively affect plant fitness by reducing seed production, and can alter the plant's developmental architecture (Salvaudon et al. 2005; Korves and Bergelson 2003, 2004).

A very important aspect of the *H. arabidopsidis* life cycle is its absolute dependence upon its host. *Hyaloperonospora arabidopsidis* can only extract nutrients from living plant cells (biotrophy) and cannot be cultured apart from its host (obligate) (Lucas et al. 1995; Koch and Slusarenko 1990; Holub et al. 1994b). This obligate biotrophic lifestyle has been documented for a number of other downy mildew species, and is thought to be characteristic of the downy mildews as a group. Contrastingly, the closest relatives of downy mildew pathogens (*Phytophthora* spp.) can be readily cultured on synthetic media (Erwin and Ribiero 1996). Additionally, *Phytophthora* pathogens employ a hemi-biotrophic lifestyle in which an initial phase of biotrophic growth is followed by a transition to necrotrophy, in which host tissue is destroyed, nutrients are assimilated from the destroyed cells, and asexual reproduction occurs (Erwin and Ribiero 1996). The differences in host dependence and pathogenic lifestyle between the downy mildews and *Phytophthora* species provide a focus for comparative genomics that comprises some of the most interesting aspects of the initial analysis of the *H. arabidopsidis* genome, described in Sect. 10.3.

### 10.1.4 Experimental Resources and Challenges for Research on *H. arabidopsidis*

*H. arabidopsidis* was initially used by several labs as a "physiological probe" to genetically define resistance genes in *Arabidopsis*. In these studies, *H. arabidopsidis* isolates were inoculated onto *Arabidopsis* ecotypes, which were then scored for resistance or susceptibility to the *H. arabidopsidis* isolates. Resistance genes were then mapped in the progeny of crosses between resistant and susceptible *Arabidopsis* ecotypes. For example, a *tour de force* of genetic analysis by Holub, Tor, and colleagues led to the genetic definition of 11 loci containing "*RPP*" genes (for resistance to *Peronospora parasitica*, as the pathogen was named at that time, (Holub et al. 1994a)). To date, 27 different *RPP* loci have been postulated (Slusarenko and Schlaich 2003).

These genetic experiments laid the groundwork for molecular cloning of the *RPP* genes using map-based approaches (Holub 1997). At present, six *RPP* genes have been described in the literature. All of these genes belong to the ubiquitous nucleotide-binding, leucine rich repeat (NB-LRR) superfamily of immune surveillance proteins. Two *RPP* genes contain a coiled-coil domain at the N-terminus (CC-NB-LRR, (McDowell et al. 1998; Bittner-Eddy et al. 2000)), while four contain a Toll, Interleukin receptor-like domain (TIR-NB-LRR, (Parker et al. 1997; van der Biezen et al. 2002; Botella et al. 1998; Sinapidou et al. 2004)). These genes, and their encoded proteins, comprise a valuable resource for studies to understand the evolution and molecular mechanisms underlying effector-triggered immunity against oomycete pathogens, as nicely discussed in (Coates and Beynon 2010).

Additionally, *H. arabidopsidis* isolates continue to be widely used as physiological probes to test whether *Arabidopsis* mutants affect responses to oomycetes (Slusarenko and Schlaich 2003). This approach has been used to map the architecture of signaling networks that underpin immunity to *H. arabidopsidis*, and has identified a broad diversity of pathways/processes with direct or indirect effects on *H. arabidopsidis* interactions, most recently including amino acid metabolism (Stuttmann et al. 2011), polarized secretion (Lu et al. 2012), and circadian regulation of immunity (Zhang et al. 2013). Although most of the emphasis in these studies has been placed on host-immune responses, more attention is currently being placed on identifying plant genes/pathways that support invasion by *H. arabidopsidis* (Lapin and Van den Ackerveken 2013). Such studies are expected to lead eventually to regulatory connections with pathogen effectors.

Contrasting with the early success of *H. arabidopsidis* as a probe to understand host immunity, the initial prospects for research on *H. arabidopsidis* pathology at the molecular level were not promising, due to the pathogen's obligate lifestyle and lack of resources for molecular experiments. However, vision and hardwork from Eric Holub, Jim Beynon, and their colleagues in the mid to late 1990s led to the development of the first tools that enabled molecular-level understanding of downy mildew pathogenesis (Coates and Beynon 2010). One such tool was a protocol for genetic crosses between *H. arabidopsidis* isolates, developed by Gunn and Holub and used to construct a mapping population that enabled genetic identification of several loci for avirulence in the segregating progeny. This formally demonstrated that *Arabidopsis* and *H. arabidopsidis* can interact according to the gene-for-gene paradigm. A linkage map using 184 AFLP markers on 81 F2 progeny was subsequently constructed (Rehmany et al. 2003). Other important tools were DNA libraries for *H. arabidopsidis* cDNAs (Bittner-Eddy et al. 2003) and large insert genomic DNA fragments (Rehmany et al. 2003) that, respectively, facilitated cloning of the first two *H. arabidopsidis* avirulence/effector proteins (*Atr13* and *Atr1*, (Rehmany et al. 2005; Allen et al. 2004)). These successes in turn provided evidence that a reference genome sequence for *H. arabidopsidis* would be feasible and valuable, which was a critical justification funding the genome initiative as outlined in Sect. 10.2.1.

The biggest challenge for *H. arabidopsidis* research is the pathogen's obligate lifestyle. Much effort has been invested in culturing other downy mildew species, with only ephemeral success (e.g., (Tiwari and Arya 1969)). To date, there have been no reports of success in culturing *H. arabidopsidis* apart from its host. The inability to grow *H. arabidopsidis* axenically, on synthetic media, essentially precludes the genetic transformation procedures that are critical for molecular biology research. Attempts have been made at transformation *in planta*, via *Agrobacterium*-mediated gene transfer, but only transient transformation has been achieved (personal communication from Guido van den Ackerveken, University of Utrecht, The Netherlands).

Other approaches have been developed to work around this impediment. To begin with, it is now common practice to express genes for pathogen-secreted proteins (i.e., effectors) *in planta*, under control of regulatory modules that render genes active in plants (e.g., the cauliflower mosaic virus 35S promoter). It is trivial to make stably transformed *Arabidopsis* using the floral dip procedure (Clough and Bent 1998). One can then assess whether and how expression of the *H. arabidopsidis* gene affects the plant (e.g., does the *H. arabidopsidis* protein trigger or suppress host immunity?). This approach is considered biologically relevant for effector proteins that are secreted to the outside or inside of plant cells because the effector transgene is configured to express the *H. arabidopsidis* protein in the processed form that it presumably adopts in the plant milieu. Indeed, this approach has arguably been the most fruitful for understanding the function of Type III effectors from bacterial pathogens (Munkvold and Martin 2009), and is now being applied to effectors from *H. arabidopsidis* and other oomycetes (Sect. 10.4.5). One major advantage of this approach, compared to studying a knockout or knockdown of the gene in the pathogen, is that it avoids problems with genetic redundancy. The biggest potential drawback of this approach is that the protein is presumably produced at a higher level than it would be in a natural interaction, and thus resultant phenotypes could be exaggerated (hypomorphic) or gains of function (neomorph). Although precedents from studies of bacterial effectors indicate that such problems are infrequent, if not negligible, the caveat must always be considered on a case-by-case basis, and it is best to focus experimentation on *Arabidopsis* lines in which the transgene is expressed at relatively low levels (Munkvold and Martin 2009).

Transient systems are also employed to study *H. arabidopsidis* gene functions in planta. The first such studies for *H. arabidopsidis* used biolistics to deliver *H. arabidopsidis* genes to plant cells, followed by assays for avirulence gene activity. This is how the avirulence proteins Atr1 and Atr13 were functionally validated (Rehmany et al. 2005; Allen et al. 2004). Another approach is to use bacterial phytopathogens as surrogates to deliver the *H. arabidopsidis* gene or protein to plant cells. Unfortunately, transient T-DNA delivery from *Agrobacterium* ("Agroinfiltration") does not work as efficiently in *Arabidopsis* as it does in *Nicotiana* species (Goodin et al. 2008; Wroblewski et al. 2005). However, Jonathan Jones' group developed an elegant strategy to exploit the Type III secretion of phytopathogenic *Pseudomonas* bacteria (Sohn et al. 2007). This strategy is based on a bacterial plasmid expression vector in which the *H. arabidopsidis* gene of interest is fused to a leader sequence from a bacterial effector (AvrRps4) that directs the chimeric protein to the Type III secretion system, through which the protein can be delivered to the interior of plant cells. This "effector detector vector" is now being used as a delivery tool for *H. arabidopsidis* effectors (Fabro et al. 2011). This approach does not require making stably transformed plants (Rentel et al. 2008). However, the system also carries caveats, including the possibility that any given effector might not be able to assume a biologically active form after passage through the Type III secretion system. Confounding effects from bacterial effectors and pathogen-associated molecular patterns must also be considered when evaluating phenotypic changes (or absence thereof).

Although the above approaches are valid for effectors that function inside plant cells, they are not useful for *H. arabidopsidis* genes encoding proteins that operate inside *H. arabidopsidis* cells. At present, the best work around for this challenge is to use a surrogate, i.e., study homologs of *H. arabidopsidis* genes in *Phytophthora* species that are more amenable (but still challenging!) for genetic transformation. Encouragingly, several *Phytophthora* species have been demonstrated to efficiently infect *Arabidopsis* (Wang et al. 2013; Schlaeppi et al. 2010; Attard et al. 2010; Rookes et al. 2008; Roetschi et al. 2001), thus enabling the investigator to manipulate the pathogen and access the experimental tools of *Arabidopsis* that vastly exceed those available in any other plant host.

## 10.2 Structure of the *H. arabidopsidis* Genome

The following sections summarize the methodology for obtaining the *H. arabidopsidis* genome sequence, and the initial insights from the genome as reported by the paper from Baxter et al. (2010).

### 10.2.1 Strategy for Sequencing

The *H. arabidopsidis* genome initiative was conceived in discussions between Brett Tyler, John McDowell, and Jim Beynon in 2003. At that time, next-generation sequencing technologies were in their infancy and Sanger sequencing was still the platform of choice for genomics. However, costs of Sanger sequencing were still high. Thus, two proposals were concurrently submitted: One, authored by Tyler and McDowell to the USDA/NSF program in Microbial Genome Sequencing, and the other, authored by Beynon, to the UK BBSRC. Fortunately, both proposals were funded on the first attempt, and encompassed the following strategy: The US group (Tyler and McDowell at Virginia Tech, and Sandra Clifton at the Washington University Genome Center) were responsible for (i) random shotgun sequencing to 8-fold depth, including assembly and machine annotation of the genome; (ii) importing the sequence into the Oomycete Community Annotation Database at the Virginia Bioinformatics Institute to enable manual annotation by the community; (iii) holding an annotation jamboree to familiarize and train community members in the use of annotation tools. The UK Group (Beynon at Horticulture Research International and Jane Rogers at the Sanger Centre were responsible for (i) fingerprinting and end sequencing of an 8,000 clone BAC library, (ii) sequencing of 30,000 ESTs, (iii) finishing of 100 BACs.

The *H. arabidopsidis* isolate Emoy2 was chosen as the source for the reference genome because it was used for construction of the BAC library and the mapping cross that were described above (Gunn et al. 2002; Rehmany et al. 2003). Because of *H. arabidopsidis*'*s* host dependency, DNA was isolated from asexual spores that were collected by rinsing from compatible plants. Libraries for sequencing were constructed with plasmid, fosmid, and BAC vectors. End sequences from these clones were obtained via Sanger technology.

Data from the US and UK were used to generate several draft assemblies. The first was publically released in April 2007, and six subsequent versions followed. The final version based only on Sanger sequences (v7.0.1) was comprised of 1.1 million input reads assembled with PCAP software (9.5-fold phred Q20 redundancy), along with 97 fully sequenced BAC inserts that were selected to encompass gaps and repetitive regions. The assembly size was 78 Mb and was comprised of 1640 contigs with a contig N50 of 45.9 Kb.

Soon after version 7.0.1 was completed, a third group joined the coalition: Jonathan Jones' group (The Sainsbury Laboratories, Norwich, UK) sequenced the Emoy2 isolate using Illumina technology to 35-fold coverage with paired-end reads. This yielded a 56.9 Mb assembly with a mean contig length of 2980 bp. Several strategies were explored to integrate the Sanger and Illumina sequences, resulting in an assembly pipeline to correct sequencing errors and span gaps in the Sanger assembly, remove possible contaminant sequences, and finally append Illumina scaffolds larger than 2 kb that were absent in the Sanger assembly.

The *H. arabidopsidis* Emoy 2 transcriptome was surveyed with conventional EST sequencing and Illumina cDNA sequencing. 13,364 unigenes were derived from 31,759 reads of a normalized cDNA library constructed using mRNA isolated from asexual spores. Illumina sequencing was performed using RNA from *Arabidopsis* leaves colonized with Emoy2.

Another key aspect of the sequencing strategy was to convene an annotation jamboree, open to all interested parties, at the Virginia Bioinformatics Institute at Virginia Tech in August 2007. This attracted 25 participants from 11 labs in the US, UK, Canada, The Netherlands, and France. The groups self-sorted according to their interests, and analyzed version 6.0 of the genome. As described below, much of the analysis involved comparisons to the previously released genomes of *Phytophthora sojae* and *Phytophthora ramorum*. Each group submitted white papers in December 2007, which comprised the basis of the paper that described the *H. arabidopsidis* genome (Baxter et al. 2010).

## 10.2.2 Genome Size and Composition

Based on precedents from other obligate pathogens (e.g., (Ochman and Moran 2001; Keeling 2004)), it was thought that the *H. arabidopsidis* genome might be relatively small compared to those from free-living relatives, due to genome compaction, including removal of genes that were rendered nonessential as the pathogens' association with the host became more intimate. However, this proved not to be the case. The size of the assembly was 82 Mb. Additionally, it soon became evident that the assembly size was an underestimate due to repeats being collapsed in the genome assembly. We conducted two types of statistical analyses of the coverage provided by the Sanger and Illumina reads (described in the Supplemental Information in (Baxter et al. 2010)), which independently yielded genome estimates of 98 and 100 Mb. The discrepancy between these estimates and the size of the assembly is likely due to approximately 13 Mb of sequence that had an average copy number of approximately three, and shared greater than 95 % identity and were therefore collapsed as single copies in the assembly. Thus, the size of the genome is most probably approximately 100 Mb. This is larger than the estimated genome sizes of *P. sojae* (65 Mb, (Tyler et al. 2006)) and *P. capcisi* (64 Mb, (Lamour et al. 2012)), comparable in size to *P. ramorum* (95 Mb, (Tyler et al. 2006)), and substantially smaller than *P. infestans* (240 Mb, (Haas et al. 2009)). Moreover, analysis of *H. arabidopsidis* gene space, described below, indicated that *H. arabidopsidis* genome has not undergone a large-scale reduction of genes relative to *Phytophthora,* i.e., has not undergone pervasive reduction. However, particular types of gene families have been reduced, as described in Sects. 10.3.1–10.3.3.

The composition of the repetitive fraction of the *H. arabidopsidis* genome was examined using databases of known repetitive elements along with de novo repeat finders. The *P. sojae* genome (v4.0 assembly) was included in the analysis to serve as a point of reference. In *H. arabidopsidis*, 42 % of the genome is comprised of repeats. The predominant class is comprised of LTR elements (20 %), followed by an unknown novel type (18 %) specific to the *H. arabidopsidis* genome. The remaining elements are largely comprised of LINES (1.65 %), small RNA genes (0.03 %), simple sequence repeats (0.38 %), and low complexity regions (0.09 %). 213 rolling circle autonomous helitrons were found in the *H. arabidopsidis* genome. These helitrons overlapped with approximately 16 gene models. By comparison, 24 % of the *P. sojae* genome is comprised of repetitive elements. The repeats exhibit similar compositions as in the *H. arabidopsidis* genome, except that the predominant repeat class is an unclassified type (12.36 %) followed by LTR elements (8.56 %).

## 10.2.3 Protein-Coding Genes

To assess how well the genome assembly captured *H. arabidopsidis* gene space, we searched for homologs of 248 core eukaryotic genes (CEGs) that are conserved amongst *A. thaliana,*

*Homo sapiens, Drosophila melanogaster, Caenorhabditis elegans, Saccharomyces cerevisiae,* and *Schizosaccharomyces pombe*, using the CEGMA pipeline combined with manual examination. 95 % of the CEGs are present in the *H. arabidopsidis* assembly, compared with 95, 95, and 98 %, respectively, in the assemblies for *P. infestans*, *P. ramorum*, and *P. sojae*. Moreover, 93 % of 31,759 EST reads align to the assembly. Together, these measures indicated that a very high percentage of *H. arabidopsidis* genes are represented in the assembly.

*Hyaloperonospora arabidopsidis* genes were predicted through ab initio approaches (Augustus, GeneZilla, and SNAP), combined with whole genome similarity searches to the NCBI nonredundant database. 14,543 genes were computationally predicted. 80 % of these are supported by ESTs and/or Illumina cDNA sequence tags, suggestive of a robust gene call. The Illumina data was particularly useful in refining the gene call. The predicted gene content of the *H. arabidopsidis* genome is similar to *P. ramorum* (15,743 genes) and lower than *P. sojae* (19,027 genes) or *P. infestans* (17,887 genes).

Orthologous proteins between the predicted proteomes of *H. arabidopsidis*, *P. sojae*, and *P. ramorum* were assigned with the InParanoid algorithm, using *A. thaliana* as the outgroup, giving 5938 'Inparalogs' between *H. arabidopsidis* and *P. ramorum*, and 6170 between *H. arabidopsidis* and *P. sojae*, reflecting the closer relationship of the latter two species. Six thousand, six hundred and eighty two predicted genes in *H. arabidopsidis* are unique, with no homologs in available *Phytophthora* genome assemblies, or matches in the NR database. As discussed below, these lineage-specific genes could provide valuable clues about adaptations that are unique to downy mildew species.

## 10.3 Insights from the Genome

At the time when the first drafts of the *H. arabidopsidis* genome were assembled and the annotation jamboree was convened, the first two *Phytophthora* genomes, from *P. sojae* and *P. ramorum* (Tyler et al. 2006), had recently been published. As mentioned above, the *Phytophthora* genus comprises the closest relative to the downy mildew clade and is generally described as a sister taxon (Cooke et al. 2000). Despite this close relationship, the downy mildews differ from *Phytophthora* species in several critical aspects of their lifestyle as described in Sect. 10.1.3. All downy mildews share an obligate biotroph life style and typically have very narrow host ranges (Lucas et al. 1995). Contrastingly, *Phytophthora* species are facultative plant parasites that can be cultured on synthetic media and in many cases have broad host ranges (Erwin and Ribiero 1996). *Phytophthora* species employ a hemi-biotrophic infection strategy, in which an initial phase of biotrophic growth is followed by a necrotrophic phase, during which plant cells are destroyed (Erwin and Ribiero 1996). It is generally believed that downy mildews evolved from a free-living, hemi-biotrophic ancestor whose life cycle resembled those of extant *Phytophthora* species (Goker et al. 2007). Thus, the availability of the *H. arabidopsidis* genome provided a novel opportunity to explore the molecular basis and evolution of obligate biotrophy in the downy mildews by comparison with the genomes of *P. sojae* and *P. ramorum*. This goal provided focus for the annotation jamboree and for several subsequent studies that formed the backbone of the *H. arabidopsidis* genome paper (Baxter et al. 2010). The following sections summarize three major "signatures of obligate biotrophy" that emerged from these comparisons.

### 10.3.1 Evolution for Stealth: Gene Families Involved in Pathogenicity are Downsized

Given the importance of downy mildews and *Phytophthora* species as pathogens, a high priority was placed on examining genes that are known to play a role in oomycete pathogenesis. A pervasive outcome from these analyses was that pathogenicity gene families are smaller in the *H. arabidopsidis* genome than in the genomes of *P. ramorum* or *P. sojae*, or in other *Phytophthora* species that have been described subsequently. This is illustrated below with several examples of pathogenicity gene reduction that are described in more detail than was possible in the original publication.

*Degradative enzymes* Gene families encoding secreted proteinases and cell wall degrading enzymes are likely to facilitate osmotrophic growth, and perhaps to contribute to cell destruction during necrotrophy. These genes are abundant in *Phytophthora* but substantially reduced in the *H. arabidopsidis* genome. For example, the *H. arabidopsidis* genome contains three genes encoding predicted endoglucanases of family 12 (*EGL12*) and three genes encoding pectin methyl esterases (*Pect*) that are predicted to be secreted into the apoplastic space to hydrolyze plant cell wall xyloglucans and pectin, respectively. By contrast, *P. sojae* has 10 *EGL12* genes and 19 *Pect* genes, and *P. ramorum* has 8 *EGL12* genes and 15 *Pect* genes. A proteomic survey of apoplastic fluid extracted from *Arabidopsis* leaves colonized by *H. arabidopsidis* confirmed that *HaEGL12-1*, *HaEGL12-2*, and *HaPect1* proteins are expressed. Furthermore, quantitative RT-PCR demonstrated that *HaEGL12-1*, *HaEGL12-2*, and *HaPect1* are induced during infection, while *HaEGL12-3*, *HaPect2*, and *HaPect3* are not. Phylogenetic analyses indicated that the genome of *H. arabidopsidis* has undergone a number of *EGL12* and *Pect* gene losses after speciation of *Phytophthora* and downy mildew lineages, while these gene families have expanded further via gene duplication in *P. sojae* and *P. ramorum*. In addition, the remaining *H. arabidopsidis* enzymes are highly divergent from the orthologous genes in *P. sojae* and *P. ramorum*. Hydrolytic enzymes that target the host cell wall can release cell wall fragments that can act as elicitors of host defense. These data suggest that in evolving a biotrophic lifestyle, *H. arabidopsidis* has lost most of the arsenal of secreted hydrolytic enzymes that were likely present in the hemi-biotrophic ancestor.

*Necrosis and ethylene-inducing (Nep1)-like proteins (NLPs)* NLPs are found in bacteria, fungi, and oomycetes, and have been implicated in evolution via horizontal gene transfer (see Sect. 10.3.3). NLPs in *Phytophthora* and *Pythium* have been shown to trigger plant cell death. Based on the timing of NLP gene expression, these genes have been postulated to play a role in cell killing during the necrotrophic phase of *Phytophthora* colonization (Gijzen and Nurnberger 2006). Thus, it is perhaps not surprising that the NLP gene family is less represented in *H. arabidopsidis*, compared to *P. sojae* and *P. ramorum*. Phylogenetic analysis demonstrates that oomycete *NLP* genes fall into 13 clades containing genes from at least two of the three species. Interestingly, *H. arabidopsidis* genes are found in only three of these clades, suggesting that the other clades were lost during evolution of the downy mildew lineage. Additionally, functional analysis indicated that HaNLP3 does not possess necrosis-inducing activity, despite its close phylogenetic relationship to NLPs from *P. sojae* and *P. infestans*. These results suggest that NLP genes in the downy mildew lineage have evolved a different function, compared to orthologs in *Phytophthora*. This was addressed further in a follow-up study described in Sect. 10.4.5.

*PAMPS/elicitors* Several oomycete pathogen-associated molecular patterns (PAMPs, (Boller and He 2009)) have been described recently. One well-characterized PAMP is produced by CBEL genes that encode proteins with a carbohydrate-binding domain (Gaulin et al. 2006). This glycoprotein is involved in organized polysaccharide deposition in the cell wall and in adhesion of the mycelium to cellulosic substrates such as the cell host surface. Only two CBEL genes are present in the *H. arabidopsidis* genome, compared to 13 in *P. sojae* and 15 in *P. ramorum*. Another class of PAMPs is comprised of so-called Elicitin genes which encode secreted proteins that are thought to bind sterols in the apoplastic space. The sterol-binding capability is thought to be important because *Phytophthora* species are sterol auxotrophs. As suggested by the name, some elicitins can induce plant cell death when transiently expressed or applied as exogenous proteins (Kamoun 2006). *Phytophthora sojae* and *P. ramorum* maintain moderately sized elicitin gene families (17 and 18, respectively). Contrastingly, only one copy is present in the *H. arabidopsidis* genome. The function of this gene is well worthy of further investigation.

*RxLR effector genes* One of the most important recent insights from molecular plant pathology research is that diverse pathogens have evolved the capacity to export effector proteins to the interior of plant cells (Hogenhout et al. 2009; Kale and Tyler 2011). Once inside, these proteins are thought to modify host regulatory pathways to make the plant host more receptive to colonization. In phytopathogenic bacteria, secretion of effectors from a Type III pilus is essential for suppression of host immune responses (Gohre and Robatzek 2008). The analog of these proteins in oomycetes is represented by RXLR effector genes, encoding proteins with a signal peptide for secretion outside of the pathogen, followed by an RXLR motif that are hypothesized to function in host cell targeting (Dou et al. 2008; Whisson et al. 2007). The signal peptide and RxLR motifs have been used in bioinformatic searches to identify large families of candidate RxLR effector genes in *Phytophthora* genomes (Jiang et al. 2008; Tyler et al. 2006; Whisson et al. 2007). Experimental characterization of candidate RXLR effector genes suggests that the bioinformatic searches have high predictive value. Moreover, functional assays suggest that suppression of host immunity is a major function of RxLR effectors, analogous to bacterial Type III effectors (see Sect. 10.4.5).

Totally, 134 high-confidence effector gene candidates (HaRxL genes) were identified using a permutation-guided HMM approach, followed by manual curation, during the annotation jamboree. This number is substantial but still lower than those for *Phytophthora* species (estimated at 350–550). The HaRxL genes appear to be under strong selection pressure to diverge. SNPs arising from heterozygosity in the assembly were observed at a five-fold higher frequency in RxLR effector candidate genes than in other genes (1 per ~2500 bp). Additionally, comparisons of the RXLR effectorome between *H. arabidopsidis* and *Phytophthora* species found matches for only a fraction of the HaRxL genes. Only 36 % display hits with an E value <1e-3 and sequence similarity >30 %). Moreover, *H. arabidopsidis* effectors generally are not located in syntenic locations relative to *Phytophthora* genomes. However, a handful of relatively conserved effectors were identified. These might represent "core" pathogenicity functions that are conserved between downy mildews and *Phytophthora*, and have been the subject of follow-up studies in the McDowell lab (Anderson et al. 2012).

In summary, these gene families, along with others implicated in pathogenicity, show a clear trend toward reduction in *H. arabidopsidis*, compared to *Phytophthora*. It is important to consider that each of these gene families can be thought of as double-edged swords. While the proteins can promote pathogenicity and/or virulence, they also have the potential to activate host defenses through various mechanisms. This is likely to be more consequential for obligate, narrow host range downy mildew pathogens than for *Phytophthora* species. It is also likely that a substantial portion of these genes promote the necrotrophic phase of growth and/or saprotrophy in *Phytophthora*, neither of which is relevant for the obligate biotroph *H. arabidopsidis*. Thus, the general reduction in pathogenicity genes in the *H. arabidopsidis* genome likely reflects evolution toward "stealth" in which genes with the potential to induce host defenses have been reduced to the bare minimum necessary for biotrophic colonization. As discussed below, similar reductions are evident in lineages of oomycetes and fungi that have independently evolved to an obligate lifestyle.

### 10.3.2 Loss of Motile Zoospores

Oomycetes are referred to colloquially as "water molds," and many oomycetes produce zoospores, for which motility is provided by flagella. This enables the spores to swim to a host or other nutrient sources. However, a number of downy mildew species do not produce zoospores; rather, they produce nonmotile, asexual conidiospores that produce an infective germ tube which directly penetrates the host and differentiates into an infectious hypha. Phylogenetic analysis suggests that the zoosporic stage has been lost several times during the evolution of downy mildew species (Beakes et al. 2012). *Hyaloperonospora arabidopsidis* represents one such nonmotile lineage, thus, it was of interest to examine whether motility-associated genes are evident in the *H. arabidopsidis* genome. The first analysis, described in the genome paper, was based on homology searches with a set of 84 genes that were chosen from *Chlamydomonas* genes that encoded structural components of flagella, intraflagellar transport proteins, and other genes associated with motility (Pazour et al. 2005). Most of these genes had clear matches in *Phytophthora* genomes. Contrastingly, no significant matches were detected in *H. arabidopsidis* for any of these genes, except for dynein and tubulin cytoskeleton genes that are critical cellular processes besides zoospores.

In a follow-up study, Judelson and colleagues examined a larger set of potential motility-associated genes, and provided more insights into the mechanisms behind the losses of these genes (Judelson et al. 2012). An expanded set of 460 genes potentially associated with flagella was compiled from comparative genomics of other eukaryotes. Two hundred fifty seven matches were found in *Phytophthora* genomes, 77 % of which are upregulated in spores coincident with biosynthesis of the flagellum. Of these, only 46 are conserved in the *H. arabidopsidis* genome, most likely because they contribute to important processes other than flagellar motility. With regard to mechanisms underlying gene loss, evidence was provided for gene deletions and for degeneration following a point mutation. The former mechanism appeared to be more prominent.

Altogether, these analyses provide a very interesting case study of how genes encoding biosynthesis of a complex organelle can be removed from the genome. It is conceivable that the loss of the zoosporic stage reflects the downy mildews' specialization to a "terrestrial niche" consisting of aerial plant organs, for which flagellar motility offers little advantage. It will be of interest to perform similar comparisons with genomes of different downy mildews that have independently lost the zoosporic stage.

### 10.3.3 Pathways for Metabolism of Inorganic Nitrogen and Sulfur Are Compromised

Considering the nutritional dependence of *H. arabidopsidis* on its host, it was of great interest to compare metabolic pathways, predicted from the genome, to those in *P. sojae* and *P. ramorum*. This was undertaken with pathway annotation using KAAS (KEGG automated annotation server), which identifies metabolic pathway genes based on sequence similarity and generates a pathway map. This approach was supplemented by manual examination of several pathways.

Somewhat surprisingly, most of the metabolic pathways predicted to be functional in *Phytophthora* are also present in *H. arabidopsidis*, suggesting that the transition to obligate biotrophy in the downy mildew group has not been accompanied by a large-scale reduction in metabolic capacity. However, several interesting deficiencies were noted. For example, *H. arabidopsidis* lacks orthologs of *Phytophthora* polyamine oxidases as well as genes required for the synthesis of arachidonic acid. In plants, polyamine oxidation-derived programmed cell death is thought to have a key role in the hypersensitive response (Yoda et al. 2006). Arachidonic acid is a polyunsaturated fatty acid that is released from germinating spores of *Phytophthora* species during plant infection and is sufficient to elicit hypersensitive cell death in potato (Ricker and Bostock 1992). Thus, the loss of polyamine oxidase and arachidonic acid in *H. arabidopsidis* might reflect selection for stealth inside the plant host, as proposed in Sect. 10.3.1 for pathogenicity genes.

The most interesting metabolic deficiencies in the *H. arabidopsidis* genome involved pathways for assimilation of inorganic nitrogen and sulfur. In *P. sojae* and *P. ramorum*, canonical enzymes for both pathways are present. Nitrate assimilation is accomplished by sequential action of nitrate and nitrite reductase, followed by a branch in which ammonium is incorporated into glutamate or glutamine. Sulfate assimilation is predicted to be carried out by a novel tri-functional enzyme (adenylsulfate kinase, ATP sulfurylase, pyrophosphotase), with subsequent steps catalyzed by phosphoadenosine reductase, sulfite reductase, and cysteine synthetase. In *H. arabidopsidis*, the genes for nitrate and nitrite reductase are absent, along with the gene for sulfite reductase. The absence of these genes suggests that *H. arabidopsidis* has evolved to assimilate nitrogen and sulfur in organic form from host-synthesized molecules (e.g., amino acids).

## 10.4 Applications from the Genome

As is typically the case with genome studies, the value of the *H. arabidopsidis* genome has been leveraged by studies, subsequent to the initial publication, that applied information from the *H. arabidopsidis* genome to a variety of research questions. Several interesting stories are summarized in the following sections.

### 10.4.1 Convergent Evolution of Obligate Biotrophy in Disparate Lineages

One of the most unique aspects of the initial description of the *H. arabidopsidis* genome is that it provided a first look at the genome of an obligate oomycete plant pathogen. At the time that the *H. arabidopsidis* paper was being completed, a similar analysis of obligate biotrophic fungal pathogen genomes (powdery mildew species from the ascomycota) was also nearing completion. The lead investigators of these initiatives became aware of each other's work and noted several common themes, suggestive of convergent evolution. The investigators then arranged for coordinate submission, and eventual publication, to highlight these convergences (Spanu and Panstruga 2012). These papers were followed by descriptions and comparative analyses of genomes from fungal rust pathogens (basidiomycota, (Duplessis et al. 2011)) and by white blister rust (*Albugo*) species that belong to the oomycetes but reside in a different lineage that independently evolved to an obligate lifestyle (Links et al. 2011; Kemen et al. 2011). Thus, these reports cover lineages that represent four independent evolutions to an obligate lifestyle on a plant host: twice in fungi and twice in oomycetes. Several striking commonalities are evident from comparison of these lineages. First, all four lineages appear to secrete large complements of effector proteins, most likely to suppress host immunity. It is thought that the ability to suppress/evade immunity enabled these pathogens to commit to the host as a stable source of nutrients. Second, genomes tend to have large fractions of repetitive elements, compared to nonobligate relatives (although this is less true for the *Albugo* species). This is thought to provide a plastic genome, in which recombination between repeats and perhaps also repeat-driven epigenetic changes can generate variation that is critical for pathogens to coevolve with their hosts (Raffaele and Kamoun 2012). Third, enzymes for assimilation of inorganic sulfur and/or nitrogen have been discarded. This presumably reflects a cost associated with maintenance of these pathways, and implies that the pathogens have evolved an efficient mechanism(s) for obtaining nitrogen and sulfur in organic, host-synthesized forms.

Altogether, the independent emergence of these three "signatures" in four different obligate lineages suggests that each adaptation is highly advantageous, if not essential, for success as an obligate, filamentous plant pathogen. This comprises one of the most exciting insights to which the *H. arabidopsidis* genome has contributed, and has inspired several commentaries that consider this subject in more depth (Spanu and Kamper 2010; McDowell 2011; Spanu 2012; Kemen and Jones 2012).

### 10.4.2 Reconstruction of Oomycete Genome Evolution

Genomes from *H. arabidopsidis* and *Phytophthora* species were nicely exploited in a pair of papers that used comparative approaches to infer several aspects of oomycete gene and genome evolution (Seidl et al. 2012a, b). One study reconstructed over 18,000 oomycete gene trees and inferred gene gains, duplications, and losses. The authors estimated that the genome of the last common ancestor (LCA) of oomycetes was comprised of about 10,000 genes. Subsequent gene family evolution was very fluid with different functional gene classes exhibiting different dynamics. The analysis suggested a high frequency of gene duplications subsequent to the divergence of the LCA of *Phytophthora* species, perhaps due to a large-scale duplication. The second study took a broader perspective in analyzing protein domain organization in 67 eukaryotic species, including *H. arabidopsidis*, *P. ramorum*, *P. infestans*, and *P. sojae* and five fungal pathogens. The most interesting aspect of this study was delineation of 773 unique combinations of protein domains in oomycete genomes, many of which could be hypothesized to play roles in pathogenesis-related processes. Such genes represent a high priority for functional characterization. Oomycetes displayed a large diversity of domain combinations compared to fungi, underscoring their ability to evolve and their pathogenic success. Together these studies provide an impressive demonstration of new applications of comparative genomics and bioinformatics to bridge the gap between genomics and experimental biology. These approaches will be even more powerful as additional oomycete genomes become available.

### 10.4.3 Identification of Horizontal Transfer of Pathogenicity Genes from Fungi

Another comparative application of the *H. arabidopsidis* genome was reported in a recent description of horizontal gene transfer (HGT) events from fungi-to-oomycete plant pathogens (Richards et al. 2011). The authors identified 33 probable fungi-to-oomycete transfer events that were present in one or more of the genomes of *P. sojae*, *P. ramorum*, *P. infestans*, and *H. arabidopsidis*. None of these genes were present in other oomycete genomes, suggesting that the genes were transferred to the last common ancestor of *Phytophthora* and downy mildew species. Many of these genes were duplicated after transfer, so that the 33 HGTs account for an estimated 329 genes in extant *Phytophthora* and *H. arabidopsidis* genomes. A strikingly disproportionate number of these genes encode putative secreted proteins, suggesting that HGT has substantially enhanced the virulence capabilities of these pathogens. One such transfer involved the NLP family described above. Another encodes a LysM-containing protein like those that have been recently implicated in sequestering chitin fragments to prevent perception by plant pattern recognition receptors (Thomma and de Jonge 2009). Interestingly, one-third of HGTs involved genes for breakdown and uptake of complex carbohydrates, and almost all of the remaining HGTs have other putative roles in nutrient acquisition. This suggests that HGTs have enhanced the pathogens' capacity for osmotrophic growth and provides intriguing clues to the understudied mystery of how oomycetes acquire nutrients from hosts.

### 10.4.4 Controversial Phylogenetic Relationships Between *Phytophthora* and Downy Mildew Species

Genome data can be used to resolve poorly defined species phylogenies. As mentioned in Sect. 10.1.1, the downy mildew species are thought to form a single, monophyletic group. *Phytophthora* species are similarly depicted as a monophyletic assemblage comprised of ten clades, numbered based on their phylogenetic proximity to each other (Blair et al. 2008). The downy mildew group and the *Phytophthora* species are typically depicted as two distinct lineages that share a common ancestor (i.e., sister taxa). However, a recent study proposed a dramatic revision of this relationship (Runge et al. 2011). The authors constructed a multi-locus data set, based on markers derived from genome data from *Phytophthora* species, *H. arabidopsidis* and from the cucurbit downy mildew pathogen *Pseudoperonospora cubensis* (Tian et al. 2011). This multi-locus dataset was designed to provide better phylogenetic resolution than had been possible with the previously limited sequence information for the downy mildew species. The resultant phylogenies placed the downy mildew species within the *Phytophthora* genus, most closely related to *Phytophthora* clades 1 and 4, which contain well-known species such as *P. infestans*, *P. nicotiane*, and *P. palmivora*. In other words, this phylogeny predicts that *P. infestans* is more closely related to *H. arabidopsidis* than it is to *P. sojae* or *P. ramorum*. Obviously, this would necessitate substantial reclassification. More importantly, models of evolution for these pathogens would have to be substantially re-evaluated. It should be noted that Seidl et al. (2012a, b) used genome data to generate a different phylogeny that placed the downy mildews outside the *Phytophthora* genus, consistent with earlier studies. This discrepancy must be resolved.

## 10.4.5 Functional Analysis of Pathogenicity Genes

The most widespread application of the *H. arabidopsidis* genome has been to use genome-enabled lists of pathogenicity genes as a starting point for functional analysis of the encoded proteins. As mentioned above, much experimental attention in the oomycete field has focused on genes that encode secreted proteins with the RXLR motif that is hypothesized to mediate host cell entry (Jiang and Tyler 2012). Bioinformatic predictions in *Phytophthora* as well as *H. arabidopsidis* yielded long lists of candidate genes, ranging from over 550 in *P. infestans* to at least 134 in *H. arabidopsidis*. Thus, a major challenge is to determine which genes encode *bona fide* effectors with biological activity that promotes virulence. This challenge has been addressed by examining gene expression patterns (are the genes induced during infection?), analysis of polymorphism (which genes are subjected to diversifying selection, indicative of a role in host interaction?), and functional tests in which individual RXLR genes are expressed, in a processed form, inside plant cells as described above. Due to the large number of candidate genes, transgene experiments typically involve transient assays using bacteria as surrogates to deliver the RXLR genes/proteins, with stable transformation used in some cases to validate results from transient delivery systems.

For effectors from *P. infestans* and *P. sojae*, *Agrobacterium*-mediated transient transformation was used to efficiently test for activation or suppression of immunity in *Nicotiana benthamiana*. As discussed above, *Agrobacterium*-mediated transient expression is not efficient in *Arabidopsis*, so Fabro and colleagues used *Pseudomonas syringae* to deliver RXLR proteins through the Type III secretion system (Fabro et al. 2011). Their system was based on the aforementioned effector detector vector to target the RXLR proteins to the Type III secretion system (Sohn et al. 2007). The authors further utilized a strain of *P. syringae* that constitutively expressed a luciferase gene. This enabled quantitative measurement of light as a proxy for bacterial growth, thereby increasing the throughput of the screen. In this system, 44 of 64 tested effectors enhanced virulence of *P. syringae*. Thus, a large proportion of *H. arabidopsidis* effectors seem to have the capacity to suppress immunity or otherwise manipulate the host to be more accommodating for the bacterium. This high capacity for immune system suppression was mirrored somewhat in functional surveys of RXLR effectors from *P. sojae* (Wang et al. 2011).

This survey was followed by a medium-throughput effort to estimate subcellular localization of 46 RXLR effector proteins from *H.*

*arabidopsidis*, based on high-throughput, transgenic expression of effectors fused to fluorescent proteins (Caillaud et al. 2011). The main point from this paper was that the nucleus and membrane networks comprise the main targets of RXLR effectors. This is the first such study for oomycete effectors, and will undoubtedly lead to similar comparisons in other oomycete pathosystems. Interestingly, other organelles that could be fertile targets for oomycete manipulation (e.g., mitochondria, chloroplasts) were not targeted. Many of the tested effectors exhibited similar localization patterns in *Arabidopsis* and *N. benthamiana*, suggesting that these effectors exploited conserved host mechanisms for protein targeting.

Another landmark effort was designed to identify candidate protein targets of *H. arabidopsidis* RXLR effectors (Mukhtar et al. 2011). The experimental platform was based on automated yeast two hybrid assays, in which 99 *H. arabidopsidis* RXLR proteins were tested, in pairwise fashion, for interaction with 8,000 *Arabidopsis* proteins representing roughly one- third of the *Arabidopsis* proteome. This screen also included 53 Type III effectors from *Pseudomonas* bacteria. These screens yielded a large number of interactions between pathogen effectors and plant proteins. Putative targets for 53 of the *H. arabidopsidis* RXLR effectors were identified. Importantly, 17 of these host proteins interacted with one or more effectors from *Pseudomonas*, suggesting that these plant proteins represent points of vulnerability in plant regulatory networks that oomycetes and bacteria have evolved independently to exploit. These interactions represent high priorities for follow up studies.

In addition to the large-scale studies, a number of recent papers take a narrower focus on the structure and/or function of individual RXLR proteins. Two such studies reported the structures of the RXLR avirulence effectors ATR1 and ATR13, which are recognized by the *Arabidopsis* NB-LRR surveillance proteins RPP1 and RPP13, respectively (Leonelli et al. 2011; Chou et al. 2011). Both of these studies identify residues that are important for recognition, and in the case of ATR13, a novel localization determinant. Another study reported functional characterization of *H. arabidopsidis* effector HaRxL96 and its homology from *P. sojae*, PsAvh163 (Anderson et al. 2012). HaRxL96 is one of the few *H. arabidopsidis* RXLR effectors with recognizable homologs in *Phytophthora* species. The two effectors were shown to activate or suppress immunity in distantly related plants (soybean, *Arabidopsis*, and *N. benthamiana*), suggesting that they might interact with a conserved host target.

Although most of the effort so far has focused on RXLR proteins, other pathogenicity gene families are now under investigation (Cabral et al. 2011, 2012). Of particular interest, given the *H. arabidopsidis* biotrophic lifestyle, is the role of NLP genes in the *H. arabidopsidis* genome. Why would an obligate biotroph maintain a multigene family for presumptive necrosis-inducing proteins? Functional analysis of *H. arabidopsidis* NLP proteins demonstrated that they are noncytolytic. Interestingly, analysis of chimeras between HaNLP and the cytolytic *P. sojae* NIP demonstrated that HaNLP3 does possess cytolytic activity, but this is masked by an exposed domain that is unique to this NLP lineage in *H. arabidopsidis*. This implies that NLPs have an important, noncytolytic role in biotrophic growth.

## 10.5 Future Perspectives

Below are thoughts about the next steps that can be taken to improve the gene inventory, to move toward a better understanding of how downy mildew pathogens and their relatives cause disease, and how this understanding can be exploited for more efficient control of oomycete diseases.

### 10.5.1 Enhanced Resources for *H. arabidopsidis* Genomics and Transcriptomics

This discussion must begin with a consideration of additional genomic infrastructure that is needed to extract full value from the initial investment in the *H. arabidopsidis* reference genome. Considering the decreasing cost and increasing efficiency of DNA sequencing and bioinformatic analysis, the most straightforward resource to develop is additional genome information. The *H. arabidopsidis* Emoy2 assembly, although adequate for many purposes, has much room for improvement and would benefit from application of new low cost, long-read technologies (e.g., Pacific Biosciences). It is also important to generate draft sequences of additional *H. arabidopsidis* isolates. Recent studies from *Phytophthora* and from fungi have clearly validated the utility of insights from deep intraspecies genome comparisons (e.g., (McDowell 2013; Cooke et al. 2012; O'Connell et al. 2012; Gan et al. 2013)).

Another resource that must be developed is better transcriptomic data for the *H. arabidopsidis-Arabidopsis* interaction. Gene expression databases that span the infection cycle in *Arabidopsis* would be invaluable for functional inferences, as demonstrated recently for *Phytophthora* pathogens (Wang et al. 2011; Whisson et al. 2007). Such resources will ideally encompass multiple combinations of *H. arabidopsidis* isolate and *Arabidopsis* ecotypes, to capture transcriptome variation on both sides of the interactions (Birch and Kamoun 2000). The obligate lifestyle of *H. arabidopsidis* is not a barrier to development of these resources, because it is trivial to extract RNA from germinated spores and from leaves over a time course following infection. The ever-improving RNAseq-based approaches should be sufficient for thorough sampling of the *H. arabidopsidis* transcriptome even from samples in which pathogen RNA comprises a very low proportion of the total sample (e.g., within the first 24 h after inoculation). However, the potential of this approach is compromised by the current necessity of analyzing "bulk" samples, i.e., RNA isolated from homogenized leaves. Transcript profiles of mixed cell populations from infected tissues reflect the averaged signal from different interaction sites, as well as a majority of cells that are not in close proximity to the pathogen. Such samples do not provide spatial and temporal information at a high level of resolution. It should be pointed out that this limitation is equally relevant for interactions involving other oomycetes, fungi, and all other pathogens with growth habits that include specialized interfaces with individual host cells. This problem is now tractable (albeit still challenging) due to breakthrough advances in laser microdissection and other technologies that enable researchers to harvest specific cells from differentiated organs (Chandran et al. 2010; Bailey-Serres 2013). Such approaches, when coupled with RNA amplification and RNAseq, will allow space- and time-resolved transcriptomic signatures that will undoubtedly provide new insights into oomycete-plant interactions.

### 10.5.2 Developing Genome and Transcriptome Resources for Diverse Downy Mildew Species

It is important to consider that every inference about downy mildew biology described above is based on one downy mildew genome. Thus, there is a critical need for genome resources from key representatives of the other 800 downy mildew species. A framework for prioritization of these species for genome analysis would ideally (1) encompass both close and distant relatives of *H. arabidopsidis*, to maximize comparative value, (2) capture unique aspects of biology that exist amongst the downy mildew species, and (3) most efficiently enable translation of genome data into useful tools for disease control. For example, close relatives of *H. arabidopsidis* (e.g., species that parasitize cultivated brassica crops) provide points of comparison from which relatively recent genome evolution can be inferred, as exemplified by (Raffaele et al. 2010). From a practical standpoint, this resource would open the door for effector-based screens for new resistance genes (see Sect. 10.5.4) that would be very useful for disease control. Given the close evolutionary distance between

*Arabidopsis* and its cultivated relatives, it is probable that *Arabidopsis* itself could serve as a facile source of resistance genes for cultivated brassica species and perhaps even more distantly related crops.

At the other end of the downy mildew phylogenetic spectrum, relative to *H. arabidopsidis*, are the downy mildews genera that cause disease on monocot hosts (e.g., *Sclerospora*, *Peronosclerospora*). These are fascinating species from both basic and applied perspectives. From a phylogenetic standpoint, these pathogens comprise an outgroup relative to the other downy mildew species and perhaps even a "missing link" between *Phytophthora* and downy mildew species (Thines 2009). These species are also very distinct in their host range-no other downy mildew species are known to cause disease on monocot hosts. In fact, none of the *Phytophthora* species can parasitize monocots. Sequenced genomes would undoubtedly provide hypotheses to address the fascinating question of why these downy mildew species are capable of causing diseases on monocots where "plant destroyers" cannot. More importantly, the monocot downy mildew species can be devastating pathogens of maize, pearl millet, and sorghum in Africa, India, and Southeast Asia, as described in Sect. 10.1.2. There is a clear need for genome-enabled tools to fight these diseases.

Many other downy mildew species are well worthy of attention, including *Plasmopora*, *Bremia*, and *Pseudoperonospora* species that cause substantial disease problems on grape, lettuce, and cucurbit crops, respectively. Transcriptomic and genomic initiatives for some species have already been initiated and led to the unexpected identification of RXLR-like effectors that vary in the consensus sequence for host cell uptake (Stassen et al. 2012, 2013; Tian et al. 2011; Savory et al. 2012a), and revealed alternative mRNA splicing as a potential source of novel RXLR effectors (Savory et al. 2012b). These studies exemplify how additional genomes will undoubtedly reveal interesting new basic biology. Of equal importance, these genomes will also provide information necessary for new strategies for downy mildew control, as outlined in Sect. 10.5.4.

### 10.5.3 Exploiting Genomics for a Better Understanding of How Downy Mildew Pathogens Cause Disease

Hopefully, this chapter has already conveyed the enormous impact that genomics has had on our understanding of oomycete pathogenicity and virulence/avirulence. In particular, the development of bioinformatic criteria for identifying candidate effector genes in sequenced genomes was a key advance. The functional studies of *H. arabidopsidis* RXLR proteins described above would not have been possible without the list of candidate genes that the *H. arabidopsidis* genome enabled. Now, the challenge is to carry out the detailed, painstaking studies that will be necessary to elucidate the mechanisms through which these RXLR effectors alter their plant targets, and most importantly, relate the functionality of those targets to plant processes that are reprogrammed to better support pathogen colonization. Such studies are already well underway for a handful of RXLR effectors from *Phytophthora* (Bozkurt et al. 2012), and it will be of great interest to compare the functions of RXLR proteins from downy mildew species and *Phytophthora* species. It will be particularly intriguing to examine whether there are lineage-specific functions that have evolved in the RXLR secretome to support the obligate lifestyle (e.g., manipulation of host metabolism). In this way, substantial strides will be made toward understanding the molecular basis of oomycete virulence and plant disease susceptibility.

As important as it is to understand RXLR effectors, a rereading of section X.3.1 will remind the reader that RXLR effectors comprise only a fraction of the predicted *H. arabidopsidis* secretome (Stassen and Van den Ackerveken 2011; Jiang and Tyler 2012). With the exception of NLP proteins, the other families of pathogenicity genes in *H. arabidopsidis* remain unstudied. This can easily be remedied using experimental approaches, both large- and small-scale, which follow the blueprints established for RXLR proteins. Another area of potentially ripe inquiry is the lineage-specific genes that are unique to downy mildew species, particularly those that appear to be secreted. Such an effort is probably premature at present, but the probable emergence of more downy mildew genomes in the near future will be of high utility for prioritizing lineage-specific downy mildew genes for functional studies.

Much of the effort invested so far in molecular plant-oomycete interactions has focused on mechanisms for induction and suppression of plant immune responses. However, evasion of plant immunity is only one of several challenges that oomycetes must overcome to successfully colonize their hosts. Another challenge is to reprogram host metabolic pathways such that infected organs become nutrient sinks, and to render host-synthesized macromolecules into forms that can be assimilated into pathogen metabolic networks to fuel growth. In this regard, the most advantageous aspect of the *H. arabidopsidis-Arabidopsis* pathosystem is the unparalleled experimental resources on the host side, which will facilitate identification of host pathways that are co-opted for this purpose. The *H. arabidopsidis* genome will complement the host resources by delineating potential transport mechanisms. For example, genome analysis has already suggested that host-derived, organic compounds are a key source of nitrogen and sulfur for the pathogen. Amino acids are an obvious potential source, and the McDowell lab is following up on this with efforts to identify host-encoded amino acid transporters that are co-opted by the pathogen to fuel its own growth.

The approaches described above will almost certainly be fruitful, however, the experimental utility of downy mildew pathosystems is limited by our inability to culture the pathogen on synthetic media. The work-arounds described in Sect. 10.1.4 are sufficient for many purposes, but there is often no good substitute for a knockout or knockdown of an endogenous gene. This is likely to be particularly true for the lineage-specific genes and for genes involved in nutrient transport and assimilation on the pathogen side of the interaction. In this regard, a disappointment from the initial genome analysis was the absence of obvious solutions for culturing *H. arabidopsidis*. For now, this remains as a major problem and a holy grail for the field. Hopefully, additional genome sequences will provide new clues to the obligations that inextricably link downy mildew pathogens with their hosts (Spanu 2006).

### 10.5.4 Exploiting Genomics for Improved Control of Downy Mildew Diseases of Crops

It is imperative that genomic and molecular dissection of plant-pathogen interactions, no matter how fundamentally oriented, must be conducted with an eye toward new solutions for control of crop diseases. Fortunately, the payoffs of oomycete genome analyses for disease management are already being recouped. The most straightforward application of downy mildew genome information is for the design of DNA-based assays that can be used for classification, diagnosis, and epidemiology (Studholme et al. 2011; Thines and Kamoun 2010). Although sufficient gene resources exist already for typing at the species level, it would be of great value to have better markers for resistance to fungicides and for pathotype (virulence/avirulence). It is now conceivable that such assays could directly test the gene that controls the trait of interest (e.g., sequence polymorphisms associated with fungicide resistance, or loss-of-function in avirulence effector genes). This will be of great value for determining the most cost-effective strategies for fungicide use and for selecting potentially resistant cultivars.

Host resistance is still widely regarded as the most cost-effective, environmentally friendly option for control of oomycete diseases. Unfortunately, history has shown clearly that oomycetes are very adept at breaking gene-for-gene resistance (Fry 2008; Kamoun 2003). We now know, from genomics, that this adaptability can be explained by large RXLR gene repertoires (providing genetic redundancy and/or compensation and thereby allowing for *Avr* gene deletions) along with genome plasticity (providing for rapid mutation, deletion, or silencing of Avr genes) (Raffaele and Kamoun 2012; Jiang and Tyler 2012). What then is the solution? In this context, the most exciting new strategy to have emerged from genomics and effector biology is to use cloned effectors as specific probes for *R* genes in germplasm screens. This approach was pioneered against *P. infestans* with highly promising results (Vleeshouwers et al. 2008, 2010; Oh et al. 2009; Rietman et al. 2012). Their approach is based on screening with effector genes, one at a time, in wild relatives of cultivated potato to look for novel *R* genes. High-throughput, *Agrobacterium*-mediated, transient assays are used to express the RXLR genes. Formation of a macroscopic HR is used as a readout for resistance. The advantage of this approach, compared to traditional, pathogen-based screens, is that it is possible to look for responses to a single effector, independently of complications from other effectors. Equally important, the screen can be extended to relatives of the crop that are nonhosts for the pathogen of interest, thereby expanding potential sources of resistance. Ideally, one can screen with effectors that are conserved in the pathogen species and essential for virulence. Any *R* genes that recognize such effectors are expected to be durable because the pathogen would suffer a substantial fitness penalty if the effector were deleted or silenced. This approach has been used to identify new genes for late blight resistance and is currently being optimized for several other oomycete and fungal pathogens. In principle, the only resources absolutely necessary for this approach are a list of RXLR effectors and a suitable system for expressing these in the germplasm to be screened. This approach has the added advantage of not requiring transgenes, as long as the screen is confined to sexually compatible relatives. Another advantage of this approach is the ability to monitor field populations of the pathogen for emergence of resistance-breaking alleles of the relevant effector gene. If such alleles emerge, their spread can be slowed by planting cultivars with different *R* genes and/or applying fungicides (Vleeshouwers et al. 2010). Thus, this effector-based approach is likely to have a strong impact on resistance breeding in the future.

Genome information is also fueling more speculative innovations for engineering resistant plants. The most widely applied technology is so-called host-induced gene silencing, in which plant transgenes are designed to synthesize double-stranded RNAs that are identical to an endogenous pathogen gene that is essential for virulence. dsRNA fragments are translocated to the inside of pathogen cells, wherein they trigger RNA interference by canonical pathways to silence the endogenous gene. This approach has been tested with promising results against a biotrophic fungal pathogen (Nowara et al. 2010). Applications to oomycete diseases are currently underway, using target genes identified based on sequenced genomes. Another approach under development is to express transgenes that block entry of RXLR effectors into plant cells (Jiang and Tyler 2012). These and other approaches are relatively high risk, and would require the large investments necessary for transgene regulation. However, they also offer a huge potential promise of broad spectrum, durable resistance against any oomycete disease. Together with effector-directed breeding, these new strategies provide excellent examples of how genome data can be translated into sustainable disease control.

**Acknowledgments** Kasia Dinkeloo provided critical reading of the manuscript. Research on plant-oomycete interactions in the McDowell lab is supported by the U.S. Department of Agriculture–Agriculture and Food Research Initiative (2004-35600-15055 and 2010-65110-20764), the National Science Foundation (ABI-1146819), and the Virginia Tech Institute for Critical Technology and Applied Sciences.

## References


Allen RL, Bittner-Eddy PD, Grenvitte-Briggs LJ, Meitz JC, Rehmany AP, Rose LE, Beynon JL (2004) Host-parasite coevolutionary conflict between *Arabidopsis* and downy mildew. Science 306(5703):1957–1960

Anderson R, Cassady M, Fee R, Deb D, Fedkenheuer K, Tyler B, McDowell J (2012) Suppression of defense responses in distantly related plants by homologous RXLR effectors from *Hyaloperonospora arabidopsidis* and *Pytophthora sojae*. Plant J 72:882–893

Attard A, Gourgues M, Callemeyn-Torre N, Keller H (2010) The immediate activation of defense responses in *Arabidopsis* roots is not sufficient to prevent *Phytophthora parasitica* infection. New Phytol 187(2):449–460

Bailey-Serres J (2013) Microgenomics: genome-scale, cell-specific monitoring of multiple gene regulation tiers. Annu Rev Plant Biol 64:293–325

Baldauf SL, Roger AJ, Wenk-Siefert I, Doolittle WF (2000) A kingdom-level phylogeny of eukaryotes based on combined protein data. Science 290(5493):972–977

Baxter L, Tripathy S, Ishaque N, Boot N, Cabral A, Kemen E, Thines M, Ah-Fong A, Anderson R, Badejoko W, Bittner-Eddy P, Boore JL, Chibucos MC, Coates M, Dehal P, Delehaunty K, Dong S, Downton P, Dumas B, Fabro G, Fronick C, Fuerstenberg SI, Fulton L, Gaulin E, Govers F, Hughes L, Humphray S, Jiang RH, Judelson H, Kamoun S, Kyung K, Meijer H, Minx P, Morris P, Nelson J, Phuntumart V, Qutob D, Rehmany A, Rougon-Cardoso A, Ryden P, Torto-Alalibo T, Studholme D, Wang Y, Win J, Wood J, Clifton SW, Rogers J, Van den Ackerveken G, Jones JD, McDowell JM, Beynon J, Tyler BM (2010) Signatures of adaptation to obligate biotrophy in the *Hyaloperonospora arabidopsidis* genome. Science 330(6010):1549–1551

Beakes GW, Glockling SL, Sekimoto S (2012) The evolutionary phylogeny of the oomycete "fungi". Protoplasma 249(1):3–19

Birch P, Kamoun S (2000) Studying interaction transcriptomes: coordinated analyses of gene expression during plant-microorganism interactions. Trends Plant Sci 5:77–82

Bittner-Eddy P, Allen R, Rehmany A, Birch P, Beynon J (2003) Use of suppressive subtrative hybridization to identify downy mildew genes expressed during infection of *Arabidopsis thaliana*. Mol Plant Pathol 4:501–508

Bittner-Eddy PD, Crute IR, Holub EB, Beynon JL (2000) RPP13 is a simple locus in *Arabidopsis thaliana* for alleles that specify downy mildew resistance to different avirulence determinants in *Peronospora parasitica*. Plant J 21(2):177–188

Blair JE, Coffey MD, Park S-Y, Geiser DM, Kang S (2008) A multi-locus phylogeny for *Phytophthora* utilizing markers derived from complete genome sequences. Fungal Genet Biol 45(3):266–277

Boller T, He SY (2009) Innate immunity in plants: an arms race between pattern recognition receptors in plants and effectors in microbial pathogens. Science 324(5928):742–744

Botella MA, Parker JE, Frost LN, Bittner-Eddy PD, Beynon JL, Daniels MJ, Holub EB, Jones JD (1998) Three genes of the *Arabidopsis RPP1* complex resistance locus recognize distinct *Peronospora parasitica* avirulence determinants. Plant Cell 10(11):1847–1860

Bozkurt TO, Schornack S, Banfield MJ, Kamoun S (2012) Oomycetes, effectors, and all that jazz. Curr Opin Plant Biol 15(4):483–492

Cabral A, Oome S, Sander N, Kufner I, Nurnberger T, Van den Ackerveken G (2012) Nontoxic Nep1-like proteins of the downy mildew pathogen *Hyaloperonospora arabidopsidis*: repression of necrosis-inducing activity by a surface-exposed region. Mol Plant Microbe Interact 25(5):697–708

Cabral A, Stassen JH, Seidl MF, Bautor J, Parker JE, Van den Ackerveken G (2011) Identification of *Hyaloperonospora arabidopsidis* transcript sequences expressed during infection reveals isolate-specific effectors. PLoS ONE 6(5):e19328

Caillaud M-C, Piquerez SJM, Fabro G, Steinbrenner J, Ishaque N, Beynon J, Jones JDG (2011) Subcellular localization of the *Hpa* RxLR effector repertoire identifies the extrahaustorial membrane-localized HaRxL17 that confers enhanced plant susceptibility. Plant J 69:252–265

Chandran D, Inada N, Hather G, Kleindt CK, Wildermuth MC (2010) Laser microdissection of *Arabidopsis* cells at the powdery mildew infection site reveals site-specific processes and regulators. Proc Natl Acad Sci USA 107(1):460–465

Chou S, Krasileva KV, Holton JM, Steinbrenner AD, Alber T, Staskawicz BJ (2011) *Hyaloperonospora arabidopsidis* ATR1 effector is a repeat protein with distributed recognition surfaces. Proc Natl Acad Sci USA 108(32):13323–13328

Clark J, Spencer-Phillips P (2000) Downy mildews. In: Encyclopedia of microbiology, vol 2, 2nd edn. Academic Press Inc., New York, pp 117–129

Clough SJ, Bent AF (1998) Floral dip: a simplified method for *Agrobacterium*-mediated transformation of *Arabidopsis thaliana*. Plant J 16(6):735–743

Coates ME, Beynon JL (2010) *Hyaloperonospora arabidopsidis* as a pathogen model. Annu Rev Phytopathol 48:329–345

Constantinescu O, Fatehi J (2002) *Peronospora*-like fungi (Chromista, *Peronospora*les) parasitic on Brassicaceae and related hosts. Nova Hedwigia 74(3–4).291–338

Cooke DE, Cano LM, Raffaele S, Bain RA, Cooke LR, Etherington GJ, Deahl KL, Farrer RA, Gilroy EM, Goss EM, Grunwald NJ, Hein I, Maclean D, McNicol JW, Randall E, Oliva RF, Pel MA, Shaw DS, Squires JN, Taylor MC, Vleeshouwers VG, Birch PR, Lees AK, Kamoun S (2012) Genome analyses of an aggressive and invasive lineage of the Irish potato famine pathogen. PLoS Pathog 8(10):e1002940

Cooke DE, Drenth A, Duncan JM, Wagels G, Brasier CM (2000) A molecular phylogeny of *Phytophthora* and related oomycetes. Fungal Genet Biol 30(1):17–32

Crute IR, Beynon J, Dangl JL, Holub EB, Mauch-Mani B, Slusarenko A, Staskawicz BJ, Ausubel FM (1994) Microbial pathogenesis of *Arabidopsis*. In: Meyerowitz EM, Somerville CR (eds) *Arabidopsis*. Cold Spring harbor Laboratory Press, Cold Spring Harbor, pp 705–748

Crute IR, Holub EB, Tor M, Brose E, Beynon JL (1993) The identification and mapping of loci in *Arabidopsis thaliana* for recognition of the fungal pathogens *Peronospora parasitica* (downy mildew) and *Albugo candida* (white blister). In: Nester EW, Verma DPS (eds) Advances in molecular genetics of plant-microbe interactions, vol 2., Current plant science and biotechnology in agricultureKluwer Academic, Dordrecht, pp 437–444

Dangl JL (1993) Applications of *Arabidopsis thaliana* to outstanding issues in plant-pathogen interactions. Int Rev Cytol 144:53–83

Dick M (2001) Straminipilous fungi: systematics of the peronosporomycetes including accounts of the marine straminipilous protists, the plasmodiophorids and similar organisms. Kluwer Academic Publishers, Dordrecht

Dou D, Kale SD, Wang X, Jiang RH, Bruce NA, Arredondo FD, Zhang X, Tyler BM (2008) RXLR-mediated entry of *Phytophthora sojae* effector Avr1b into soybean cells does not require pathogen-encoded machinery. Plant Cell 20(7):1930–1947

Duplessis S, Cuomo C, Lin YC, Aerts A, Tisserat N (2011) Obligate biotrophy features unraveled by the genomic analysis of rust fungi. Proc Natl Acad Sci USA 14:9166–9171

Erwin DC, Ribiero OK (1996) *Phytophthora* diseases worldwide. APS Press, St. Paul

Fabro G, Steinbrenner J, Coates M, Ishaque N, Baxter L, Studholme DJ, Korrner E, Allen RL, Piquerez SJM, Rougon-Cardoso A, Greenshields D, Lei R, Badel JL, Caillaud M-C, Sohn K-H, Van den Ackerveken G, Parker JE, Beynon J, Jones JDG (2011) Multiple candidate effectors from the oomycete pathogen *Hyaloperonospora arabidopsidis* suppress host plant immunity. PLoS Pathog 7(11):e1002348

Fry W (2008) *Phytophthora infestans*: the plant (and *R* gene) destroyer. Mol Plant Pathol 9(3):385–402

Gan P, Ikeda K, Irieda H, Narusaka M, O'Connell RJ, Narusaka Y, Takano Y, Kubo Y, Shirasu K (2013) Comparative genomic and transcriptomic analyses reveal the hemibiotrophic stage shift of *Colletotrichum* fungi. New Phytol 197(4):1236–1249

Gaulin E, Drame N, Lafitte C, Torto-Alalibo T, Martinez Y, Ameline-Torregrosa C, Khatib M, Mazarguil H, Villalba-Mateos F, Kamoun S, Mazars C, Dumas B, Bottin A, Esquerre-Tugaye MT, Rickauer M (2006) Cellulose binding domains of a *Phytophthora* cell wall protein are novel pathogen-associated molecular patterns. Plant Cell 18(7):1766–1777

Gäumann E (1918) Ueber die formen der *Peronospora parasitica* (Pers.) Fries. 35(1.Abt.): Beih Bot Zentralblatt 35:395–533

Gijzen M, Nurnberger T (2006) Nep1-like proteins from plant pathogens: recruitment and diversification of the NPP1 domain across taxa. Phytochemistry 67(16):1800–1807

Gisi U (2002) Chemical control of downy mildews. In: Spencer-Phillips P, Gisi U, Lebeda A (eds) Advances in downy mildew research. Kluwer Academic Publishers, Dordrecht, pp 119–160

Gohre V, Robatzek S (2008) Breaking the barriers: microbial effector molecules subvert plant immunity. Annu Rev Phytopathol 46:189–215

Goker M, Voglmayr H, Riethmuller A, Oberwinkler F (2007) How do obligate parasites evolve? A multi-gene phylogenetic analysis of downy mildews. Fungal Genet Biol 44(2):105–122

Goodin MM, Zaitlin D, Naidu RA, Lommel SA (2008) *Nicotiana benthamiana*: its history and future as a model for plant-pathogen interactions. Mol Plant-Microbe Interact 21(8):1015–1026

Gunn ND, Byrne J, Holub EB (2002) Outcrossing of two homothallic isolates of *Peronospora parasitica* and segregation of avirulence matching six resistance loci in *Arabidopsis thaliana*. In: Spencer-Phillips P, Gisi U, Lebeda A (eds) Advances in downy mildew research. Kluwer Academic Publishers, Dordrecht, pp 185–188

Haas BJ, Kamoun S, Zody MC, Jiang RH, Handsaker RE, Cano LM, Grabherr M, Kodira CD, Raffaele S, Torto-Alalibo T, Bozkurt TO, Ah-Fong AM, Alvarado L, Anderson VL, Armstrong MR, Avrova A, Baxter L, Beynon J, Boevink PC, Bollmann SR, Bos JI, Bulone V, Cai G, Cakir C, Carrington JC, Chawner M, Conti L, Costanzo S, Ewan R, Fahlgren N, Fischbach MA, Fugelstad J, Gilroy EM, Gnerre S, Green PJ, Grenville-Briggs LJ, Griffith J, Grunwald NJ, Horn K, Horner NR, Hu CH, Huitema E, Jeong DH, Jones AM, Jones JD, Jones RW, Karlsson EK, Kunjeti SG, Lamour K, Liu Z, Ma L, Maclean D, Chibucos MC, McDonald H, McWalters J, Meijer HJ, Morgan W, Morris PF, Munro CA, O'Neill K, Ospina-Giraldo M, Pinzon A, Pritchard L, Ramsahoye B, Ren Q, Restrepo S, Roy S, Sadanandom A, Savidor A, Schornack S, Schwartz DC, Schumann UD, Schwessinger B, Seyer L, Sharpe T, Silvar C, Song J, Studholme DJ, Sykes S, Thines M, van de Vondervoort PJ, Phuntumart V, Wawra S, Weide R, Win J, Young C, Zhou S, Fry W, Meyers BC, van West P, Ristaino J, Govers F, Birch PR, Whisson SC, Judelson HS, Nusbaum C (2009) Genome sequence and analysis of the Irish potato famine pathogen *Phytophthora infestans*. Nature 461(7262):393–398

Hogenhout SA, Van der Hoorn RA, Terauchi R, Kamoun S (2009) Emerging concepts in effector biology of plant-associated organisms. Mol Plant Microbe Interact 22(2):115–122

Holub E, Brose E, Tör M, Clay C, Crute IR, Beynon JL (1994a) Phenotypic and genotypic variation in the interaction between *Arabidopsis thaliana* and *Albugo candida*. Molec Plant Microbe Interact 8:916–928

Holub EB (1997) Organisation of resistance genes in *Arabidopsis*. In: Crute IR, Holub EB, Burdon JJ (eds) The gene-for-gene relationship in plant-parasite interactions. CAB International, New York, pp 5–26

Holub EB (2008) Natural history of *Arabidopsis thaliana* and oomycete symbioses. Eur J Plant Pathol 122(1):91–109

Holub EB, Beynon JL (1996) Symbiology of mouse ear cress (*Arabidopsis thaliana*) and oomycetes. Adv Bot Res 24:228–273

Holub EB, Beynon JL, Crute IR (1994b) Phenotypic and genotypic characterization of interactions between isolates of *Peronospora parasitica* and accessions of *Arabidopsis thaliana*. Mol Plant-Microbe Interact 7:223–239

Jeger Gilijamse, Bock Frinking (1998) The epidemiology, variability and control of the downy mildews of pearl millet and sorghum, with particular reference to Africa. Plant Pathol 47(5):544–569

Jiang RH, Tripathy S, Govers F, Tyler BM (2008) RXLR effector reservoir in two *Phytophthora* species is dominated by a single rapidly evolving superfamily with more than 700 members. Proc Natl Acad Sci USA 105(12):4874–4879

Jiang RH, Tyler BM (2012) Mechanisms and evolution of virulence in oomycetes. Annu Rev Phytopathol 50:295–318

Judelson HS, Shrivastava J, Manson J (2012) Decay of genes encoding the oomycete flagellar proteome in the downy mildew *Hyaloperonospora arabidopsidis*. PLoS ONE 7(10):e47624

Kale SD, Tyler BM (2011) Entry of oomycete and fungal effectors into plant and animal host cells. Cell Microbiol 13(12):1839–1848

Kamoun S (2003) Molecular genetics of pathogenic oomycetes. Eukaryot Cell 2(2):191–199

Kamoun S (2006) A catalogue of the effector secretome of plant pathogenic oomycetes. Annu Rev Phytopathol 44:41–60

Keeling PJ (2004) Reduction and compaction in the genome of the apicomplexan parasite *Cryptosporidium parvum*. Dev Cell 6(5):614–616

Kemen E, Gardiner A, Schultz-Larsen T, Kemen AC, Balmuth AL, Robert-Seilaniantz A, Bailey K, Holub E, Studholme DJ, MacLean D, Jones JDG (2011) Gene gain and loss during evolution of obligate parasitism in the white rust pathogen of *Arabidopsis thaliana*. PLoS Biol 9(7):e1001094

Kemen E, Jones JD (2012) Obligate biotroph parasitism: can we link genomes to lifestyles? Trends Plant Sci 17(8):448–457

Koch E, Slusarenko A (1990) *Arabidopsis* is susceptible to infection by a downy mildew fungus. Plant Cell 2(5):437–445

Korves T, Bergelson J (2004) A novel cost of *R* gene resistance in the presence of disease. Am Nat 163(4):489–504

Korves TM, Bergelson J (2003) A developmental response to pathogen infection in *Arabidopsis*. Plant Physiol 133(1):339–347

Lamour KH, Mudge J, Gobena D, Hurtado-Gonzales OP, Schmutz J, Kuo A, Miller NA, Rice BJ, Raffaele S, Cano LM, Bharti AK, Donahoo RS, Finley S, Huitema E, Hulvey J, Platt D, Salamov A, Savidor A, Sharma R, Stam R, Storey D, Thines M, Win J, Haas BJ, Dinwiddie DL, Jenkins J, Knight JR, Affourtit JP, Han CS, Chertkov O, Lindquist EA, Detter C, Grigoriev IV, Kamoun S, Kingsmore SF (2012) Genome sequencing and mapping reveal loss of heterozygosity as a mechanism for rapid adaptation in the vegetable pathogen *Phytophthora capsici*. Mol Plant Microbe Interact 25(10):1350–1360

Lapin D, Van den Ackerveken G (2013) Susceptibility to plant disease: more than a failure of host immunity. Trends Plant Sci 18(10):546–554

Latijnhouwers M, de Wit PJGM, Govers F (2003) Oomycetes and fungi: similar weaponry to attack plants. Trends in Microbiol 11(10):462–469

Leonelli L, Pelton J, Schoeffler A, Dahlbeck D, Berger J, Wemmer DE, Staskawicz B (2011) Structural elucidation and functional characterization of the *Hyaloperonospora arabidopsidis* effector protein Atr13. PLoS Pathog 7:e1002428

Lindau G (1901) Hilfsbuch für das Sammein Parasitischer Pilze. Bornträger Verlag, Berlin

Links M, Holub E, Jiang R, Sharpe A, Hegedus D, Beynon E, Sillito D, Clarke W, Uzuhashi S, Borhan M (2011) De novo sequence assembly of Albugo candida reveals a small genome relative to other biotrophic oomycetes. Trends Microbiol 12(1):503

Lu YJ, Schornack S, Spallek T, Geldner N, Chory J, Schellmann S, Schumacher K, Kamoun S, Robatzek S (2012) Patterns of plant subcellular responses to successful oomycete infections reveal differences in host cell reprogramming and endocytic trafficking. Cell Microbiol 14(5):682–697

Lucas J, Hayter J, Crute I (1995) The downy mildews: host specificity and pathogenesis. In: Singh U, Singh R (eds) Pathogenesis and host specificity in plant diseases. Permagon, UK, pp 217–234

McDowell JM (2011) Genomes of obligate plant pathogens reveal adaptations for obligate parasitism. Proc Natl Acad Sci USA 108(22):8921–8922

McDowell JM (2013) Genomic and transcriptomic insights into lifestyle transitions of a hemi-biotrophic fungal pathogen. New Phytol 197(4):1032–1034

McDowell JM, Dhandaydham M, Long TA, Aarts MG, Goff S, Holub EB, Dangl JL (1998) Intragenic recombination and diversifying selection contribute to the evolution of downy mildew resistance at the *RPP8* locus of *Arabidopsis*. Plant Cell 10(11):1861–1874

Mims CW, Richardson EA, Holt BF III, Dangl JL (2004) Ultrastructure of the host-pathogen interface in *Arabidopsis thaliana* leaves infected by the downy mildew *Hyaloperonospora parasitica* (vol 82, p 1001, 2004). Can J Botany-Revue Can De Botanique 82(10):1545

Mukhtar MS, Carvunis AR, Dreze M, Epple P, Steinbrenner J, Moore J, Tasan M, Galli M, Hao T, Nishimura MT, Pevzner SJ, Donovan SE, Ghamsari L, Santhanam B, Romero V, Poulin MM, Gebreab F, Gutierrez BJ, Tam S, Monachello D, Boxem M, Harbort CJ, McDonald N, Gai L, Chen H, He Y, Vandenhaute J, Roth FP, Hill DE, Ecker JR, Vidal M, Beynon J, Braun P, Dangl JL (2011) Independently evolved virulence effectors converge onto hubs in a plant immune system network. Science 333(6042):596–601

Munkvold KR, Martin GB (2009) Advances in experimental methods for the elucidation of *Pseudomonas syringae* effector function with a focus on AvrPtoB. Mol Plant Pathol 10(6):777–793

Nowara D, Gay A, Lacomme C, Shaw J, Ridout C, Douchkov D, Hensel G, Kumlehn J, Schweizer P (2010) HIGS: host-induced gene silencing in the obligate biotrophic fungal pathogen *Blumeria graminis*. Plant Cell 22(9):3130–3141

O'Connell RJ, Thon MR, Hacquard S, Amyotte SG, Kleemann J, Torres MF, Damm U, Buiate EA, Epstein L, Alkan N, Altmuller J, Alvarado-Balderrama L, Bauser CA, Becker C, Birren BW, Chen Z, Choi J, Crouch JA, Duvick JP, Farman MA, Gan P, Heiman D, Henrissat B, Howard RJ, Kabbage M, Koch C, Kracher B, Kubo Y, Law AD, Lebrun MH, Lee YH, Miyara I, Moore N, Neumann U, Nordstrom K, Panaccione DG, Panstruga R, Place M, Proctor RH, Prusky D, Rech G, Reinhardt R, Rollins JA, Rounsley S, Schardl CL, Schwartz DC, Shenoy N, Shirasu K, Sikhakolli UR, Stuber K, Sukno SA, Sweigard JA, Takano Y, Takahara H, Trail F, van der Does HC, Voll LM, Will I, Young S, Zeng Q, Zhang J, Zhou S, Dickman MB, Schulze-Lefert P, Ver Loren van Themaat E, Ma LJ, Vaillancourt LJ (2012) Lifestyle transitions in plant pathogenic *Colletotrichum* fungi deciphered by genome and transcriptome analyses. Nat Genet 44(9):1060–1065

Ochman H, Moran NA (2001) Genes lost and genes found: evolution of bacterial pathogenesis and symbiosis. Science 292(5519):1096–1099

Oh SK, Young C, Lee M, Oliva R, Bozkurt TO, Cano LM, Win J, Bos JI, Liu HY, van Damme M, Morgan W, Choi D, Van der Vossen EA, Vleeshouwers VG, Kamoun S (2009) In planta expression screens of *Phytophthora infestans* RXLR effectors reveal diverse phenotypes, including activation of the *Solanum bulbocastanum* disease resistance protein Rpi-blb2. Plant Cell 21(9):2928–2947

Parker JE, Coleman MJ, Szabo V, Frost LN, Schmidt R, van der Biezen E, Moores T, Dean C, Daniels MJ, Jones JDG (1997) The *Arabidopsis* downy mildew resistance gene *Rpp5* shares similarity to the toll and interleukin-1 receptors with *N* and *L6*. Plant Cell 9:879–894

Parker JE, Holub EB, Frost LN, Falk A, Gunn ND, Daniels MJ (1996) Characterization of eds1, a mutation in *Arabidopsis* suppressing resistance to *Peronospora parasitica* specified by several different *RPP* genes. Plant Cell 8(11):2033–2046

Pazour GJ, Agrin N, Leszyk J, Witman GB (2005) Proteomic analysis of a eukaryotic cilium. J Cell Biol 170(1):103–113

Raffaele S, Farrer RA, Cano LM, Studholme DJ, MacLean D, Thines M, Jiang RH, Zody MC, Kunjeti SG, Donofrio NM, Meyers BC, Nusbaum C, Kamoun S (2010) Genome evolution following host jumps in the Irish potato famine pathogen lineage. Science 330(6010):1540–1543

Raffaele S, Kamoun S (2012) Genome evolution in filamentous plant pathogens: why bigger can be better. Nat Rev Microbiol 10(6):417–430

Rehmany AP, Gordon A, Rose LE, Allen RL, Armstrong MR, Whisson SC, Kamoun S, Tyler BM, Birch PR, Beynon JL (2005) Differential recognition of highly divergent downy mildew avirulence gene alleles by *RPP1* resistance genes from two *Arabidopsis* lines. Plant Cell 17(6):1839–1850

Rehmany AP, Grenville LJ, Gunn ND, Allen RL, Paniwnyk Z, Byrne J, Whisson SC, Birch PR, Beynon JL (2003) A genetic interval and physical contig spanning the *Peronospora parasitica* (At) avirulence gene locus *ATR1Nd*. Fungal Genet Biol 38(1):33–42

Reignault P, Frost LN, Richardson H, Daniels MJ, Jones JD, Parker JE (1996) Four *Arabidopsis* RPP loci controlling resistance to the Noco2 isolate of *Peronospora parasitica* map to regions known to contain other RPP recognition specificities. Mol Plant Microbe Interact 9(6):464–473

Rentel MC, Leonelli L, Dahlbeck D, Zhao B, Staskawicz BJ (2008) Recognition of the *Hyaloperonospora parasitica* effector ATR13 triggers resistance against oomycete, bacterial, and viral pathogens. Proc Natl Acad Sci USA 105(3):1091–1096

Richards TA, Soanes DM, Jones MDM, Vasieva O, Leonard G, Paszkiewicz K, Foster PG, Hall N, Talbot NJ (2011) Horizontal gene transfer facilitated the evolution of plant parasitic mechanisms in the oomycetes. Proc Natl Acad Sci USA 108(37):15258–15263

Ricker KE, Bostock RM (1992) Evidence for release of the elicitor arachidonic acid and its metabolites from sporangia of *Phytophthora infestans* during infection of potato. Physiol Mol Plant Pathol 41(1):61–72

Rietman H, Bijsterbosch G, Cano LM, Lee HR, Vossen JH, Jacobsen E, Visser RG, Kamoun S, Vleeshouwers VG (2012) Qualitative and quantitative late blight resistance in the potato cultivar *Sarpo Mira* is determined by the perception of five distinct RXLR effectors. Mol Plant Microbe Interact 25(7):910–919

Roetschi A, Si-Ammour A, Belbahri L, Mauch F, Mauch-Mani B (2001) Characterization of an Arabidopsis-Phytophthora Pathosystem: resistance requires a functional PAD2 gene and is independent of salicylic acid, ethylene and jasmonic acid signalling. Plant J 28(3):293–305

Rookes JE, Wright ML, Cahill DM (2008) Elucidation of defence responses and signalling pathways induced in

*Arabidopsis thaliana* following challenge with *Phytophthora* cinnamomi. Physiol Mol Plant Pathol 72(4'Äì6):151–161

Runge F, Telle S, Ploch S, Savory E, Day B, Sharma R, Thines M (2011) The inclusion of downy mildews in a multi-locus-dataset and its reanalysis reveals a high degree of paraphyly in *Phytophthora*. IMA Fungus 2(2):163–171

Salvaudon L, Haraudet V, Shykoff JA, Koella J (2005) Parasite-host fitness trade-offs change with parasite identity: genotype-specific interactions in a plant-pathogen system. Evolution 59(12):2518–2524

Savory EA, Adhikari BN, Hamilton JP, Vaillancourt B, Buell CR, Day B (2012a) mRNA-Seq analysis of the *Pseudoperonospora cubensis* transcriptome during cucumber (*Cucumis sativus* L.) infection. PLoS ONE 7(4):e35796

Savory EA, Zou C, Adhikari BN, Hamilton JP, Buell CR, Shiu SH, Day B (2012b) Alternative splicing of a multi-drug transporter from *Pseudoperonospora cubensis* generates an RXLR effector protein that elicits a rapid cell death. PLoS ONE 7(4):e34701

Schlaeppi K, Abou-Mansour E, Buchala A, Mauch F (2010) Disease resistance of *Arabidopsis* to *Phytophthora brassicae* is established by the sequential action of indole glucosinolates and camalexin. Plant J 62(5):840–851

Seidl MF, Van den Ackerveken G, Govers F, Snel B (2012a) Reconstruction of oomycete genome evolution identifies differences in evolutionary trajectories leading to present-day large gene families. Genome Biol Evol 4(3):199–211

Seidl MF, Wang RP, Van den Ackerveken G, Govers F, Snel B (2012b) Bioinformatic inference of specific and general transcription factor binding sites in the plant pathogen *Phytophthora infestans*. PLoS ONE 7(12):e51295

Sinapidou E, Williams K, Nott L, Bahkt S, Tor M, Crute I, Bittner-Eddy P, Beynon J (2004) Two TIR-NB-LRR genes are required to specify resistance to *Peronospora parasitica* isolate Cala2 in *Arabidopsis*. Plant J 38(6):898–909

Slusarenko A, Mauch-Mani B (1991) Downy mildew of *Arabidopsis thaliana* caused by *Peronospora parasitica*: a model system for the investigation of the molecular biology of host-pathogen interactions. In: Hennecke H, Verma DPS (eds) Advances in molecular genetics of plant-microbe interactions, vol 1. Kluwer Academic Publishers, The Netherlands, pp 280–283

Slusarenko A, Schlaich N (2003) Downy mildew of *Arabidopsis thaliana* caused by *Hyaloperonospora parasitica* (formerly *Peronospora parasitica*). Mol Plant Pathol 4:159–170

Sohn KH, Lei R, Nemri A, Jones JD (2007) The downy mildew effector proteins ATR1 and ATR13 promote disease susceptibility in *Arabidopsis thaliana*. Plant Cell 19(12):4077–4090

Soylu E, Soylu S (2003) Light and electron microscopy of the compatible interaction between *Arabidopsis* and the downy mildew pathogen *Peronospora parasitica*. J Phytopathol 151:300–306

Spanu P, Kamper J (2010) Genomics of biotrophy in fungi and oomycetes–emerging patterns. Curr Opin Plant Biol 13(4):409–414

Spanu PD (2006) Why do some fungi give up their freedom and become obligate dependants on their host? New Phytol 171(3):447–450

Spanu PD (2012) The genomics of obligate (and nonobligate) biotrophs. Annu Rev Phytopathol 50:91–109

Spanu PD, Panstruga R (2012) Powdery mildew genomes in the crosshairs. In: 2nd International Powdery Mildew Workshop and 3rd New Phytologist Workshop, in Zurich, Switzerland, February 2012. New Phytol 195(1):20–22

Spencer-Phillips P (1997) Function of fungal haustoria in epiphytic and endophytic infections. Adv Bot Res 24:309–333

Stassen J, den Boer E, Vergeer PW, Andel A, Ellendorff U, Pelgrom K, Pel M, Schut J, Zonneveld O, Jeuken MJ, Van den Ackerveken G (2013) Specific in planta recognition of two GKLR proteins of the downy mildew *Bremia lactucae* revealed in a large effector screen in lettuce. Mol Plant Microbe Interact 26(11):1259–70

Stassen JH, Seidl MF, Vergeer PW, Nijman IJ, Snel B, Cuppen E, Van den Ackerveken G (2012) Effector identification in the lettuce downy mildew *Bremia lactucae* by massively parallel transcriptome sequencing. Mol Plant Pathol 13(7):719–731

Stassen JH, Van den Ackerveken G (2011) How do oomycete effectors interfere with plant life? Curr Opin Plant Biol 14(4):407–414

Studholme DJ, Glover RH, Boonham N (2011) Application of high-throughput DNA sequencing in phytopathology. Annu Rev Phytopathol 49:87–105

Stuttmann J, Hubberten HM, Rietz S, Kaur J, Muskett P, Guerois R, Bednarek P, Hoefgen R, Parker JE (2011) Perturbation of *Arabidopsis* amino acid metabolism causes incompatibility with the adapted biotrophic pathogen *Hyaloperonospora arabidopsidis*. Plant Cell 23(7):2788–2803

Szabo LJ, Bushnell WR (2001) Hidden robbers: the role of fungal haustoria in parasitism of plants. Proc Natl Acad Sci USA 98(14):7654–7655

Thines M (2009) Bridging the gulf: *Phytophthora* and downy mildews are connected by rare grass parasites. PLoS ONE 4(3):e4790

Thines M, Kamoun S (2010) Oomycete-plant coevolution: recent advances and future prospects. Curr Opin Plant Biol 13:427–433

Thomma BPHJ, de Jonge R (2009) Fungal LysM effectors: extinguishers of host immunity? Trends Microbiol 17(4):151–157

Tian M, Win J, Savory E, Burkhardt A, Held M, Brandizzi F, Day B (2011) 454 Genome sequencing of *Pseudoperonospora cubensis* reveals effector proteins with a QXLR translocation motif. Mol Plant Microbe Interact 24(5):543–553

Tiwari MM, Arya HC (1969) *Sclerospora graminicola* axenic culture. Science 163(3864):291–293

Tyler BM, Tripathy S, Zhang X, Dehal P, Jiang RH, Aerts A, Arredondo FD, Baxter L, Bensasson D, Beynon JL, Chapman J, Damasceno CM, Dorrance AE, Dou D, Dickerman AW, Dubchak IL, Garbelotto M, Gijzen M, Gordon SG, Govers F, Grunwald NJ, Huang W, Ivors KL, Jones RW, Kamoun S, Krampis K, Lamour KH, Lee MK, McDonald WH, Medina M, Meijer HJ, Nordberg EK, Maclean DJ, Ospina-Giraldo MD, Morris PF, Phuntumart V, Putnam NH, Rash S, Rose JK, Sakihama Y, Salamov AA, Savidor A, Scheuring CF, Smith BM, Sobral BW, Terry A, Torto-Alalibo TA, Win J, Xu Z, Zhang H, Grigoriev IV, Rokhsar DS, Boore JL (2006) *Phytophthora* genome sequences uncover evolutionary origins and mechanisms of pathogenesis. Science 313(5791):1261–1266

van der Biezen EA, Freddie CT, Kahn K, Parker JE, Jones JD (2002) *Arabidopsis RPP4* is a member of the *RPP5* multigene family of TIR-NB-LRR genes and confers downy mildew resistance through multiple signalling components. Plant J 29(4):439–451

Vleeshouwers V, Raffaele S, Vossen J, Champouret N, Oliva R, Segretin ME, Rietman H, Cano LM, Lokossou A, Kessel G, Pel MA, Kamoun S (2010) Understanding and exploiting late blight resistance in the age of effectors. Ann Rev Phytopathol 47:507–531

Vleeshouwers VG, Rietman H, Krenek P, Champouret N, Young C, Oh SK, Wang M, Bouwmeester K, Vosman B, Visser RG, Jacobsen E, Govers F, Kamoun S, Van der Vossen EA (2008) Effector genomics accelerates discovery and functional profiling of potato disease resistance and *Phytophthora infestans* avirulence genes. PLoS ONE 3(8):e2875

Voglmayr H (2003) Phylogenetic relationships of *Peronospora* and related genera based on nuclear ribosomal ITS sequences. Mycol Res 107(Pt 10):1132–1142

Voglmayr H, Riethmuller A, Goker M, Weiss M, Oberwinkler F (2004) Phylogenetic relationships of *Plasmopara*, *Bremia* and other genera of downy mildew pathogens with pyriform haustoria based on Bayesian analysis of partial LSU rDNA sequence data. Mycol Res 108(Pt 9):1011–1024

Wang YAN, Bouwmeester K, van de Mortel JE, Shan W, Govers F (2013) A novel *Arabidopsis*–oomycete pathosystem: differential interactions with *Phytophthora capsici* reveal a role for camalexin, indole glucosinolates and salicylic acid in defence. Plant, Cell Environ 36(6):1192–1203

Wang YC, Wang QQ, Han CZ, Ferreira AO, Yu XL, Ye WW, Tripathy S, Kale SD, Gu BA, Sheng YT, Sui YY, Wang XL, Zhang ZG, Cheng BP, Dong SM, Shan WX, Zheng XB, Dou DL, Tyler BM (2011) Transcriptional programming and functional interactions within the *Phytophthora sojae* RXLR effector repertoire. Plant Cell 23:2064–2086

Whisson SC, Boevink PC, Moleleki L, Avrova AO, Morales JG, Gilroy EM, Armstrong MR, Grouffaud S, van West P, Chapman S, Hein I, Toth IK, Pritchard L, Birch PR (2007) A translocation signal for delivery of oomycete effector proteins into host plant cells. Nature 450(7166):115–118

Wroblewski T, Tomczak A, Michelmore R (2005) Optimization of *Agrobacterium*-mediated transient assays of gene expression in lettuce, tomato and *Arabidopsis*. Plant Biotechnol J 3(2):259–273

Yoda H, Hiroi Y, Sano H (2006) Polyamine oxidase is one of the key elements for oxidative burst to induce programmed cell death in tobacco cultured cells. Plant Physiol 142(1):193–206

Zhang C, Xie Q, Anderson RG, Ng G, Seitz NC, Peterson T, McClung CR, McDowell JM, Kong D, Kwak JM, Lu H (2013) Crosstalk between the circadian clock and innate immunity in *Arabidopsis*. PLoS Pathog 9(6):e1003370

# Index



R. A. Dean et al. (eds.), *Genomics of Plant-Associated Fungi and Oomycetes: Dicot Pathogens*,
DOI: 10.1007/978-3-662-44056-8, © Springer-Verlag Berlin Heidelberg 2014

Printed in the USA
CPSIA information can be obtained
at www.ICGtesting.com
LVHW070341201123
764347LV00011B/989

9 783662 440551